Building the World

Building the World

An Encyclopedia of the
Great Engineering
Projects in History

Volume 1

FRANK P. DAVIDSON AND
KATHLEEN LUSK BROOKE

Greenwood Press
Westport, Connecticut • London

Library of Congress Cataloging-in-Publication Data

Building the world : an encyclopedia of the great engineering projects in history / by
Frank P. Davidson and Kathleen Lusk Brooke.

     p. cm.

  Includes bibliographical references and index.

  ISBN 0–313–33354–8 (set: alk. paper)—0–313–33373–4 (vol 1: alk. paper)—0–313–33374–2
(vol 2: alk. paper)

  1. Industrial engineering. I. Title: Encyclopedia of the great engineering projects in
history. II. Davidson, Frank Paul, 1918– III. Lusk Brooke, Kathleen.

  T56.I44 1997

  620—dc22   2005037902

British Library Cataloguing in Publication Data is available.

Library of Congress Catalog Card Number: 2005037902

ISBN: 0–313–33354–8 (set)

        0–313–33373–4 (vol 1)

        0–313–33374–2 (vol 2)

First published in 2006

Greenwood Press, 88 Post Road West, Westport, CT 06881
An imprint of Greenwood Publishing Group, Inc.
www.greenwood.com

Printed in the United States of America

The paper used in this book complies with the
Permanent Paper Standard issued by the National
Information Standards Organization (Z39.48–1984).

10 9 8 7 6 5 4 3 2 1

CONTENTS

Volume 1

Contents

Volume 2

Map:

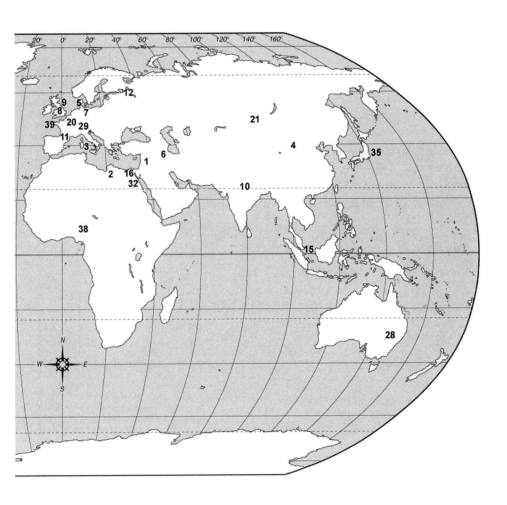

DOCUMENTS OF AUTHORIZATION

The editors have endeavored in each chapter to provide the original authorizing documents. However, in those cases where no identifiable document authorized the project, the process of authorization is recounted through the recollections and remarks of individuals who witnessed or knew of the various stages of the decision process. In some cases the authorization was not by document but a combination of symbolic actions and verbal explanations. For example, the plan for founding Baghdad was drawn on the sand by Caliph al-Mansur and later described by reputable historians. For Charlemagne, his decrees were recorded in the annals of the Royal Frankish Court. As such, they are considered a reliable and authoritative record.

PERMISSIONS

The editors wish to state that they have been diligent in seeking permission from every author and information source quoted or cited in these two volumes. In the few cases where we have been unable to obtain explicit permission, the reader should be assured that we have exerted considerable time and good-faith effort to research the sources, identify copyright holders, and obtain the desired permissions.

INTERNET REFERENCES

The editors wish to advise those who utilize the Internet references provided in the chapters and the bibliography, that we have made every effort to confirm the validity of each website as of this edition's publication date.

If you would like to contact the editors, you can write to Greenwood Press or e-mail any member of the editorial team:

Frank P. Davidson
Kathleen Lusk Brooke
Cherie Potts
Richard Hantula
E-mail: buildingtheworld@aol.com

PREFACE

This first encyclopedia of great engineering projects is a pioneering and necessarily selective presentation of macroengineering as a key element in the building, maintenance, and expansion of civilization. While readers will be familiar with a number of these wonders of the world, this is the first time their founding documents have been made available in one publication. We have all seen photos of the Taj Mahal, but how many of us have read the real estate agreement?

The scope of the current volumes is admittedly vast—from the wide-ranging achievements of ancient China, Greece, and Arabia to modern satellite communications systems and voyages to the moon. The selection criteria for inclusion hinged on the survival of the founding document in each case (the editors ardently hope that archaeologists will eventually unearth an original decree that ordered the building of the Great Pyramid). However, this encyclopedia is intended to be much more than just these documents. Our goal is to provide the reader with a greater understanding of the role such structures have played not just in the transformation of the physical world, but in the growth and development of societies and cultures. Thus, we have included lengthy examinations of the cultural background of the project, the planning and building stages, and the historical importance of the structure. We also provide sources for further exploration with references to books, articles, and Web sites. Where relevant, we have included cultural works inspired by these marvels of engineering; Verdi's *Aida* was written as a tribute to the Suez Canal and is now available on the Internet. One can hear the great opera while reading about the equally dramatic story of the Suez Canal itself.

What, precisely, do we mean by *macroengineering*? The word connotes the largest and most complex technical projects that can be accomplished in any given period of history. It follows that such enterprises characteristically involve an immense call on resources of all kinds: manpower, finance, materials, energy, and statecraft. And there are substantial impacts on the environment, society, culture, and the economy. There is necessarily an element of organizational and technical complexity.

The editors have had personal connections with large engineering projects. Frank P. Davidson was the American cofounder, in 1957, of the Channel Tunnel Study Group. In 1970, he was appointed head of the Macro-Engineering Research Group of the School of Engineering at the Massachusetts Institute of Technology (MIT), a post he held for more than a quarter century. Kathleen Lusk Brooke was one of the youngest Fulbright Scholars in history. This enabled her to spend a year of study in Iceland—precisely as the island of Surtsey emerged out of a volcanic eruption. In 1972 she was awarded a Harvard Graduate Prize Fellowship, which included a year of travel. Her itinerary took her to 41 countries, including Afghanistan, India, Japan, Singapore, and most of Europe. From 1974 to 1976, Davidson and Lusk Brooke taught "Failure of Human Systems" at the School of Engineering at MIT, followed by tandem courses on "Failure" and "Success" at Harvard University's Radcliffe Institute.

The audience for *Building the World* is, inevitably, diverse. Students of national and regional history, engineering and technology, law, architecture, public art, and historic preservation will discover information relevant to their specialties, while professionals in these and related fields will find useful and interesting structured information. Government leaders, diplomats, and planners will join social scientists in pondering past accomplishments as a prologue to the proliferation of large-scale technical projects that stretch the mind (one could cite Robert Goddard's design for maglev trains operating at supersonic velocities). Scholars will, at last, have ready access to the text of documents that launched some of the greatest achievements of human history.

If there is one connecting thread among works as different as the Grand Canal of China and COMSAT, it is the combination of public and private resources. Technologies now exist that could supply plentiful energy on a scale never before contemplated; travel and transport could be designed to offer both greater safety and speed; medical care, water supplies, and communication and education systems could be optimized by knowledgeable interdisciplinary cooperation. *Building the World* suggests precedents for the powerful opportunities offered by macroengineering to transform and improve the future.

ACKNOWLEDGMENTS

We owe great thanks to Kevin J. Downing of Greenwood Publishing Group, who has worked closely with us to develop a work of significance, and to Dr. Jeffrey K. Stine, curator of engineering, National Museum of American History, Smithsonian Institution, Washington, DC, who first suggested Greenwood as our publisher.

This book owes much to the guidance, over a period of years, of John W. Landis, D.Sci., president of the International Association of Macro-Engineering Societies (IAMES), and of Uwe Kitzinger, CBE, chairman of IAMES and former president of Templeton College, Oxford.

We must give special thanks to an expert in his field. Professor Pieter Huisman, former director of flood control at the Rijkswaterstaat and a retired professor at Delft Technical University. Dr. Huisman is the author of leading works on Dutch engineering history, and we benefited greatly from his unique expertise in a review of chapter 5 on the dikes of the Netherlands. Others who have contributed counsel and encouragement include Philippe Bernard, former head of the Humanities Division, Ecole Polytechnique; Janet R. Caristo-Verrill, chairman, Boston chapter of The American Society for Macro-Engineering (TASME); Professor Emeritus Ernst G. Frankel of the Massachusetts Institute of Technology; Cordell W. Hull, Esq., retired (Bechtel Corporation); Dr. Betty-Ann Kevles, former Charles Lindbergh Visiting Scholar, Smithsonian Institution; Dr. Andrew C. Lemer, president of the American Society for Macro-Engineering and staff member of the National Research Council; Dr. David Marks and Dr. Fred Moavenzadeh of the Massachusetts Institute of Technology; the late Professor

Manabu Nakagawa, former dean of the Faculty of Economics at Hitotsubashi University, Tokyo; Dr. Roy J. Nirschel, president, and Stephen K. White, AIA, dean of the School of Architecture, Art, and Historic Preservation and director of the Center for Macro Projects and Diplomacy at Roger Williams University, Bristol, Rhode Island.

For excellent and artful preparation of documents, for initial copyediting, for supporting research and identification of illustrations, and for design and execution of the publication copy:

Cherie Potts (president, WordWorks), manager of production and editorial adviser

For initial research and writing on the history of each project, and for insightful commentary:

Dr. Richard Hantula (president, HES), senior writer and permissions editor

For documentary research, ancient and modern; for translation assistance; and for other beneficial advice:

Rucker Alex	Justin Lake
Christine Alexander	George H. Litwin
Alan Altshuler	Edward J. Lusk
Evgenii Victorovich Anisimov	Kenneth Mandel
Ali Asani	Timothy Moi
Christine Boedler	Mya Mya Mon
Judith A. Boyajian	Gülrü Necipoglu
Christoph-Friedrich von Braun	Gloria Ferrari Pinney
Marilynn G. Burmeister	Earl R. Potts, Jr.
Nicholas H. Davidson	Francis Raud
Lucien Deschamps	Joan Reiter
Stéphane Dieu	Alberto Renault
Ruth Dorfman	Elisabeth Richard-Rossignol
John Evans	Rachael Rusting
Susanne Fairclough	Khalid Saeed
Gary Fix	Frederick Salvucci
Clive Foss	George Sanborn
Dieter Garbade	Renata Silvano
Thierry Gaudin	Gary C. Solar
Alexander Gorlov	Sarayu Srinivasan
Ella Gorlov	Anne Thayer
M. A. Hamid	Mead Treadwell
John P. Hennigan	Yanni Tsipis
Lin-lin Hu	Nicholas Vaczek
Barbara Kehoe	Norio Yamamoto
Hasan-Uddin Khan	Jan Ziolkowski
Seth Kramer	

NOTES TO THE READER

There are several characteristics of macroengineering works that may be of interest to the reader when considering and exploring *Building the World*.

LONG TIME FRAME

Spanning decades, and sometimes centuries, macro projects have very long time frames. Engineers, architects, constructors, developers, government and legislative regulators, financiers, and those who live in contiguity to the project may need to take different approaches to the customary ways we think of beginnings and ends. So we invite the reader to consider how the projects presented in *Building the World* were managed over the time span each encompassed, comparing the approaches and systems used. And, in recognition of the long time frame of each project, we point out to the reader that the dates of the documents presented are distributed along the time span of the projects. Some documents mark the inception; others come at some midpoint. Because macro projects often develop strength through momentum, critical funding and/or design agreements can actually come after the start of building.

PLANNING AND PROFIT

Because of their long time frames, the costs of macro projects can escalate due to unforeseen delays. For example, if the Channel Tunnel linking France and

England had been built when designed in 1959, the cost would have been $100 million instead of the $15 billion at completion in 1994. On the other hand, careful planning and precise specifications can result in on-time and on-budget success. Gustave Eiffel spent twice as long planning his tower as building it; as a result, the project was finished early. The Hoover Dam was on time and under budget. The Suez Canal was efficiently conceived and built; the Panama Canal, initiated by the same entrepreneur, was not well planned and required several changes of leadership. There is a correlation between the depth of planning and the degree of success.

MANY VOICES

Like a large stone cast into a pond, a macro project creates waves that affect its galvanizing cause. Macro projects are vast in the many resources they require—natural, material, and human. Some projects are actual cities: examples presented here are Cyrene (630 B.C.); Baghdad (762 A.D.); St. Petersburg (1703); Washington, DC (1787); Singapore (1819); Brasilia (1956); and Abuja (1976). Arnold Toynbee, in *Cities of Destiny*, explored how culture can reach heights given certain conditions. Bridging continents, uniting regions in transport or energy distribution, macro projects can have a unifying effect. Hundreds and thousands of people are employed; millions are affected.

ONE VISION

Even though they involve many people, macro works are often ignited by the vision of a single individual. In earlier times, perhaps it was the emperor, the king, the high priest, or the czar. In more recent times, it can be a legislator or a passionate leader who belives ardently in the vision. As the reader examines the macro projects presented here, it is worthwhile to look for the champion and the way in which his or her vision is transmitted and shared, for instance, DeWitt Clinton for the Erie Canal, and Duff Roblin for the Red River Floodway that saved Winnipeg from inundation and disaster in 1997.

ENVIRONMENTAL CONCERNS

Environmental structural change is a primary consideration of macroengineering. For example, dams create hydroelectric power and increase water resources for people and agriculture; however, salinity can later become an environmental problem, as seen in the High Dam at Aswan or the Snowy Mountains hydroelectric power project. The reader might note the progression of environmental agreements in macro projects—from the earliest times, when such provisions were not included, to more recent endeavors such as the Tennessee Tombigbee Waterway, described in Dr. Jeffrey K. Stine's classic work *Mixing the Waters*. In many current projects, design is often influenced by environmental concerns.

TECHNICAL AND SCIENTIFIC INNOVATIONS

In *How Big and Still Beautiful? Macro-Engineering Revisited,* Meador and Parthé posited three areas of impact to be considered in developing decision support systems for macro projects: cultural impact potential, public/private resource requirements, and technology development difficulty. Macroengineering projects present technical and scientific challenges because of their size and scope, often requiring the development of new technologies. While technical and scientific issues for macro projects can be challenging, they are also a source of innovation. As the reader examines the projects presented here, tracing the innovations that resulted from macro endeavors can be rewarding.

COOPERATIVE ADVANTAGE

While one pillar of business and economic strategy is competitive advantage, macro projects also illustrate the power of cooperation. In *Macro* (1983), Frank P. Davidson articulated the idea of macroengineering and how this field influences the dynamics of systems. Mobilization of macro endeavors requires shared purpose and cooperation, as suggested by Litwin, Bray, and Lusk Brooke in *Mobilizing the Organization: Bringing Strategy to Life.* L. Sprague de Camp, in *The Ancient Engineers*, observed that when humans learned to farm, communities settled in one place long enough to develop tools, ideas, and expertise for building projects beyond the scope of one family or one village, hence giving time and space for public works. Edward O. Wilson, in *Sociobiology*, traced the evolution of the instinct to devote action to the well-being of those beyond one's own immediate circle. Perhaps as the reader explores the macro projects presented in *Building the World*, the hand of evolution might emerge and even point the way to future works that can improve and unite the world.

FOR FURTHER REFERENCE

Davidson, Frank P. *Macro: Big Is Beautiful*. London: Anthony Blond, 1986.
———. *Macro: A Clear Vision of How Science and Technology Will Shape Our Future*. New York: William Morrow, 1983.
de Camp, L. Sprague. *The Ancient Engineers*. London: Ballantine Books, Random House, 1960.
Fox, J. Ronald, and Donn B. Miller. *Challenges in Managing Large Projects*. Fort Belvoir, VA: Defense Acquisition University Press, 2006.

Litwin, George, John J. Bray, and Kathleen Lusk Brooke. *Mobilizing the Organization: Bringing Strategy to Life*. London: Prentice Hall, 1996.

Meador, C. Lawrence, and Arthur C. Parthé Jr. "Managing Macro-Development Policy, Planning and Control System Implications." pp. 78–108. In *How Big and Still Beautiful? Macro-Engineering Revisited*, edited by Frank P. Davidson, C. Lawrence Meador, and Robert Salkeld. Boulder, CO: Westview Press, 1980.

Morris, Peter W. G. *The Management of Projects*. London: Thomas Telford, 1994.

Stine, Jeffrey K. *Mixing the Waters: Environment, Politics, and the Building of the Tennessee-Tombigbee Waterway*. Akron, OH: University of Akron Press, 1993.

Toynbee, Arnold, ed. *Cities of Destiny*. London: Thames and Hudson, 1967.

Wilson, Edward O. *Sociobiology*. Cambridge, MA: Belknap Press, 1975.

INTRODUCTION

AUTHORIZATION: THE *SINE QUA NON* OF MACRO-ENGINEERING

This compendium presents texts of laws and edicts—and in rare cases, accounts of utterances—that authorized some of the largest engineering projects in history. The publication, believed to be the first of its kind, is illustrative rather than comprehensive. It is offered for the use of historians of technology, specialists in legal history, scholars interested in the interaction of law and technology, and the general public.

In identifying texts that constituted decisive approval for large-scale technical enterprises, the editors have had to take into consideration the occasional murkiness of procedures for indicating *yea* or *nay*. One might expect that archives of absolute monarchies would be replete with edicts, decrees, rescripts, and *firmans*. Indeed, such is frequently the case. But in monolithic societies too, authority tends to be exercised, on a day-to-day basis, by subordinate officials and even by committees! Thus we learned that the key initial step in the founding of St. Petersburg—the 1703 decision to fortify the bastion of St. Peter and St. Paul—was taken by a military committee. This was confirmed by no less a personage than the czar himself, Peter the Great, in his own diary, from which we have quoted the relevant authorizing passage.[1]

The word *authorization* is not entirely free of ambiguity. Concepts of legitimacy vary with time and context, and a comment that in fact incites thousands of people to take a series of momentous steps may be far different in character from what the lawyers among us might have expected and approved. And where the best evidence for an otherwise inexplicable paradigmatic change no longer exists—or perhaps never did exist in written form—shall we be permitted to cite literature and legend? In a few instances, we have done so; the very remoteness of prehistoric public events challenges us to probe more deeply into the essential meaning of authorization. In our own times as well, a decisive impetus may be imparted to the "decision matrix" by an individual or group outside the loop of officialdom.

Authorization cannot be separated from the nexus of causation: an order not obeyed, an authorization without implementation, would not qualify for citation in these pages. An entire volume could be devoted to the topic of failed authorizations. Charlemagne is said to have issued orders to dig a canal to connect the Rhine and the Danube. Work commenced and a section was completed, but the enterprise was beyond the means and management capacity of the Holy Roman emperor. It was not until postwar Europe that the impetus toward infrastructural unity revived the scheme after an interval of more than a thousand years, and Charlemagne's design became a reality.

Then, too, a project can be completed and promptly fail. The great ship *Vasa*, the pride of the Swedish navy, sank in Stockholm Harbor within minutes of its launch in the seventeenth century. The government of Clement Atlee in postwar Britain was jeopardized by the collapse of its carefully nurtured and fully authorized scheme to develop groundnuts as a cash crop for one of its former colonies in Africa.[2] And we could mention Chernobyl, the *Challenger*, and Bhopal.

Lawyers will be quick to point out the insufficiency, in most cases, of a single document as an adequate reflection of the process of authorization. Where supreme power is vested in an individual, however, a single sentence, even in personal correspondence, may be quite enough. As a relatively modern case in point, we have cited Czar Alexander's terse note[3] expressing his impatience at the hesitancy of his bureaucracy to proceed with the Trans-Siberian Railway. In the twentieth century, Albert Einstein's note to Franklin D. Roosevelt may have been the decisive advice that launched the Manhattan Project to build an atomic bomb; Einstein's stature as a scientist drove and focused the processes of decision.[4]

Readers with a taste for the history of technology will not be surprised to learn that a century earlier England and France had concluded a treaty to build a tunnel across the channel between the two countries. After a year of impressive boring, much of it under the sea, Prime Minister Gladstone yielded to the fears of the adjutant general of the British army, Sir Garnet Wolseley, and withdrew the authorization previously given. The decision process, *grosso modo*, might have been choreographed by Gilbert and Sullivan.

A few years ago, a small group of experienced engineers and investors studied locations for artificial islands to be built along the eastern seaboard of the United States.[5] They concluded that $10 billion would be needed for the construction program, and that this sum could be justified and obtained. However, no one would invest in projects that might never be authorized, and it was not deemed feasible to raise the enormous sums—perhaps one-quarter of a billion dollars—that might be absorbed by legal fees required to apply and negotiate for the requisite authorizations. (Meanwhile, more than 60 artificial islands had been built offshore of Japan.)

This book is deliberately oriented toward large projects, but without in any sense denigrating the equal validity of small and medium-size enterprises. Kathleen Murphy established the statistics in a landmark work,[6] demonstrating that more and more of the world's economic activity, even in developing countries, is performed in engineering ventures costing upward of $100 million each. More recently, the book *Big Science*[7] has detailed the proliferation of huge laboratories, science parks, and science cities. The strands of legislative and executive fiat may not provide total understanding of macroengineering history and its evolution and implications, but without awareness of the formal and informal processes of decision, it is not possible to trace with confidence the chain of causation that leads, step by step, from conceptualization through design, prototype building and testing, assessment, and ultimately authorization and construction. If we have concentrated on one step of an interactive process, it is in no sense out of a wish to devalue the significance of other aspects. But to arrive at a sense of the system dynamics of macroengineering, we cannot afford to neglect the inescapable role of law and authority. For those motivated to probe further this packet of complexities, we recommend careful perusal of *La Dynamique des Grands Projets,* the thesis that earned Jean-Claude Huot of Montreal the coveted *Doctorat d'Etat Es-Sciences* (Lyon, France, 1988).

If limitations of time, as well as the framework of two volumes—not a whole library—have constricted our search, there does appear, nonetheless, to be one lesson that arises clearly from the available data. On a few pages, which we have reproduced in full, a decree signed by Louis XIV granted a concession to build the Canal des Deux Mers, which still links the Atlantic Ocean and the Mediterranean Sea.[8] Today's macro projects routinely require dozens of pages merely to list and enumerate the requisite authorizations! Long before he became U.S. secretary of state, George P. Shultz reported on Sohio's efforts to secure authorization for a pipeline designed to move oil inland from the West Coast:

> Sohio began in January 1975 the process of securing necessary permits and governmental approvals: a total of approximately 700 permits were required from about 140 local, state, federal or private agencies. On March 13, 1979, fifty months later, the decision was reached to abandon the project. In the interim, Sohio had spent $50 million and managed to secure only 250 of the 700 permits.[9]

In a lecture presented at the 1992 North American Tunneling Conference, Jack K. Lemley, then in charge of engineering activities of *Eurotunnel,* observed,

It is not uncommon today for the approval, permitting, and licensing process to consume more time and cost more, in terms of financial and professional resources, than the actual design and construction of the facility. This is a disgrace.[10]

A glance at the existing complexities of authorization, as manifested in the documents that follow, suggests that our society—and not in North America alone—must somehow improve its Hamlet-like approach to the macro opportunities as well as the macro hazards posed by contemporary science and technology. This publication highlights an issue that is nontrivial: if North America and its trading partners still expect to define and develop "enterprises of great pith and moment," (Shakespeare, *Hamlet*, Act III, Scene 1) if *GrandsTravaux* designed to facilitate a meaningful and secure future for this beleaguered planet are to have their day in court, then we must have a regulatory process that safeguards both the freedom of debate and the option to say *yea* as well as *nay*. Perhaps this is why engineer-scholars such as John W. Landis have recommended the establishment of a National Commission on Macro-Engineering.

NOTES

1. From E. S. Gorlov, *The Founding of St. Petersburg as Recounted in the Journal (Diary Notes) of Tsar Peter the Great*. MS in Structural Engineering (May 1956), The Institute of Railroad and Transportation Engineering. Pub. Art. 1777, part 1, p. 76. Translated and adapted from the archaic Russian script of the eighteenth century.

For information on the choice of the site for Russia's new capital city, see Evgenii Victorovich Anisimov, *The Reforms of Peter the Great: Progress through Coercion in Russia*, trans. John T. Alexander (Armonk, NY: M. E. Sharpe, 1993). Another account is accessible in the chapter "The Founding of St. Petersburg" in Robert K. Massie, *Peter the Great: His Life and Work* (New York: Knopf, 1980).

2. Alan Wood, *The Groundnut Affair* (London: Bodley Head, 1950). See also Annual Reports of the Overseas Food Corporation (London: H. M. Stationery Office, 1948 *et seq.*).

3. On July 12, 1890, the czar "wrote in the margin of Baron Korf's report on the Yellow Peril the historic words: 'Necessary to proceed at once to the construction of the line!'" (Eric Newby, *The Big Red Train Ride* [New York: St. Martin's Press, 1978], 67). On March 17, 1891, an Imperial Rescript was addressed by the czar to his son Nicholas, His Imperial Highness the Grand Duke Czarevitch: "YOUR IMPERIAL HIGHNESS I have given the order to build a continuous line of railway across Siberia! I desire you to lay the first stone at Vladivostok" (ibid., 68).

4. The sequence of events is amusingly recounted in Joseph H. Lehmann, *All Sir Garnet: A Life of Field-Marshal Lord Wolseley* (London: Jonathan Cape, 1964), 296–98.

5. In 1989 a cogent brochure, "Island Complex Offshore North Atlantic," was issued by Nigel Chattey Associates, Irvington-on-Hudson, New York. An earlier brochure, "The ICONN-ERIE PROJECT," detailing a scheme to modernize the Erie Canal and provide a deepwater port on an artificial island offshore of New York, was published by Chattey in 1977.

6. Kathleen J. Murphy, *Macroproject Development in the Third World* (Boulder, CO: Westview Press, 1983).

7. Peter Galison and Bruce Hevly, eds., *Big Science: The Growth of Large-Scale Research* (Stanford, CA: Stanford University Press, 1992).

8. M. de La Lande, "Edit d'Octobre 1666 pour le Canal de Languedoc," in *Des Canaux de Navigation et Spécialement du Canal de Languedoc* (Paris: Chez la Veuve Desaint, 1778), 115–18. For a well-illustrated history of the canal (also known as Le Canal du Midi), see Odile de Roquette-Buisson and Christian Sarramon, *The Canal du Midi* (Toulouse, France: Editions Rivages, Marseilles, and Technal, 1983).

9. See George P. Shultz, "The Abrasive Interface," in *Business and Public Policy*, edited by John T. Dunlop (Cambridge, MA: Harvard University Press, 1980), 17.

10. Lemley's lecture, "Leadership Management and the Future of Tunneling in North America," was reported in *AUA News*, October 1992.

1
Solomon's Temple

Israel

DID YOU KNOW . . . ?

➤ The temple was conceived by King David but built by King Solomon circa 960 B.C.
➤ Construction materials included gold, silver, stone, iron, brass, cedar trees, and stone.
➤ The temple was a rectangular structure 20 cubits wide, 60 cubits long, and 30 cubits high (1 cubit equaled 17.5 inches or 44 centimeters).
➤ It was the first permanent structure of the Hebrew nation, which had been nomadic until that time.
➤ Its construction included early examples of a materials-delivery system, go-ahead agreements, a plan for rotating staff, outside contractors, international partnerships, and prefabrication.
➤ The temple was built and destroyed at least five times.

The Temple of Solomon, built around 960 B.C., was not only the first permanent structure of the Hebrew nation, it embodied the tradition expressed by Isaiah: "a house of prayer for all nations" (56:7). It was still highly regarded in the Middle Ages when the Knights Templar, or "Poor Knights of Christ and of the Temple of Solomon," became one of the great military orders in the twelfth century.

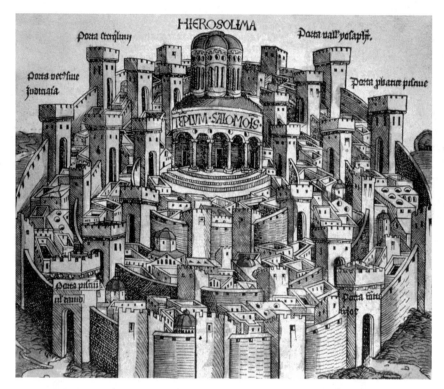

Jerusalem and the Temple of Solomon, woodcut by Melchior Wolgemuth, 1493.
© The Art Archive / Bibliothèque Mazarine, Paris / Marc Charmet.

HISTORY

The ancient Hebrews built a series of great temples in Jerusalem to provide a focal point for their religion. The first was erected in a grand, intricately ornamented style by King Solomon. It was intended to serve as a permanent center of worship, replacing the movable shrine known as the tabernacle. Solomon's Temple was destroyed around 586 B.C. by the Babylonians, and the ark was lost. Some seven decades later the temple was rebuilt, in a simpler manner and without the ark. Beginning around 20 B.C., this "Second Temple" was refurbished and expanded by Herod the Great. Herod's edifice was the temple visited by Jesus in the New Testament. It was destroyed by the Romans in A.D. 70.

The Bible is the principal source of information about Solomon's Temple, of which not a trace remains. The document reproduced here contains a description of the temple's construction. The biblical account suggests that the project was years in the making, beginning with preparatory work under Solomon's father, King David. David wanted to create a grand house for the Lord but was prevented from building the temple by ongoing wars, according to the story presented in 1 Kings. (A slightly different slant is given in 1 Chronicles 22:8: David shed so much blood in his military campaigns that

God forbade him to build the temple.) David did, however, gather substantial materials for the construction job, including gold, silver, stone, iron, brass, and cedar trees, and specifically instructed Solomon to erect the temple, a design for which David supplied.

The structure was built on Mount Moriah—the Temple Mount, now the site of the Dome of the Rock mosque. As the text reproduced here reports, construction began "in the fourth year of Solomon's reign" (1 Kings 6:1). Solomon, who launched a number of major construction projects involving forced labor during his reign, assigned tens of thousands of his people to the job. Hiram, the king of Tyre in Phoenicia, or Lebanon, helped provide materials, and an important part of the labor force on the temple project consisted of skilled Phoenician workers. Solomon paid Hiram 20,000 cors of wheat and 20 cors of fine oil (a cor was about 230 liters, equivalent to 60.7 gallons in liquid measure or 6.5 bushels in dry measure). Solomon also gave Hiram 20 cities to help compensate for all the cedar trees, fir trees, and gold supplied by Tyre.

The tabernacle is said to have served as the model for the considerably larger temple. The temple proper, a rectangular structure of hewn stone, was 20 cubits wide, 60 cubits long, and 30 cubits high. It did not need to be enormous to fulfill its function of housing the Ark of the Covenant, which was kept in a small inner sanctuary, or "holy of holies," measuring 20 cubits on each side. People assembled in two large courtyards surrounding the temple. The inner court is said to have measured 200 cubits by 100 cubits; within it were a bronze altar and a huge bronze basin, or *sea*, 10 cubits wide and 5 cubits high, which held water for ablutions. The inner court was contained within a much larger "great court."

CULTURAL CONTEXT

Solomon's Temple is unique among ancient temples in that it did not have idols. This is significant because in ancient times the common practice was to house idols in grand structures so that the spirits of the gods could enter and be present. However, the Hebrews believed (and still do) in one deity present everywhere who therefore could not be contained in an idol, so Solomon's Temple had no idols. But Solomon's Temple was not empty. Inside the temple was the Law, the agreement between the Hebrews and God (sometimes referred to as the Ark of the Covenant).

In building a temple, the Hebrews were following and extending the traditional Near East custom of building temples to hold the symbols of their civilization. The Phoenicians built temples. The Egyptians built structures to house arks or small chests for important documents. Thus the Ark of the Covenant can be seen as part of a regional tradition. In Greece, the earliest example of a temple is the Heraeum at Olympia, which was believed to have been erected in the tenth century B.C.

Cutting Down Cedars for Construction of the Temple by Gustave Dore. Courtesy of the Library of Congress.

In order to begin constructing the temple, Solomon needed a partner, and the Phoenicians were a natural choice. The Phoenicians had built temples and palaces at Tyre, and their king, Hiram, already had built a palace for King David, Solomon's father, who first thought of the temple. The close relationship between King Hiram of Tyre and King David and David's son Solomon may be one of the first records of a strategic construction partnership among nations. It may also be one of the first examples of the use of an outside expert contractor. And while the Hebrews lacked construction know-how and needed outside expertise, they did not need to import the labor to construct the building.

PLANNING

Shortly after construction of the temple was authorized, Solomon requested that Hiram supply him with "timber of cedar" and "timber of fir" (1 Kings 5:8). But how could they deliver these building supplies to the site? The easiest way was to float the lumber along the eastern Mediterranean coast from Lebanon to a designated pickup spot near the building site. Thus was devised one of the first materials-delivery plans in recorded history.

In an excellent example of seamless planning, the drop-off point was also the pickup location for the food supplies by which Solomon repaid Hiram for the timber. In 1 Kings 5:10–11 it is noted that when the Phoenicians delivered lumber, they picked up "twenty thousand measures of wheat . . . and twenty measures of pure oil," which Solomon gave to Hiram "year by year."

While Hiram was giving Solomon lumber, and Solomon was giving Hiram food, God was giving Solomon wisdom. The outcome was "peace between Hiram and Solomon, and they two made a league together" (1 Kings 5:12). This cooperation prolonged and deepened the peace between the two nations.

Although the Phoenicians had expertise and know-how, the Hebrews supplied labor via conscription: 30,000 men were divided into three troops of

An artist's rendition of the original temple. Courtesy of the Library of Congress.

10,000 each. The troops were rotated, each sent to Lebanon for one month on, two months off. When one troop came back, a fresh one was ready to begin. Among the builders, agreements were made regarding how many would carry heavy burdens, how many would hew trees in the mountains, and even how many managers were needed. This sensible and humane personnel plan produced not only rested, healthy workers but also workers who were not removed from their daily routines and jobs for long periods.

BUILDING

The construction phase was initiated by a go-ahead agreement. In 1 Kings 6:1–10, there is lengthy discussion of the building process. But verses 11–14 state that "the word of the Lord came to Solomon saying, concerning this house which thou art building, if thou wilt walk in my statutes, and execute my judgments, and keep all my commandments to walk in them . . . then I will perform my word unto thee, which I spake unto David thy father, I will dwell among the children of Israel and will not forsake my people Israel. So Solomon built the house and finished it." As in present times, it is common for large-scale projects to transition from a planning phase to a building phase with such a go-ahead agreement. Here is an early example of an approval agreement to authorize the commencement of building.

In the document authorizing the Temple of Solomon, there is one of the first known instances of prefabrication. In 1 Kings 6:7 it states, "And the house, when it was in building, was built of stone made ready before it was brought thither: so that there was neither hammer nor axe nor any tool of iron heard in the house, while it was in building." The intent was to avoid disturbing the sacred place with the noise of hammering, so the sections of the temple were prefabricated elsewhere and installed at the site.

The document presented here is filled with details about the construction. Why? Because the Bible or the Torah is not just a religious book but also a technical book. There was at that time no better repository of know-how, inasmuch as universities and engineering institutes did not yet exist. The Bible was the first attempt at an encyclopedic history covering military, religious, social, and technical events. Thus the know-how recorded in the Bible was passed on.

Ancient civilizations were themselves technological. We do not think of them as such because they seem low-tech to our modern eyes. But in those days much of the technology recorded in the Bible was the high tech of the time. So ancient civilizations were just as excited about their projects and structures as we are now with our technologies for exploring space or building high-speed trains.

In addition to its function as a place of public worship, the temple also served an important function as a fortress. By the time of Pompey's assault in 63 B.C., the temple had become almost impregnable, fortified on its weakest northern exposure by towers and a barricade ditch. Twenty-six years later, Herod too attacked from the north.

The history of Solomon's Temple must, of course, also include its rebuilding. In 586 B.C., the temple was burned to the ground, destroyed by order of King Nebuchadnezzar (2 Kings 20:8; Jer. 52:12 ff.). Seventy years later, a successor temple was finished and dedicated. In 165 B.C., three years after the destruction of the temple by Antiochus IV, Judas Maccabeas rededicated the holy house, ordered new interior furnishings, and erected a new altar (1 Macc. 4:41 ff.).

In the 18th year of his reign (20–19 B.C.), Herod obtained the reluctant consent of his subjects to rebuild the temple and improve its courts. It took 18 months and a thousand priests trained for this special purpose to rebuild the temple, and the courts surrounding the temple required another eight years. The full restoration lasted 80 years, until just before the break with Rome. That break was capped by the destruction of the temple yet again by the soldiers of Titus in A.D. 70.

IMPORTANCE IN HISTORY

The building of Solomon's Temple was a symbol of the transition of the Hebrews from a nomadic to a settled society. Biblical traditions state that houses of prayer were to be open for all nations: "for mine house shall be called an house of prayer for all people" (Isa. 56:7). There are very few instances of such explicit openness in the history of macro projects. It is significant that the Temple of Solomon was to be open to all.

The Temple of Solomon resulted from a series of innovations in design, construction, staffing, and management. Its achievement was the result of deliberate research and intellectual as well as administrative teamwork on an international level.

FOR FURTHER REFERENCE

Books and Articles

Garber, Paul Leslie. "A Reconstruction of Solomon's Temple." Cambridge, MA: Archaeological Institute of America, 1952. Stand-alone monograph reprinted from the original article published in *Archaeology* 5, no. 3 (1952): 165–72. Includes discussion of a model made by E. G. Howland.

Sagiv, Tuvia. "What if This Isn't the Western Wall." *Jerusalem Report,* November 23, 1998. About a possible third temple location.

Waldman, Neil. *The Two Brothers: A Legend of Jerusalem.* New York: Athenaeum Books for Young Readers, 1997.

Waterman, Leroy. "The Damaged 'Blueprints' of the Temple of Solomon." Chicago: University of Chicago Press, 1943. Zielinski, A. *Tajemnice polskich templariuszy* [The Polish Knights Templar]. Warsaw: Dom Wydawn, Bellona, 2003.

Internet

For information and links exploring Solomon's Temple, see http://www.templemount. org/.

For Phoenician influence on the design and building of the Temple of Solomon, including an bibliography, see http://phoenicia.org/temple.html.

Music

Norman, Jessye. "The Holy City," by Stephen Adams. On *Sanctus: Sacred Songs.* With the Ambrosia Singers and the Royal Philharmonic Orchestra, Sir Alexander Gibson, conductor. Philips 400019, 1992.

Documents of Authorization

THE BIBLE (KING JAMES VERSION)

1 Kings, Chapter 5

[1]   And Hiram king of Tyre sent his servants unto Solomon; for he had heard that they had anointed him king in the room of his father: for Hiram was ever a lover of David.

[2]   And Solomon sent to Hiram, saying,

[3]   Thou knowest how that David my father could not build an house unto the name of the LORD his God for the wars which were about him on every side, until the LORD put them under the soles of his feet.

[4]   But now the LORD my God hath given me rest on every side, so that there is neither adversary nor evil occurrent.

[5]   And, behold, I purpose to build an house unto the name of the LORD my God, as the LORD spake unto David my father, saying, Thy son, whom I will set upon thy throne in thy room, he shall build an house unto my name.

[6]   Now therefore command thou that they hew me cedar trees out of Lebanon; and my servants shall be with thy servants: and unto thee will I give hire for thy servants according to all that thou shalt appoint: for thou knowest that there is not among us any that can skill to hew timber like unto the Sidonians.

[7]   And it came to pass, when Hiram heard the words of Solomon, that he rejoiced greatly, and said, Blessed be the LORD this day, which hath given unto David a wise son over this great people.

[8]   And Hiram sent to Solomon, saying, I have considered the things which thou sentest to me for: and I will do all thy desire concerning timber of cedar, and concerning timber of fir.

[9]   My servants shall bring them down from Lebanon unto the sea: and I will convey them by sea in floats unto the place that thou shalt appoint me, and will cause them to be discharged there, and thou shalt receive them: and thou shalt accomplish my desire, in giving food for my household.

[10]   So Hiram gave Solomon cedar trees and fir trees according to all his desire.

[11]   And Solomon gave Hiram twenty thousand measures of wheat for food to his household, and twenty measures of pure oil: thus gave Solomon to Hiram year by year.

[12]   And the LORD gave Solomon wisdom, as he promised him: and there was peace between Hiram and Solomon; and they two made a league together.

[13]   And king Solomon raised a levy out of all Israel; and the levy was thirty thousand men.

[14]   And he sent them to Lebanon ten thousand a month by courses: a month they were in Lebanon, and two months at home: and Adoniram was over the levy.

[15]   And Solomon had threescore and ten thousand that bare burdens, and fourscore thousand hewers in the mountains;

[16]   Beside the chief of Solomon's officers which were over the work, three thousand and three hundred, which ruled over the people that wrought in the work.

[17]   And the king commanded, and they brought great stones, costly stones, and hewed stones, to lay the foundation of the house.

[18]   And Solomon's builders and Hiram's builders did hew them, and the stonesquarers: so they prepared timber and stones to build the house.

1 Kings, Chapter 6

[1]   And it came to pass in the four hundred and eightieth year after the children of Israel were come out of the land of Egypt, in the fourth year of Solomon's reign over Israel, in the month Zif, which is the second month, that he began to build the house of the LORD.

[2]   And the house which king Solomon built for the LORD, the length thereof was threescore cubits, and the breadth thereof twenty cubits, and the height thereof thirty cubits.

[3]   And the porch before the temple of the house, twenty cubits was the length thereof, according to the breadth of the house; and ten cubits was the breadth thereof before the house.

[4]   And for the house he made windows of narrow lights.

[5]   And against the wall of the house he built chambers round about, against the walls of the house round about, both of the temple and of the oracle: and he made chambers round about:

[6]   The nethermost chamber was five cubits broad, and the middle was six cubits broad, and the third was seven cubits broad: for without in the wall of the house he made narrowed rests round about, that the beams should not be fastened in the walls of the house.

[7]   And the house, when it was in building, was built of stone made ready before it was brought thither: so that there was neither hammer nor axe nor any tool of iron heard in the house, while it was in building.

[8]   The door for the middle chamber was in the right side of the house: and they went up with winding stairs into the middle chamber, and out of the middle into the third.

[9]   So he built the house, and finished it; and covered the house with beams and boards of cedar.

[10]   And then he built chambers against all the house, five cubits high: and they rested on the house with timber of cedar.

[11]   And the word of the LORD came to Solomon, saying,

[12]   Concerning this house which thou art in building, if thou wilt walk in my statutes, and execute my judgments, and keep all my commandments to walk in them; then will I perform my word with thee, which I spake unto David thy father:

[13]   And I will dwell among the children of Israel, and will not forsake my people Israel.

[14]   So Solomon built the house, and finished it.

[15]   And he built the walls of the house within with boards of cedar, both the floor of the house, and the walls of the ceiling: and he covered them on the inside with wood, and covered the floor of the house with planks of fir.

[16]   And he built twenty cubits on the sides of the house, both the floor and the walls with boards of cedar: he even built them for it within, even for the oracle, even for the most holy place.

[17]   And the house, that is, the temple before it, was forty cubits long.

[18]   And the cedar of the house within was carved with knobs and open flowers: all was cedar; there was no stone seen.

[19]   And the oracle he prepared in the house within, to set there the ark of the covenant of the LORD.

[20]   And the oracle in the forepart was twenty cubits in length, and twenty cubits in breadth, and twenty cubits in the height thereof: and he overlaid it with pure gold; and so covered the altar which was of cedar.

[21]   So Solomon overlaid the house within with pure gold: and he made a partition by the chains of gold before the oracle; and he overlaid it with gold.

[22]   And the whole house he overlaid with gold, until he had finished all the house: also the whole altar that was by the oracle he overlaid with gold.

[23]   And within the oracle he made two cherubims of olive tree, each ten cubits high.

[24]   And five cubits was the one wing of the cherub, and five cubits the other wing of the cherub: from the uttermost part of the one wing unto the uttermost part of the other were ten cubits.

[25]   And the other cherub was ten cubits: both the cherubims were of one measure and one size.

[26]   The height of the one cherub was ten cubits, and so was it of the other cherub.

[27]   And he set the cherubims within the inner house: and they stretched forth the wings of the cherubims, so that the wing of the one touched the one wall, and the wing of the other cherub touched the other wall; and their wings touched one another in the midst of the house.

[28]   And he overlaid the cherubims with gold.

[29]   And he carved all the walls of the house round about with carved figures of cherubims and palm trees and open flowers, within and without.

[30]   And the floor of the house he overlaid with gold, within and without.

[31]   And for the entering of the oracle he made doors of olive tree: the lintel and side posts were a fifth part of the wall.

[32]   The two doors also were of olive tree; and he carved upon them carvings of cherubims and palm trees and open flowers, and overlaid them with gold, and spread gold upon the cherubims, and upon the palm trees.

[33]   So also made he for the door of the temple posts of olive tree, a fourth part of the wall.

[34]   And the two doors were of fir tree: the two leaves of the one door were folding, and the two leaves of the other door were folding.

[35]   And he carved thereon cherubims and palm trees and open flowers: and covered them with gold fitted upon the carved work.

[36]   And he built the inner court with three rows of hewed stone, and a row of cedar beams.

[37]   In the fourth year was the foundation of the house of the LORD laid, in the month Zif:

[38]   And in the eleventh year, in the month Bul, which is the eighth month, was the house finished throughout all the parts thereof, and according to all the fashion of it. So was he seven years in building it.

From the Bible, King James version, 1 Kings, chapters 5 and 6.

2
The Founding of Cyrene

Libya

DID YOU KNOW . . . ?

➤ Cyrene, situated several miles south of the Mediterranean Sea in Libya, was founded in 630 B.C. following consultation with the Oracle at Delphi.
➤ Cyrene was the first of five flourishing cities known as the Pentapolis of Cyrenaica.
➤ Cyrene virtually disappeared around A.D. 630 and now exists only as ruins and the site of archaeological excavations near the village of Shahhat.
➤ It became a UNESCO World Heritage cultural site in 1987.
➤ Cyrene's constitution was based on demographic profiling.
➤ Eratosthenes, a Cyrenian mathematician, determined the earth's circumference.

Of the more than 1,000 colonies established by ancient Greece, Cyrene was outstanding in its fostering of an intellectual climate conducive to the discovery and encouragement of genius. It was in Cyrene that the earth's circumference was first determined. It was in Cyrene that number-theory research on the development of prime numbers evolved. Curiously, however, Cyrene's success as a new city owed much to the accident that its soil nurtured a popular and effective aphrodisiac. The mysterious herb silphium (or sylphium) grew only there, and was believed to enhance amatory powers. No wonder that soon in this land would develop Cyreniacs, followers of a philosophy that held the highest goal of life to be pleasure. No wonder the new city became so popular!

Temple of Zeus. Courtesy of Shutterstock.

HISTORY

The ancient city of Cyrene, situated several miles south of the Mediterranean Sea in Libya, is largely forgotten today, its site now the location of striking ruins, archaeological excavations, and the obscure village of Shahhat. In its heyday, however, Cyrene was a significant economic and cultural force.

The city was founded about 630 B.C. It was established by design to relieve a drought-caused food shortage that afflicted the residents of the Greek island of Thera, now called Santorini. According to an account of the origins of Cyrene by the Greek historian Herodotus, the Theran colonists were fulfilling a command by the Oracle of Apollo at Delphi to found a city in Libya. The settlers picked a spot in an eroded stream valley about 1,800 feet (550 meters) above sea level. The city was protected from desert winds by high ground on its southern flanks and grew around a spring where the colonists created a sanctuary to Apollo.

Under the dynasty inaugurated by Battos, the leader of the Theran colonists, Cyrene eventually prospered, founding additional towns in its region (Cyrenaica). Cyrene's fortunes took flight under its third king, Battos the Happy. Greeks came from every part of Libya to join the settlement, taking advantage of the Cyrenians' offer to share in their new land.

The Cyrenians agreed to accept Therans as citizens, in accordance with their ancestral custom. The Theran decree presented here arranged for the conscription of settlers who would establish the new colony. Any recalcitrants were subject to the death penalty. The Assembly allowed for the possibility of failure: if

Temple of Apollo. Courtesy of Shutterstock.

in five years the settlement failed to prosper, the colonizers would be permitted to return home to reclaim their citizenship rights in Thera.

After some two centuries under Battos and his successors, Cyrene continued for several decades as a republic before coming under the control of Alexander the Great and then, in 323 B.C., Ptolemaic Egypt. The city, which boasted a school of medicine and a vigorous intellectual life, enjoyed particular fame as a cultural nexus during the Egyptian period. Following several centuries of Roman rule, Cyrene declined, ravaged by earthquakes and foreign invasion, notably the Arab conquest of A.D. 642.

Among major figures who were born or lived in Cyrene were Aristippus (born ca. 435 B.C., died ca. 366 B.C.), the founder of the Cyreniac school of philosophy, which advanced Epicurean-like ideas; the distinguished poet and scholar Callimachus (born ca. 305 B.C., died ca. 240 B.C.); the famous geographer Eratosthenes (born ca. 276 B.C., died ca. 194 B.C.); and, much later, the Christian thinker and poet Synesius (born ca. A.D. 370, died ca. A.D. 413). The Greek lyric poet Pindar (born ca. 522 B.C., died ca. 438 B.C.) sang of the prowess shown by Cyrene's rulers in the Greek games.

Extensive archaeological excavations have revealed a portion of Cyrene's former grandeur. In addition to the temple of Apollo, other ruins that have been identified are the remains of the marketplace, the acropolis, the theater, and a huge temple of Zeus.

CULTURAL CONTEXT

Although Thera's move to Cyrene may have been occasioned by a drought (and an oracular decree), such moves were a sign of the times. And Greeks were not the only ones on the move. The Phoenicians sailed and mapped vast areas. One might compare the ancient colonizing activity of the region to that of the seventeenth and eighteenth centuries, when European countries began to expand globally.

A difference between the ancient Greek method of colonizing and the way the seventeenth-century Europeans chose to do so can be defined in one word: *ownership*. When the Dutch colonized, the new site became not only land of the company but land of the Dutch kingdom. The Greek mode of colonizing was to create independence, so when Syracuse or Cyrene became established by the Greeks, these colonies were independent. That factor of independence/dependence is tellingly depicted in the founding document of Cyrene presented here. While there is a pledge to support the colony for five years, no mention is made of taxes or duties owed.

PLANNING

Planning to found a new colony involves people, and the document presented here is clear about the methods and provisions used at that time. Citizenship was an issue—that is, to what land would the colonists belong? In the document of authorization, citizenship was guaranteed for the new residents of the colony of Cyrene who took the oath as Therans and were assigned to a tribe. Adding to the guarantee, the agreement was recorded on marble and placed in the shrine of Apollo, and also written on a tablet or *stele* that the colonists carried with them from Thera to Cyrene. So detailed is the document describing the citizenship agreements that it even notes who is going to pay for the stone and the engraving! These were not people to leave details to chance.

Once it was established that one's status as a citizen of Thera would not be lost or jeopardized, it may have been easier to recruit residents willing to make the journey. Apparently, it was not that easy, however, because the document also details the process of conscription—one son per family. In some accounts, the brothers of Theran families were said to have drawn lots.

The document authorizing the new city's founding guarantees, "If they do not successfully establish the settlement and the Therans are incapable of giving it assistance, and they are pressed by hardship for five years, from that land they shall depart, without fear, to Thera, to their own property, and they shall be citizens." So Cyrene may be one of the earliest documentations of insurance provisions legislated into the founding of a new enterprise. In this case, the anticipated failures were numerous: the settlement might be unsuccessful, Thera might run out of money to support the colony, life might prove to be too hard to continue the experiment. A testing period of five years' duration was a plausible amount of time that might reasonably be required to launch a new city.

While guarantees were offered, and positive inducements and protections were evident, there were also threats: "Any man who, if the city sends him, refuses to sail, will be liable to the death penalty and his property shall be confiscated." If the possibilities of the opportunity did not induce people to take the chance, the penalties for refusing were persuasive. Further strictures were equally convincing: "The man harboring him or concealing him [i.e., those who would not go although conscripted], whether he be a father aiding his son or a brother aiding his brother, is to suffer the same penalty as the man who refuses to sail."

The planning document describes a haunting image. The whole society came together—men, women, boys, girls—to make wax figures and burn them, saying, "The person who does not abide by this sworn agreement but transgresses it shall melt away and dissolve like the images—himself, his descendants and his property."

But threats are no way to induce inspired action, so the Therans concluded their send-off with the promise that "those who abide by the sworn agreement—those sailing to Libya and those staying in Thera—shall have an abundance of good things, both themselves and their descendants." With that promise, the new colony was launched.

BUILDING

In the planning phase, the Therans sailed only to Platea, an island off the coast of Libya. Finally they made the move to the mainland at Aziris, where they remained for six years, when the Libyans told them there were better places to establish themselves and offered to guide the way. Any traveler will warn: always be wary of guides who may have their own reasons for providing a particular route. The Libyan guides deliberately paced the journey so that the Therans missed the most beautiful district in the country, Irasa, because

Aristippus, the founder of the Cyreniac school of philosophy. Courtesy of the Library of Congress.

that area was concealed by the cloak of night, thus bringing to pass the night march, another interesting factor in the founding of Cyrene. Eventually, however, the Libyans led the settlers to an area where it was said that the "sky leaked." It turned out to be quite true. While the drought in Thera had prompted the move to Cyrene, the new land's moisture would become the fountainhead of their success, from crops of grain to healthy sheep, legendary horses, and the mysterious herb silphium.

Building flourished as well. As Cyrenaica, with its five cities, grew into a sprawling region called the Pentapolis (*penta*, five; *polis*, town), residences and commercial centers were constructed, and the city became a focal point for major public art. The Therans built a sanctuary and temple to Apollo, as well as a theater to the west of the sanctuary. They also built the largest Doric temple in Africa in the sixth century B.C., dedicated to Zeus, which is often compared to the Temple of Zeus at Olympia.

Not all the building was physical. A society was also built, with a social strata based on immigration history. When Battus III became ruler, he enlisted the aid of Demonax, a mainland Greek, to help write a new constitution. Demonax sorted the Cyrenaics into three classes of citizenship: those who hailed from Thera, those who came from Crete and the Peloponnese, and those who came from the islands in the Aegean. It might be one of the first evidences in history of demographic profiling.

In 321 B.C., Egypt occupied the region but eventually lost control to Rome in 96 B.C. During Roman times, building became a transformational activity that Romanized the culture and its public art. With so much water flowing from the land (just as the Oracle at Delphi had promised), the Romans built fountains and baths. Roman artists and sculptors fashioned statues that Romanized the previous edifices of the sanctuary and Temple of Apollo. The Romans also transformed the theater west of the sanctuary, enlarging it to hold gladiatorial fights. True to their engineering tradition, the Romans also built other infrastructure improvements.

But even those strong structures could not withstand the great floods that repeatedly engulfed the region throughout ancient and even modern times. In 1987, UNESCO gave Cyrene a World Heritage designation so the United Nations could intervene with a flood-protection plan that continues in effect to the present day.

IMPORTANCE IN HISTORY

The measurement of the earth's circumference, the mechanics of doubling a cube, research into prime numbers, the advancement of geometry, an early map of the stars (tallied at 675 at the time), and one of the first time lines of the history of the world—all of these advances would not exist today if not for Eratosthenes, who was born in Cyrene. In addition to his mathematical skills, Eratosthenes was trained by Callimachus, the second director of the famous Library of Alexandria, and in 240 B.C. Eratosthenes succeeded his tutor in this role. That legendary library contained scrolls, documents, papyrus, and all the extant knowledge of the time. The great poet Apollonia was also a resident of Cyrene. The inspiration of Apollo was so pervasive that Cyrene's harbor was named after the deity.

It could be said that the cultural climate of Cyrene was as important as (and perhaps even more important than) the agricultural bounty of the region. Great thinkers were born here, and others migrated to this center of thinking, philosophy, and learning.

FOR FURTHER REFERENCE

Books and Articles

Calame, Claude. *Myth and History in Ancient Greece: The Symbolic Creation of a Colony.* Translated by Daniel. W. Berman. Princeton, NJ: Princeton University Press, 2003. Original book published in French. An intermingling of history and myth, with six pages of bibliography.

Freeman, Kathleen. *Greek City-States.* New York: Norton, 1950. Discussion of Cyrene and the evolution of the Greek city-states; includes maps.

Hope, Valerie M., and Eireann Marshall, eds. *Death and Disease in the Ancient City.* London and New York: Routledge, 2000. This collection of essays includes a study of Cyrene, as well as studies of the towns and marshes of the ancient world, death and epidemic disease in Athens, Greek ideas about the city, and disease in the fifth century B.C.

Lipsey, Roger. *Have You Been to Delphi? Tales of the Ancient Oracle for Modern Minds Including an Afterword with Lobsang Lhalungpa on Tibetan Oracles.* Albany: State University of New York Press, 2001.

Parkins, Helen M., ed. *Roman Urbanism: Beyond the Consumer City.* London and New York: Routledge, 1997. Includes a chapter on Roman Cyrene by Eireann Marshall, as well as studies of mobility and social change and research on the first elite consumers.

Wood, Michael. *The Road to Delphi: The Life and Afterlife of Oracles.* New York: Farrar, Straus, and Giroux, 2003. Information about the Oracle at Delphi.

Internet

For Cyrene and cultural sites in the Libyan Arab Jamahiriya, see http://whc.unesco.org/sites/190.htm, and http://www.geocities.com/Athens/8744/unesco.htm.

For the Library of Congress Country Study on Libya, see http://workmall.com/wfb2001/libya/libya_history_cyrenaica_and_the_greeks.html.

For Herodotus, see http://www.fordham.edu/halsall/ancient/630cyrene.html.

For illustrations of Cyrene, see http://www.galenfrysinger.com/cyrene_libya.htm.

To see an image of silphium (engraved on an ancient coin), see http://www.livius.org/ct-cz/cyrenaica/cyrenaica.html.

Music

Brody, Martin. *Earth Studies Sound Recording.* Recorded live at the Duncan Theatre, January 12–13, 1996. Janice Felty, mezzo-soprano; William Hite, tenor; James

Maddalena, baritone; Demetrius Klein Dance Company; Mary Forcase, narrator. Sound cassette. Cambridge, MA: Harvard University, Loeb Music Record Coll., Call No. AC 33342. Track II Cyrene, Ode of Pindar, Greece, 5th Center BC Published by S.I.: s.n., 1996.

Documents of Authorization

God. Good Fortune. Damis, son of Bathykles, made the motion. As to what is said by the Therans, Kleudamas son of Euthykles, in order that the city may prosper and the People of Cyrene enjoy good fortune, the Therans shall be given the citizenship according to that ancestral custom which our forefathers established, both those who *founded* Cyrene from Thera and those at Thera who *remained*—just as Apollo granted Battos and the Therans who founded Cyrene good fortune if they abided by *the* sworn agreement which our ancestors concluded with them when they sent out the colony according to the command of Apollo *Archagetes.* With good fortune.

It has been resolved by the People that the Therans shall continue to enjoy equal citizenship in Cyrene in the same way (as of old). There shall be sworn by all Therans who are domiciled in Cyrene the same oath which the others once swore, and they shall be assigned to a tribe and a phratry and nine Hetaireiai. This decree shall be written on a stele of marble and placed on the ancestral shrine of Apollo Pythios; and that sworn agreement also shall be written down on the stele which was made by the colonists when they sailed to Libya *with* Battos from Thera to Cyrene.

As to the expenditure necessary for *the stone* or for the engraving, let the Superintendents of the Accounts *provide* it from Apollo's revenues.

THE SWORN AGREEMENT OF THE SETTLERS

Resolved by the Assembly: Since Apollo spontaneously told Battos and the Therans *to colonize* Cyrene, it has been decided by the Therans to send Battos off to *Libya,* as Archagetes *and* as King, with the Therans to sail *as his Companions.* On equal and fair terms shall they sail *according to family* ,with one son to be conscripted, and from the (other) Therans those who are free-born shall sail. If they (the colonists) establish the settlement, *kinsmen* who sail later to *Libya* shall be entitled to *citizenship* and offices and *shall be allotted* portions of the land *which has no owner.* But if they do not successfully establish the settlement and *the Therans* are incapable of giving it assistance, and they are pressed by hardship for five years, from that land *shall* they depart, without fear, to Thera, to their own property, and they shall be citizens. Any man who, if the city sends him, refuses to sail, will be liable to the death-penalty and his property shall be confiscated. The man harboring him or concealing him, whether he be a father (aiding his) son or a brother his brother, is to suffer the same penalty as the man who refuses to sail. On these conditions, a sworn agreement was made by those who stayed there and by those who sailed to found the colony, and they invoked curses against those transgressors who would not abide by it—whether they were those settling in Libya or those who remained. They made waxen images and burnt them, calling down (the following) curse, everyone having assembled together, men, women, boys, girls: "The person who does not abide by this sworn agreement but transgresses it shall melt away and dissolve like the images—himself, his descendants and his property; but those who abide by the sworn agreement—those sailing to Libya *and* (those) *staying* in Thera—shall have an abundance of good things, both *themselves* [and] *their descendents.*"

From: E. Badian and Robert Sherk, eds. *The Roman Empire: Augustus to Hadrian (Translated Documents of Greece and Rome).* (Cambridge: Cambridge University Press, 1988) 27, 40, 224–26.

The Roman Aqueduct at Segovia. Courtesy of Corbis.

3
The Aqueducts of Rome

Italy

DID YOU KNOW . . . ?

➤ The first Roman aqueduct was built in 313 B.C. by the army during peace-time; aqueducts continued to be built over the next five centuries.

➤ 11 aqueducts serving Rome supplied over 1.5 million cubic yards (1.1 million cubic meters) of water per day, or about 200 gallons (750 liters) per person per day. Compare this to the 1975 average per capita consumption of water in the United States of 150 gallons (563 liters) per day.

➤ Roman engineers and Greek and Syrian experts constructed the aqueducts using thousands of workers drawn from among official military and government workers, pickup laborers, and slaves who dragged heavy loads, sometimes using pulleys or even treadmills.

➤ At its peak, ancient Rome had a population of one million people, which would have been impossible without aqueducts.

➤ More fresh water was available to ancient Romans than to present-day New Yorkers.

➤ Without the aqueducts, there might not have been a Roman empire.

The water-supply and wastewater systems of ancient Rome were by far the best in antiquity. The Romans were realists, more interested in physics than in metaphysics. The aqueducts were one of their greatest achievements in physics. When Frontinus was the chief of water resources, Rome contained more than one million people. Thanks to the infrastructure of the aqueducts, Rome expanded and maintained a healthy population capable of developing and defending an empire.

HISTORY

The ancient Romans were not the first to use aboveground and underground conduits, or aqueducts, to transport water over long distances. But the systems constructed to supply Rome and cities under its sway with water were achievements in hydraulic engineering of a scope unprecedented at the time and unsurpassed for well over a millennium. The capital city's aqueducts, built over five centuries beginning with the Appia in 312 B.C., delivered water over distances approaching 60 miles (100 kilometers). According to one estimate, 11 aqueducts serving Rome around A.D. 300 supplied over 1.5 million cubic yards (1.1 million cubic meters) of water a day, or about 200 gallons (750 liters) per person. Sextus Julius Frontinus, a former governor of Britain who was named Rome's water commissioner in A.D. 97, wrote a classic two-volume treatise on the city's water supply system contrasting the "indispensable" aqueducts to "the idle Pyramids or the useless, though famous, works of the Greeks."

Reproduced here are two documents: (1) a portion of Frontinus's work elucidating the history of aqueduct construction at Rome up to his time, and (2) the

The Aqua Marcia. Courtesy of the Library of Congress.

extant portions of an edict issued by Emperor Augustus on the construction and maintenance of the aqueduct at Venafrum (today's Venafro).

According to Frontinus, the Appian aqueduct was 11,190 paces long (a pace was 5 Roman feet, or about 1.5 meters), and at some points the arches of the New Anio aqueduct rose as high as 109 Roman feet (more than 32 meters).

CULTURAL CONTEXT

The world's earliest cultures can be identified by their crucial ties to water. In the Middle East, all the successful civilizations lay between the Tigris and Euphrates Rivers. In Asia, most major settlements are adjacent to a good water supply. In later history, capitals were moved for water. The selection of Brasília as the capital of Brazil was based entirely on its easy access to water.

Rome, too, was founded near a great river. But then why did the Romans build aqueducts? Rome was not the first to do so. The Greeks had earlier built aqueducts, and the Egyptians had explored hydraulics. But Rome perfected the process.

As is often the case with engineering advancements, it was failure that produced success. Even before the aqueducts were built, ancient Rome depended on the Tiber River, which, even at its best, was muddy. And when Rome built its first sewer, the Cloaca Maxima, all the refuse, garbage, and human waste was dumped into the Tiber. As one can imagine, it was not long before the need for clean water became urgent. That need crested during the Second Samnite War, when it was feared that the water supply was in danger of being poisoned.

These factors—inadequacy compounded by endangerment of the present water supply—forced the engineers of Rome to get to work fast. The first aqueduct was a simple underground channel. The technology quickly improved, resulting in 11 major systems piping water into Rome and its surrounding areas to service agriculture and human needs for drinking water. In addition, lavish public baths and 20-person latrines with heated marble appointments abounded, as well as 1,000 fountains throughout the city; even huge arenas were temporarily flooded to accommodate mock naval battles that rivaled the gladiatorial contests. One could say that water enabled Rome to be a city of both bread and circuses.

How much of the fabled water of the spring-fed Aqua Marcia—the favorite aqueduct of many Romans, who spoke of its nectar the way oenophiles speak of vintage wine—reached the masses? Not much, according to Juvenal, a Roman satirist who wrote at the same time as Frontinus. Juvenal described Rome as a city crowded with tenements; very few people could afford to live in a single-family *domus* with private plumbing. Walking down the street could be hazardous, according to Juvenal, who cautioned fair readers to watch where they stepped lest waste from chamber pots that had been emptied into the street meet their feet, or to watch their heads because the same refuse might be tossed into the street by those who lived on the upper floors of the tenements.

PLANNING

Knowing where to dig was an art, not a science. Springs could be found by walking into the fields and looking for lakes, of course. One can imagine the surveyor, or *librator* (in Latin), following the advice of Vitruvius, an engineer who wrote 10 books on architecture in which are extensive chapters on assaying wet ground and springs. Vitruvius advised that water might be found if the observer laid down on the ground and peered forward to see any evidence of water gently rising in the mist. Vitruvius even suggests the most comfortable position—chin resting on hands. Another method was to dig a hole and place a bowl of bronze in it; in the morning, if moisture had condensed in the bowl, prospects were good. The land also provided hints by the kind of vegetation that grew there, so surveyors became horticulturalists as well.

However, not just any water was desired (although less pure sources were tapped via aqueducts for the purpose of irrigation); the water had to be wholesome and redolent with properties of health. That could best be found, claimed Vitruvius, by looking at the complexions of the folk living near the fields. Clear skin suggested clear water. A final method was to simply ask the locals; the Aqua Virgo was named after a little girl who told water surveyors where the ground was often damp in her village.

But too damp also meant trouble. Marshy swamps, although filled with water, were the wrong places for springs, and they had to be drained to prevent disease. Romans had already drained the Pontine marshes before 600 B.C. The Roman conquerors of the Volsci failed to pay due attention to the drainage canals, resulting in the resurgence of malaria. Several emperors tried to reverse the situation, which was not finally resolved until 1931 under Mussolini.

BUILDING

Roman engineers, with the help of Greek and Syrian experts, constructed the water systems using masses of workers drawn from among official military and government workers, pickup laborers, and slaves who dragged heavy loads, sometimes using pulleys or even treadmills. One chore was to deal with the mud in the Tiber—the primary reason the aqueducts and clean water were needed in the first place. Layers of walls, designed by Roman constructors, held the Tiber's soft sides in place and reduced the amount of mud.

Having cleaned up the Tiber, the Romans moved on to the problem of bringing the new spring water into the city. The *librator* had to look for a good route from the source to the destination, preferably sloping gently downward. Once identified, the surveyor marked the route with wooden stakes. Specialized tools were needed, such as the *groma*, to determine angles and crossroad measurements. The source and destination points had to be measured and calculated using a *dioptra* to take readings. The *dioptra* had a limited scope, so measurements had to be taken every 120 feet (37 meters). When the length was determined and the elevation and slope calculated, the surveyor could estimate the rate of falling of the water, and the route was thus finalized.

Work camps were established up and down the construction route so crews could build simultaneously; small camps of men dotted the landscape, cooking, living, and working along the length of the planned aqueduct. When the route brought an aqueduct through a town, it gave unemployed locals an opportunity to join the work crews.

The work camps enabled an important part of the aqueduct construction to be tested along the route to make sure the path of the water would be secure. Every 60 feet (19 meters), shafts were built, usually made of stone to withstand the force of the water. Just as with the banking of the Tiber, workmen and slaves, sometimes with the aid of draft animals, pulled heavy boulders. If the banks turned out to be too soft or marshy, wooden supports were added. But if the shafts had to be placed in a stony section, burly hewers were employed. The whole process was inspected by the surveyor who used an instrument called a *chorobate* to check the slope. Sometimes the flow of water called for an arch (a shape the Romans

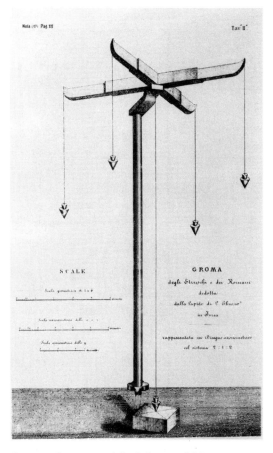

Groma. Courtesy of the Library of Congress.

perfected). The arches, although strong, were the weakest part of the construction and later often needed repair or rework. Once the shaft, the supports, and any arches were built, the final step was to line the shaft with concrete. Vitruvius gives the recipe: 1 part chalk + 2−3 parts sand + 20% water. But to make really good concrete the Roman way, it was recommended that the sand be volcanic (pozzolanic), which was known for its superior strength.

Water created a veritable industry in Rome. It is likely that the Roman army itself had an engineering corps. It is certain that there were water officials, of whom Frontinus is the most famous. There were surveyors, who must have been trained and licensed. Instruments were made, and concrete and other building materials produced. The work crews that gathered in settlements soon generated local commerce. There were even lawyers who specialized in water issues, because water was so valuable that theft was common.

The legal and illegal pathways of water offer a fascinating look at human nature, which seems relatively unchanged from the days of the toga to the present. When Frontinus took office, he was shocked to discover how many people had persuaded or even bribed those in authority to look the other way as they tapped the pipes and brought water into their own homes. Legal experts were required to adjudicate easements and rights of way. Lawyers may have handled personal-injury claims when pots of sewage, which Juvenal satirizes, landed on the heads of people strolling by the *insulae* or tenement buildings. While Rome's water system may have been a success, there were enough failures to spawn a parallel industry of legal intervention.

Frontinus revealed the extent to which water had become a true industry, one over which he ruled. In contrast, Augustus's edict (reproduced as the second Document of Authorization) tells us about the darker side of the water industry—the failures that required repair, the easements needed when an aqueduct crossed a landowner's property, the lawsuits. In section 5 of the edict, Augustus discusses the sale of water, decreeing that water "shall not be diverted in any other way than by *lead* pipes up to 50 feet from the water main." As we now know, lead is dangerous to humans. An archaeological dig at the end of the nineteenth century found a lead pipe that had been used to convey water to the Roman Forum. The 5,742 feet (1,750 meter)-long pipe had 513,130 pounds (232,752 kilograms) of lead. And there were thousands of such pipes. So, can we assume that Rome fell as a result of the presence of lead? Surprisingly not, because Roman water contained a high percentage of calcium, the deposits of which coated the lead pipes and probably prevented lead from leaking into the water supply. Lead seeps out in standing water, but flowing water retards the leakage. Confirmation of low levels of lead in the skeletal remains of ancient Romans (lead being traceable in bones) is now being verified by scientific tests.

While lead may not have been a problem with the Roman aqueducts, fire was. In a city with so much water, one would assume that fires were quickly extinguished. Strangely, however, water was not commonly found in homes; most residents used their homes for sleeping and went out to bathe and draw water. So fires occurred frequently in ancient Rome. Juvenal talks about the dangers of fire, especially in the tenements. During Nero's reign, whole sections of Rome were destroyed by fire.

The zenith of water entertainment had to be the *naumachia*, staged naval battles on artificial lakes. The *naumachia* were public celebrations, similar to gladiatorial fights, where the victor was sometimes freed. When Domitian was emperor, he ordered special piping laid under the floor of the Coliseum, then ordered the entire arena flooded for a *naumachia*. But even the Coliseum proved too small for Domitian's hobby, so he dug an artificial lake near the Tiber and surrounded it with seats from which he could view the elaborate naval staging.

IMPORTANCE IN HISTORY

In their time, the aqueducts were part of the most technically advanced water-delivery system the world had ever seen. And because the availability of water is critical to the advance and expansion of any human settlement, the early achievement of the Romans continues to inspire innovation.

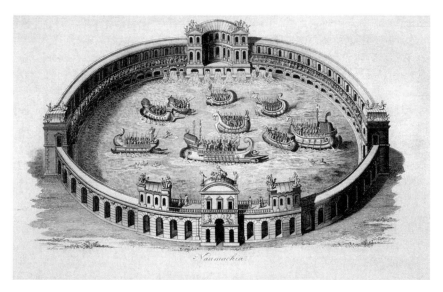

Copper plate etching showing a Roman arena with *naumachia*, a simulated naval engagement staged as an exercise or for entertainment. © Time Life Pictures / Mansell / Time Life Pictures / Getty Images.

The Imperial Valley in California evolved as the result of a brilliant legal system devised by the Chaffey brothers, who in 1900 figured out a way to (1) guarantee farmers in California desert land that they would have reliable supplies of water, and (2) assure the government that public rights would be fully respected. The result became the greatest agricultural breadbasket of North America.

The decision to relocate the capital city of Brazil was made almost entirely on the evidence of a good water supply, when the capital was moved from Rio de Janeiro to Brasília.

In a period when security considerations are front and center, it is relevant to recall that the building of the first Roman aqueducts was, in part, a response to the threat of poisoning a river water supply that could not be placed under adequate surveillance.

It has been suggested that fresh water running to the Atlantic from the Amazon River be transported across the south Atlantic in a plastic conduit to irrigate and develop the Sahara Desert. It might, however, be economical—and more logical—to exploit, for this purpose, the under-utilized river and lake resources of Africa itself.

FOR FURTHER REFERENCE

Books and Articles

Blackman, Deane R., and A. Trevor Hodge, eds. *Frontinus' Legacy: Essays on Frontinus' De Aquis Urbis Romae*. Ann Arbor: University of Michigan Press, 2001.

The California Water Atlas. Sacramento: State of California, 1978.

Durant, Will, and Ariel Durant. "Cesar and Christ." Chap. 15 in *The Story of Civilization.* New York: Simon and Schuster, 1944.

Evans, Harry B. *Water Distribution in Ancient Rome: The Evidence of Frontinus.* Ann Arbor: University of Michigan Press, 1994. Frontinus, Sextus Julius. *Frontinus: De Aquaeductu Urbis Romae.* Edited by R. H. Rodgers. Cambridge: Cambridge University Press, 2004.

———. *Strategems and the Aqueducts of Rome.* Translated by Charles Bennett. Cambridge, MA: Harvard University Press, 1961.

Hansen, R. D. "Water and Waste Water in Imperial Rome." *Water Resources Bulletin* 19 (1983): 263–69. Also available online at http://www.waterhistory.org/histories/rome/.

Hodge, A. T. "Lead Pipes and Lead Poisoning." AJA 85 (1981): 486–91.

———. *Roman Aqueducts and Water Supply.* London: Duckworth, 2002.

Juvenal, Decimus Julius. *Satire.* Translated by Jerome Mazzaro. Ann Arbor: University of Michigan Press, 1965.

Shaeffer, John R., and Stevens, Leonard A. *Future Water.* New York: Morrow & Co., 1983.

Internet

For a brief overview, see http://www.dl.ket.org/latin3/mores/aqua/.

For the use of water in times of war, and as a regional water supply, see the Raker Bill of 1913 at http://www.sfmuseum.org/hetch/hetchy10.html.

On the Roman army and its engineering corps, see http://www.roman-empire.net.

Film and Television

Secrets of Lost Empires: Roman Baths. (TV program, 2/22/02) Boston: WGBH TV. This WGBH Nova series, and an interview with Peter Aicher, classicist and consultant to Nova, can be found, on PBS's Web site, which includes a section called "Construct an Aqueduct" by Dennis Gaffney. See http://www.pbs.org/wgbh/nova/lostempires/roman.

Music

Respighi, Ottorino. *Fountains of Rome.* Conducted by Malcolm Sargent and Eugene Goossous. Everest Records CD #9018, 1995.

Documents of Authorization—I

For 441 years from the foundation of the City, the Romans were satisfied with the use of such waters as they drew from the Tiber, from wells, or from springs. Esteem for springs still continues, and is observed with veneration. They are believed to bring healing to the sick, as, for example, the springs of the Camenae, of Apollo, and of Juturna. But there now run into the City: the Appian aqueduct, Old Anio, Marcia, Tepula, Julia, Virgo, Alsietina, which is also called Augusta, Claudia, New Anio.

In the consulship of Marcus Valerius Maximus and Publius Decius Mus, in the 30th year after the beginning of the Samnite War, the Appian aqueduct was brought into the City by Appius Claudius Crassus, the Censor, who afterwards received the surname of "the Blind," the same man who had charge of constructing the Appian Way from the Porta Capena as far as the City of Capua. As colleague in the censorship Appius had Gaius Plautius, to whom was given the name of "the Hunter" for having discovered the springs of this water. But since Plautius resigned the censorship within a year and six months, under the mistaken impression that his colleague would do the same, the honor of giving his name to the aqueduct fell to Appius alone, who, by various subterfuges, is reported to have extended the term of his consulship, until he should complete both the Way and this aqueduct. The intake of the Appia is on the Lucullan estate, between the seventh and eighth milestones, on the Praenestine Way, on a crossroad, 780 paces to the left. From its intake to the Salinae at the Porta Trigemina, its channel has a length of 11,190 paces, of which 11,130 paces run underground, while above ground 60 paces are carried on substructures and, near the Porta Capena, on arches. Near Spes Vetus, on the edge of the Torquatian and Epaphroditian Gardens, there joins it a branch of Augusta, added by Augustus as a supplementary supply. This branch has its intake at the sixth milestone, on the Praenestine Way, on a crossroad, 980 paces to the left, near the Collatian Way. Its course, by underground channel, extends to 6,380 paces before reaching the Twins. The distribution of Appia begins at the foot of the Publician Ascent, near the Porta Trigemina, at the place designated as the Salinae.

Forty years after Appia was brought in, in the 481st year from the founding of the City, Manius Curius Dentatus, who held the censorship with Lucius Papirius Cursor, contracted to have the waters of what is now called Old Anio brought into the City, with the proceeds of the booty captured from Pyrrhus. This was in the second consulship of Spurius Carvilius and Lucius Papirius. Then two years later the question of completing the aqueduct was discussed in the Senate on the motion of the praetor. At the close of the discussion, Curius, who had let the original contract, and Fulvius Flaccus were appointed by decree of the Senate as a board of two to bring in the water. Within five days of the time he had been appointed, one of the two commissioners, Curius, died; thus the credit of achieving the work rested with Flaccus. The intake of Old Anio is above Tibur at the 20th milestone outside the Gate, where it gives a part of its water to supply the Tiburtines. Owing to the exigence of elevation, its conduit has a length of 43,000 paces. Of this, the channel runs underground for 42,779 paces, while there are above ground substructures for 221 paces.

One hundred and twenty-seven years later, that is in the 608th year from the founding of the City, in the consulship of Servius Sulpicius Galba and Lucius Aurelius Cotta, when the conduits of Appia and Old Anio had become leaky by reason of age, and water was also being diverted from them unlawfully by individuals, the Senate commissioned Marcius, who at that time administered the law as praetor between citizens, to reclaim and repair these conduits; and since the growth of the City was seen to demand a more bountiful supply of water, the same man was charged by the Senate to bring into the City other waters so far as he could. He restored the old channels and brought in a third supply, more wholesome than these, which is called Marcia after the man who introduced it. We read in Fenestella, that 180,000,000 sesterces were granted to Marcius for these works, and since the term of his praetorship was not sufficient for the completion of the enterprise, it was extended for a second year. At that time the Decemvirs, on consulting the Sibylline Books for another purpose, are said to have discovered that it was not right for the Marcian water, or rather the Anio (for tradition more regularly mentions this) to be brought to the Capitol. The matter is said to have been debated in the Senate, in the consulship of Appius Claudius and Quintus Caecilius, Marcus Lepidus acting as spokesman for the Board of Decemvirs; and three years later the matter is said to have been brought up again by Lucius Lentulus, in the consulship of Gaius Laelius and Quintus Servilius, but on both occasions the influence of Marcius Rex carried the day; and thus the water was brought to the Capitol. The intake of Marcia is at the 36th milestone on the Valerian Way, on a crossroad, 3 miles to the right as you come from Rome. But on the Sublacensian Way, which was first paved under the Emperor Nero, at the 38th milestone, within 200 paces to the left [a view of its source may be seen]. Its waters stand like a tranquil pool, of deep green hue. Its conduit has a length, from the intake to the City, of 61,710 1/2 paces; 54,247 1/2 paces of underground conduit; 7,463 paces on structures above ground, of which, at some distance from the City, in several places where it crosses valleys, there are 463 paces on arches; nearer the City, beginning at the seventh milestone, 528 paces on substructures, and the remaining 6,472 paces on arches.

The Censors, Gnaeus Servilius Caepio and Lucius Cassius Longinus, called Ravilla, in the year 627 after the founding of the City, in the consulate of Marcus Plautus Hypsaeus and Marcus Fulvius Flaccus, had the water called Tepula brought to Rome and to the Capitol, from the estate of Lucullus, which some persons hold to belong to Tusculan territory. The intake of Tepula is at the tenth milestone on the Latin Way, near a crossroad, 2 miles to the right as you proceed from Rome. From that point it was conducted in its own channel to the City.

Later in the second consulate of the Emperor Caesar Augustus, when Lucius Volcatius was his colleague, in the year 719 after the foundation of the City, [Marcus] Agrippa, when able, after his first consulship, took another independent source of supply, at the 12th milestone from the City on the Latin Way, on a crossroad 2 miles to the right as you proceed from Rome, and also tapped Tepula. The name Julia was given to the new aqueduct by its builder, but since the waters were again divided for distribution, the name Tepula remained. The conduit of Julia has a length of 15,426 1/2 paces; 7,000 paces on masonry above ground, of which 528 paces next the City, beginning at the seventh milestone, are on substructures, the other 6,472 paces being on arches. Past the intake of Julia flows a brook, which is called Crabra. Agrippa refrained from taking in this brook either because he had condemned it, or because he thought it ought to be left to the proprietors at Tusculum, for this is the water which all the estates of that district receive in turn, dealt out to them on regular days and in regular quantities. But our watermen, failing to practice the same restraint, have always claimed a part of it to supplement Julia, not, however, thus increasing the actual flow of Julia, since they habitually exhausted it by diverting its waters for their own profit. I therefore shut off the Crabra brook and at the Emperor's command restored it entirely to the Tusculan proprietors, who now, possibly not without surprise, take its waters, without knowing to what cause to ascribe the unusual abundance. The Julian aqueduct, on the other hand, by reason of the destruction of the branch pipes through which it was secretly plundered, has maintained its normal quantity even in times of most extraordinary drought. In the same year, Agrippa repaired the conduits of Appia, Old Anio, and Marcia, which had almost worn out, and with unique forethought provided the City with a large number of fountains.

The same man, after his own third consulship, in the consulship of Gaius Sentius and Quintus Lucretius, twelve years after he had constructed the Julian aqueduct, also brought Virgo to Rome, taking it from the estate of Lucullus. We learn that June 9 was the day that it first began to flow in the City. It was called Virgo, because a young girl pointed out certain springs to some soldiers hunting for water, and when they followed these up and dug, they found a copious supply. A small temple, situated near the spring, contains a painting which illustrates this origin of the aqueduct. The intake of Virgo is on the Collatian Way at the eighth milestone, in a marshy spot, surrounded by a concrete enclosure for the purpose of confining the gushing waters. Its volume is augmented by several tributaries. Its length is 14,105 paces. For 12,865 paces of this distance it is carried in an underground channel, for 1,240 paces above ground. Of these 1,240 paces, it is carried for 540 paces on substructures at various points, and for 700 paces on arches. The underground conduits of the tributaries measure 1,405 paces.

I fail to see what motive induced Augustus, a most sagacious sovereign, to bring in the Alsietinian water, also called Augusta. For this has nothing to commend it,—is in fact positively unwholesome, and for that reason is nowhere delivered for consumption by the people. It may have been that when Augustus began the construction of his Naumachia, he brought this water in a special conduit, in order not to encroach on the existing supply of wholesome water, and then granted the surplus of the Naumachia to the adjacent gardens and to private users for irrigation. It is customary, however, in the district across the Tiber, in an emergency, whenever the bridges are undergoing repairs and the water supply is cut off from this side of the river, to draw from Alsietina to maintain the flow of the public fountains. Its source is the Alsietinian Lake, at the 14th milestone, on the Claudian Way, on a crossroad, 6 miles and a half to the right. Its conduit has a length of 22,172 paces, with 358 paces on arches.

To supplement Marcia, whenever dry seasons required an additional supply, Augustus also, by an underground channel, brought to the conduit of Marcia another water of the same excellent quality, called Augusta from the name of its donor. Its source is beyond the springs of Marcia; its conduit, up to its junction with Marcia, measures 800 paces.

After these aqueducts, Gaius Caesar, the successor of Tiberius, in the second year of his reign, in the consulate of Marcus Aquila Julianus and Publius Nonius Asprenas, in the year 791 after the founding of the City, began two others, inasmuch as the seven then existing seemed insufficient to meet both the public needs and the luxurious private demands of the day. These works Claudius completed on the most magnificent scale, and dedicated in the consulship of Sulla and Titianus, on the 1st of August in the year 803 after the founding of the City. To the one water, which had its sources in the Caerulean and Curtian springs, was given the name Claudia. This is next to Marcia in excellence. The second began to be designated as New Anio, in order the more readily to distinguish by title the two Anios that had now begun to flow to the City. To the former Anio the name of "Old" was added.

The intake of Claudia is at the 38th milestone on the Sublacensian Way, on a crossroad, less than 300 paces to the left. The water comes from two very large and beautiful springs, the Caerulean, so designated from its appearance, and the Curtian. Claudia also receives the spring which is called Albudinus, which is of such excellence that, when Marcia, too, needs supplementing, this water answers the purpose so admirably that by its addition there is no change in Marcia's quality. The spring of Augusta was turned into Claudia, because it was plainly evident that Marcia was of sufficient volume by itself. But Augusta remained, nevertheless, a reserve supply to Marcia, the understanding being that Augusta should run into Claudia only when the conduit of Marcia would not carry it. Claudia's conduit has a length of 46,606 paces, of which 36,230 are in a subterranean channel, 10,176 on structures above ground; of these last there are at various points in the upper reaches 3,076 paces on arches; and near the City, beginning at the 17th milestone, 609 paces on substructures and 6,491 on arches.

The intake of New Anio is at the 42nd milestone on the Sublacensian Way, in the district of Simbruvium. The water is taken from the river, which, even without the effect of rainstorms, is muddy and discolored, because it has rich and cultivated fields adjoining it, and in consequence loose banks. For this reason, a settling reservoir was put in beyond the inlet of the aqueduct, in order that the water might settle there and clarify itself, between the river and the conduit. But even despite this precaution, the water reaches the City in a discolored condition whenever there are rains. It is joined by the Herculanean brook, which has its source on the same Way, at the 38th milestone, opposite the springs of Claudia, beyond the river and the highway. This is naturally very clear, but loses the charm of its purity by admixture with New Anio. The conduit of New Anio measures 58,700 paces, of which 49,300 are in an underground channel, 9,400 paces above ground on masonry; of these, at various points in the upper reaches are 2,300 paces on substructures or arches; while nearer the City, beginning at the seventh milestone, are 609 paces on substructures, 6,491 paces on arches. These are the highest arches, rising at certain points to 109 feet.

With such an array of indispensable structures carrying so many waters, compare, if you will, the idle Pyramids or the useless, though famous, works of the Greeks!

From Sextus Julius Frontinus, *The Aqueducts of Rome*, book 1, 7–8. In *Frontinus: The Stratagems and The Aqueducts of Rome*, with an English translation by Charles E. Bennett, the translation of the aqueducts being a revision of that of Clemens Hershell, edited by Mary B. McElwain. Cambridge, MA: Harvard University Press and London: William Heinemann, 1950, 339–47.

Documents of Authorization—II

EDICT OF AUGUSTUS ON THE AQUEDUCT AT VENAFRUM

1) Edict of Emperor Caesar Augustus. . . .

2) . . . in the name of the people of Venafrum . . . it shall be right and permissible. . . .

3) In regard to channels, conduits, sluices, and springs . . . have been made, built, or constructed above or below the water level for the purpose of building or repairing the aqueducts; or in regard to any other work which has been performed above or below the water level for the purpose of building or repairing the said aqueduct: it is ordered that whatever of the following operations have been done in the past are to continue in effect in the same manner, and workmen are to remake, to replace, to restore, or to repair in the same manner regardless of the number of times, and are to lay culverts and pipes of all sizes, to make openings therein, and to do any other work necessary to construct the aqueduct. There shall be no destruction or removal, however, of that wall or any part of that wall whereby any tract or field has been enclosed on the estate, which is, or is said to be, the property of Quintus Sirinius, son of Lucius, of the tribe Terentina, and on the estate which is, or is said to be, the property of Lucius Pompeius Sulla, son of Marcus, of the tribe Terentina, through which tract or under which tract the conduits of the said aqueduct pass, except as is necessary for the repair or the inspection of the conduit. No privately owned structures shall be constructed thereon, however, whereby passage, flow, or conduction of the water can be impeded. . . .

4) It is ordered that there shall be a cleared space of 8 feet vacant on the right and the left sides of this watercourse and around those structures built to carry the water; and that to the citizens of Venafrum or to a person acting in their name . . . there shall be granted a right of way through that space for the building and the repair of the said aqueduct or of the structures pertaining to it, provided that this right is exercised only in legitimate pursuits. And it shall be right and permissible that whatever materials are necessary for the said construction

or repair shall be conveyed, carried, or transported by the shortest route possible to the aqueduct, and, together with whatever is removed from the aqueduct, shall be dumped, as uniformly as is practicable, within the 8-foot space on the right and the left sides, provided that promise is made on oath that restitution will be made for damages likely to be inflicted through those activities. It is ordered that the colonists of Venafrum shall have jurisdiction and authority over all the said matters so constituted, provided that the owner of any property, field, or tract through which the said water ordinarily passes, flows, or is conducted is not himself denied passage over it because of the said activity; and provided that his power to pass, to convey, or to drive directly from one part of his property to another is unimpaired; and provided that no one through whose fields the said water is conducted is permitted to damage the aqueduct, to steal or to divert the water, or to do anything whereby the ability of the said water to flow or to be conducted directly into Venafrum is diminished.

5) In regard to the water which comes, flows, or is conducted into the town of Venafrum: it is ordered that authority and power to allot and to distribute the said water by sale, or to impose and to determine the fee therefor, shall be entrusted to the duumvir or the duumvirs of this colony placed in charge of this task by a decree of a majority of the decurions of the town, provided that not less than two-thirds of the decurions are present when the said decree is passed; and by the decree of the decurions, which has been passed in the manner as has been described above, he shall have the right and the authority to establish a regulation therefor. Furthermore, the said water, allotted and distributed in such manner, and concerning which such decree has been passed, shall not be diverted in any other way than by lead pipes up to 50 feet from the watermain; and the said pipes or mains shall be laid or located only under ground, and such ground shall have the legal status of a public road, way, or boundary; and the said water shall not be piped through private property without the consent of its owner. If the duumvirs, placed in charge by a decree of the decurions, which has been passed in the manner described above, establish any regulation for the protection of the said water and for the protection of the structures, which have been or shall be created . . . for the conduction or the distribution of the said water, the said regulation is ordered to be valid and confirmed. . . .

6) . . . as to a colonist or an inhabitant . . . the person to whom the matter has been entrusted, in accordance with a decree of the decurions, as has been explained above, when he brings suit he shall be granted a recuperatory action by the peregrine praetor to assess 10,000 sesterces for each case and he shall summon for the investigation no more than ten witnesses for each side, provided that the complainant and the defendant have the same right to reject recuperators, as will be their right and privilege by the terms of the law established to govern private suits.

4
The Grand Canal

China

DID YOU KNOW . . . ?

➤ The Grand Canal is the longest continuous construction project in history.
➤ At 1,118 miles (1,800 kilometers), the canal is the longest man-made river of the ancient world.
➤ The canal's elevation varies from sea level to 138 feet (42 meters) above sea level.
➤ It is estimated that more than five million people worked to build the canal, including many women.
➤ The canal made it possible to sustain Beijing as the capital of China.
➤ It was the first major multipurpose, large-scale infrastructure project in history.
➤ It had the first chambered lock in history (A.D. 983).
➤ Straightening the canal was Kublai Khan's greatest engineering feat.

The longest man-made river in the ancient world, the Grand Canal of China has been under intermittent construction since 600 B.C. The country's major rivers flow generally from west to east, and to connect the economic and agricultural centers of the south with the political and military centers in the north, a gigantic inland waterway—the Grand Canal, or Da Yunhe—took shape over a period of more than a millennium. The program announced in 2002 for a whole network of north–south canals on the model of the Grand Canal, which runs from Beijing in the north to Hangzhou in the south, will be the most costly engineering project in the history of the world.

Several barges travel along the Grand Canal near Suzhou, 2002. Courtesy of Shutterstock.

HISTORY

The Chinese have built canals since antiquity to control flooding and irrigate farmland, and also to provide transportation. "There are no undertakings of utility and invention for which the Chinese are more celebrated," observed the eighteenth-century British diplomat Lord Macartney, "than the numberless communications by water through the interior of their country" (Harrington, 12). The Grand Canal was the longest canal in the ancient world, a network of artificial waterways and canalized rivers and lakes, and today it remains the world's oldest major canal still in operation.

The Grand Canal's builders confronted huge engineering challenges, among them differences in elevation, varying from about sea level at Hangzhou to as high as 138 feet (42 meters), and the barrier presented by the Yellow (Huang) River, which the canal crosses en route. The Yellow River, traditionally called "China's sorrow," has made life miserable for millions of Chinese over the centuries; its floods, dramatic changes of course, and immense silt loads complicated building and maintaining the Grand Canal.

The first four imperial commands cited here relate to a spurt of building in the late sixth and early seventh centuries A.D., during the Sui dynasty. It involved the reconstruction and linking of segments built in the fifth century B.C. and also extended the canal south to Hangzhou and north to the Wei River and Tianjin. Major reconstruction, including rerouting, was authorized and initiated by Mongol emperors in the thirteenth and fourteenth centuries, particularly under Kublai Khan. A new effort was made by Kublai Khan and entrusted

The fifty-three-arch bridge spanning the Grand Canal (nearly 1,000 miles long) near Soo-chow, China, 1900. Courtesy of the Library of Congress.

to Water Resources Director Guo Shoujing, who used springs; today's Kunming Lake at the Summer Palace originated as a regulating reservoir for Guo's canal network, which, under the name Tonghui Canal (Channel of Communicating Grace), extended on to Tongzhou.

The ensuing Ming dynasty improved the canal, broadening and deepening some portions and adding locks and dams. Much of the canal fell into disrepair in the nineteenth and early twentieth centuries; a large-scale renovation and modernization program was carried out in the mid-twentieth century.

CULTURAL CONTEXT

Providing everyone with food and water was the key to maintaining peace in the Chinese empire, and that was the primary motive for building the Grand Canal. Drought-stricken northern China was always in need of irrigation and more potable water, while the south had plenty. The Grand Canal was constructed to connect these two regions.

There was another reason for building this massive inland waterway: to link the centralized government and administration of all of the provinces and districts of the vast country of China. The canal was possibly the first major multipurpose infrastructure project in history, helping to shape the development of China in many ways. It was a secure interior communications link guaranteeing the cohesiveness of a growing Chinese empire, facilitated flood control and irrigation, and helped assure the indispensable delivery of grain to the military and political centers in the north. It fostered the creation of a huge bureaucracy to administer and maintain the canal. It also brought a measure of national security because it offered a safe alternative to the hazardous seacoast route infested by pirates.

An eighteenth-century Chinese painting on silk of Emperor Yang Ti on his boat on the Grand Canal, China. © The Art Archive / Bibliothèque Nationale, Paris.

The availability of a common written language made it feasible for a Chinese ruler to plan and communicate instructions for the building of a canal system that would link regions that previously lacked a common written language. In 221 B.C. the first Quin emperor standardized writing in China. Before this, each area of China had its own peculiar writing system. This so-called Quin script became the official script and was required for all state and governmental affairs. It made possible the government of China.

PLANNING

Once an idea sprang up in the mind of an emperor of ancient China (or Son of Heaven, as emperors were called), there was little planning, and no requests for proposals or bidding on contracts. Action took place quickly following even a minor imperial utterance. When Emperor Yang Jian gave a verbal order to build a canal in A.D. 584, construction commenced immediately.

The Sui dynasty perceived the need and utility of connecting the canals. The motivation was power and food: the Sui emperors had established Loyang as the capital, and that required a connection from Loyang to the fertile areas surrounding the Yangtze and its valleys of grain. During the Sui era (A.D. 581–618), grain began to flow to northern China.

Human-resource arrangements were simple, as the order by Emperor Yang Guang (A.D. 608) reveals: "Recruit thousands of men and women from north of the Yellow River to build the Yongji Canal." The decree is evidence of imperial clout: it may be the first instance of women working on construction projects. Estimates of the number of people forced to work on the Sui's canal projects run as high as five million.

Part of planning involved measuring the land and terrain elevations and calculating anticipated changes in the water levels of the canals. It appears that the Chinese created the first lock, and it was a Chinese engineer, Chaio Wei-yo, who invented the chambered lock in A.D. 983. The Grand Canal was finally straightened, rebuilt, and completed in 1300 after Kublai Khan appointed the great Chinese mathematician Gou Shoujing as water-resources director.

BUILDING

Unlike Rome, which built its infrastructure with labor from its huge army, thereby keeping the troops busy during peacetime by improving the water systems and roads, the Chinese relied on the recruitment of peasant labor according to recommendations of experts who were appointed and trusted by the emperor. This was the style of a famous line of emperors beginning with the legendary Genghis Khan.

Genghis Khan's son Ogodei relied on his resident expert, Liu Conglu, who was responsible for renovating and repairing ancient sections of the Grand Canal. In the text presented here, the clock turns back to A.D. 1235, where we read a conversation between the boss and his project manager, Liu Conglu. The emperor relies so heavily on his expert engineer that even the decree quotes Liu Conglu as the reason for the action: "Liu Conglu suggested recently, 'I have led 200 skilled laborers following the schedule and finishing and renovating the loose and cracking parts of the Lugou Canal on time because without the renovation, the canal would fail to stand any flood tide' . . . If further renovation comes, the required labor and tools should be made ready for dispatch. You can keep 50 skilled laborers from the original 200 for emergency." This passage shows how human resources were trained and employed in units headed by hydraulic-engineering experts.

When Kublai Khan became emperor, he directed that grain be moved northward to his new capital of Dadu. The southern valleys of the Yangtze were lush and fertile, with an abundant harvest to send north. However, Kublai's initial success in commanding that grain be delivered to Dadu ultimately resulted in the failure that changed the shape of the Grand Canal.

His initial success? The emperor received 816,000 tons of grain annually, with the major share—537,000 tons—coming from the south. The failure part? When grain from the Yangtze region arrived by sea or by the canal, it still had to be transported about 20 miles (32 kilometers) to reach Dadu. The local draft animals, instead of helping plow the farmers' fields, were conscripted to haul grain, and many died along the route. To his credit, Kublai Khan turned this failure into a stunning success by ordering the completion of the seventh and final section of the Grand Canal, straightening the route to make it more efficient, and enabling all of China to communicate with the new center of political and cultural life in Dadu, which the world knows today as Beijing.

There is some evidence that Kublai's water-resources director, Gou Shoujing, was valued not only for what he had (knowledge of hydraulic engineering) but also for what he did not have. Unlike some of the court sycophants, Gou Shoujing was not a member of the old court. He had no political alliances, but instead was dedicated to improving water resources for the country. For these reasons Kublai Khan made him water-resources director. Soon grain barges were docking right outside the palace.

It was not just the emperors who prospered. Businesses near the canal prospered, trade expanded, and centers of economic activity sprang up along the route. Not only goods, but music, crafts, and culture were exchanged along this inland artery. It was the aquatic equivalent of the Silk Road.

IMPORTANCE IN HISTORY

An American engineer was inspired by the Grand Canal. In 1796, he wrote "The Treatise on the Improvement of Canal Navigation." In the essay, he said that China became a great nation primarily because of the Grand Canal. This engineer was such an admirer of the achievement that he lobbied to build the first canal in New York, perhaps inspiring the eventual building of the Erie Canal, before he died in 1815. But this engineer, Robert Fulton, is more famously known for building the first successful steamboat in 1807.

Unlike many ancient and medieval macro projects, the Grand Canal is still in daily use—unusual for a structure begun in 600 B.C. In 1958, 240,000 workers began dredging and widening the canal to create sufficient depth and width to accommodate ships of the megatransport category. The goal was that by the 1980s the Grand Canal could be used by ships of 1,000 tons.

China has applied to UNESCO for World Heritage status for the Grand Canal. China's director of state administration of cultural heritage, at the 28th session of the World Heritage Committee in Suchow, suggested that the Grand Canal, as the world's longest artificial river and the central north–south waterway of China, is a great engineering wonder and should be granted special status and preserved for the ages.

In 2002, China decided to reverse the flow of water in the Grand Canal by creating a south-to-north water diversion, its purpose being to bring water from the moist and agriculturally rich south to the more arid north. The Chinese Ministry of Water Resources announced the project, triggering a barrage of proposals from foreign and Chinese pump manufacturers and machinery specialists. The estimated cost for the first phase is $22 billion. The western route, the largest of the three planned, will be completed later for an estimated $36 billion. According to current calculations, experts judge this megaproject to be the largest of its kind ever planned and undertaken.

FOR FURTHER REFERENCE

Books and Articles

Ch'len, Ssu Ma (Kuang). *The Grand Scribe's Records: The Basic Annals of Pre-Han China.* Bloomington: Indiana University Press, 1994 (English language).

Delfs, R. "Arteries of the Empire." *Far Eastern Economic Review,* March 15, 1990, 28–29.

Harrington, Lyn. *The Grand Canal of China.* Chicago: Rand McNally, 1967.

Leonard, Jane Kate. *Controlling from Afar: The Dauguang Emperor's Management of the Grand Canal Crisis, 1824–1826.* Ann Arbor: Center for Chinese Studies, University of Michigan, 1996.

Needham, Joseph. *The Development of Iron and Steel Technology in China.* London: Newcomen Society for the Study of History of Engineering and Technology, Science Museum, 1958.

———. *Science and Civilisation in China.* Cambridge: Cambridge University Press, 1954.

New China News Ltd. and New China Pictures Co. *The Grand Canal of China.* Hong Kong: South China Morning Post, New China News, 1984.

Reischauer, Edwin O. *Ennin's Travels in T'ang China.* New York: Ronald Press Company, 1955.

Temple, Robert. *The Genius of China.* New York: Simon and Schuster, 1986. With an introduction by Joseph Needham.

Yao, Han-yüan. *The Grand Canal: An Odyssey.* Edited by Liao Pin. Beijing: Foreign Languages Press, 1987.

Internet

In an illustrated presentation of landmarks, this discussion calls the Grand Canal the "artificial Nile," which did for China what the real Nile did for Egypt; See http://library.thinkquest.org/20443/grandcanal.html.

Ever wondered who built the first canal in human history? The Egyptians, in 4000 B.C. For a time line of all the major canals from the beginnings to 1749, see http://home.eznet.net/~dminor/Canalto1749.html.

For more on the influence of the Grand Canal on America's development of canals, see http://members.tripod.com/~american_almanac/canal.htm.

For a brief illustrated guide to the Grand Canal, see http://www.chinapage.com/canal.html.

For information on Genghis Khan, who may have made a greater contribution than hitherto believed to the population of the Mongol domain, see the genetic study indicating that eight percent of the men living in that region now carry his chromosomes: http://news.nationalgeographic.com/news/2003/02/0214_030214_genghis.html.

For an excellent gallery of images of the Grand Canal, see http://www.altavista.com/image/results?pg=qdstype=simage&imgset=2&q=The%20Grand%20Canal%20of%20China&Avkw=aaps.

For a map of the Silk Road and more, see http://www.silkroadproject.org/.

Music

Ma, Yo Yo, and the Silk Road Ensemble. *Silk Road Journeys: When Strangers Meet* (Sony Classical 089782, 2002) and *The Silk Road: A Musical Caravan* (Smithsonian Folkways Recording SFW40438, 2002). While the Silk Road was an overland route from China across Asia to the Middle East and Europe, it might be possible that the inland waterway of the Grand Canal was also a route of exchange of the art, music, and culture of the times.

Documents of Authorization

A selection of imperial decrees from the Sui dynasty and the Mongol period relating to extension and renovation of the canal.

I. Sui Dynasty, founded by Yang Jian in 581

In 584, Emperor Yang Jian gave the order: "Build the canal from the Wei Shui in Daxing (Xi'an) to the Yellow River to enable grain transport on an inland waterway."

In 587, Emperor Yang Jian ordered: "Dig the canal of Shanyang Du in Yangzhou to create a waterway."

In 605, Emperor Yang Guang gave the decree: "Recruit thousands of men and women from every county in the south of the Yellow River to build the Tongji Canal, and lead the water of the Gu Shui and Luo Shui in Xi Yuan into the Yellow River, and lead the water of the Yellow River in Ban Zhu into the Huai."

In 608, Emperor Yang Guang ordered again: "Recruit thousands of men and women from all the counties in the north of the Yellow River to build the Yongji Canal, to reach the Yellow River in the south through the Qin Shui and to connect Zhu Chun in the north."

From Ruifang Liu, *The History of Chinese Emperors* ([Beijing ?]: National Defense University Publishing House, n.d.) 304–5.

II. Mongol rule and the Yuan Dynasty: Genghis (Chinggis) Khan was succeeded by his son Ögödei, referred to here as Yuan Taizong. Genghis Khan's grandson Kublai (Khubilai) was the Yuan emperor Hubilie.

In August 1235, Yuan Taizong gave the edict: "Liu Conglu suggested recently, 'I have led 200 skilled laborers following the schedule and finishing renovating the loose and cracking parts of the Lugou Canal on time because without the renovation, the canal would fail to stand any flood tide. Now I am asking your majesty to give an order to stop some selfish people from stealing water from the canal to irrigate their own fields.' I assign Liu Conglu to be in charge, and to make sure there is no water stealing. Violators will be considered criminals, sentenced to two years' imprisonment and punished with 70 heavy stick beatings. And if further renovation comes, the required labor and tools should be made ready for dispatch. You can keep 50 skilled laborers from the original 200 for emergency. Now that I have given the assignment, make sure to inquire and investigate the situation of the canal very often, and report annually. Any neglect of duty will be punished as crimes."

From Lian Song, comp., *The History of Yuan* (Shanghai: Chinese Book Bureau, 1976), 64: 1593.

In 1291, water conservancy director Guo Shoujing implemented Hubilie's imperial orders to undertake large-scale water conservancy projects because the emperor (Hubilie) accepted his suggestion: "Dredge Tongzhou to the Dadu River, lead the muddy water to irrigate, draw clear water from the old sluiceway, north from Shen spring in Baifu village of Changpin county, to turn southwest, passing the river of Shuang Ta, You He, Yi Mu, and You Chuan, to enter the capital through Xi Shui Gate. Waters in the south will be merged into a cistern or reservoir. The canal will run southeast through Wen Ming Gate, reach Tongzhou Gao Li Village in the east and go into the Bai River. Its total length will be 164 *li* and 140 *bu*. There will be 12 places of clear water, 310 in length, 20 sluice gates in ten different places. The canal will save water and provide a waterway, which will really benefit people."

The building of this canal started in the spring of 1292 and was completed in the fall of 1293, and it was granted by the emperor the name Tonghui.

From Lian Song, comp., *The History of Yuan* (Shanghai: Chinese Book Bureau, 1976), 64:1588–89.

The following are imperial decrees, selected from several dynasties, relating to extensions and preservation of the canal.

The Tang dynasty was long gone, and the Sung dynasty was withering, when a conqueror from Mongolia became the new Son of Heaven. Kublai Khan, now emperor of North China (Cathay), in 1260 chose a site for his capital. Khanbalik—the khan's city—had been the capital of a small kingdom, and would eventually be named Peking.

In years of fighting, the Mongol armies had pushed the Sung emperors into South China (Manchi). At the lower end of the Grand Canal, the Sungs had transformed a fishing village on the Chientang River into the lovely city of Hangchow.

Kublai Khan was full of plans and energy. South China had to be conquered by his generals, for the North needed its tribute grain. The granaries his troops had burned had to be rebuilt. The Grand Canal had to be restored to carry the grain north, and the roads built to move his armies quickly to suppress any rebellion.

Kublai was wary in choosing his officials from the foreigners at his court. But he trusted one Chinese hydrologist, Kuo Shou-ching, as his Minister of Water Conservation. Even his enemies admitted, "The craftsmen of Cathay are the most skillful by far, beyond those of any other nation."

Kublai's trust was rewarded. Kuo Shou-ching completely rebuilt the Grand Canal, rerouted the northern half, and extended it into the khan's new city. He channeled creeks through Khanbalik to provide water for the citizens and artificial lakes for the emperor. Kuo brought water from the Jade Spring at the north, through several small regulating lakes into a moat around the city walls. A conduit from the rampaging Yungting River on the west helped to tame that stream, and brought more water into the moat.

The moat drained east through Tunghui Canal, the final lap of the Grand Canal of China. Now grain barges could dock inside the imperial city.

From Lyn Harrington, *The Grand Canal of China* (New York: Rand McNally, 1967).

The work of combining the sections of the Grand Canal was done by the Sui dynasty (+581 to +618) when the need to link the capital at Loyang with the key economic area of the lower Yangtze valley became imperative. In the last decades of the +13th century, under the Yuan emperors, the same need continued, but as the capital was now Peking, a vast remodelling of the canal was carried out, so that it finally extended from Hangchow in the south to the furthest northern parts of the North China plain. To visualise this major work in its final stages, it is only necessary to remember that it covered 10 degrees of latitude, which would be comparable to a canal extending from New York to Florida. Its total length attained nearly 1,100 miles. Its summit, reached when skirting the mountains of Shantung, was some 138 ft. above sea level.

On Lu Kou canal—part of the Grand Canal, was built by the Yuan emperor, decades before Hu Pi Lieh.

In August 1235, Yuan T'ai Tsung gave the edict: "Ts'ung-lu Liu suggested recently 'I have led two hundreds of skilled labor following the schedule and finishing renovating the loose and cracking parts of the Lu Kou canal on time because without the renovation, the canal would fail to stand any flood tide. Now I am asking your majesty to give order to stop some selfish people from stealing water of the canal to irrigate their own fields.' I assign Ts'ung-lu Liu to be in charge, and to make sure of not water stealing. Violators will be considered criminals, sentenced 2 years imprisonment and punished with 70 heavy stick beatings. And if further renovation comes, the required labor and tools should be made ready for dispatch. You can keep 50 skilled labor from the original two hundreds for emergency. Now that I have given the assignment, make sure to inquire and investigate the situation of the canal very often, and report annually. Any neglect of duty will be punished as crimes."

From Lian Song, comp., *The History of Yuan* (Shanghai: Chinese Book Bureau, 1976), 64:1593.

On Thung Hui Canal

In 1291, water conservancy director Shou-ching Kuo implemented Hu Pi Lieh's imperial orders to undertake large-scale water conservancy projects because the emperor (Hu Pi Lieh) accepted his suggestion: "dredge Thung Chou to Ta Tu river, lead the muddy water to irrigate, draw clear water from the Old sluiceway, north from Shen Spring in Pai Fu village of Ch'ang P'in county, to turn southwest, passing the river of Shuang T'a, Yu He, Yi Mu and Yu Ch'uan, to enter the capital through Hsi Shui Men. Waters in the south will be merged into a cistern or reservoir. The canal will run southeast through Wen Ming Men, reach Thung Chou Kao Li Chuang in the east and go into Pai He river. Its total length will be 164 Li & 140 Pu. There will be 12 places of clear water, 310 in length, 20 sluice gates in 10 different places. The canal will save water and provide waterway, which will really benefit people." The building of this canal started in the Spring of 1292, completed in the Fall of 1293, and it was granted a name by the emperor Thung Hui.

From Lian Song comp., *The History of Yuan* (Shanghai: Chinese Book Bureau, 1976), 64:1588–89.

In 588, Emperor Yang Chien ordered in his imperial instruction: "we must listen to public opinions, select among the choices of common people and consider their interests. Thus, we can estimate the failure or success of our administration. Today, I want to go on a tour of inspection of the waterways in Huai and near the sea to get to know the customs and scenery there."

From Ssu Ma Kuang, comp., *Tzu-chi t'ung chien* [Comprehensive mirror for aid in government] (N.p.: Yueh Lu Books Association, n.d.), 180:342. See related book: Ssu Ma Ch'len (aka Kuang), *The Grand Scribe's Records: The Basic Annals of Pre-Han China.* Bloomington: Indiana University Press, 1994 (English language).

In 1273, Emperor Hu Pi-lieh ordered: "the director of the Agriculture Department assign some minister to teach people and persuade them to ensure the accomplishments in Agriculture."

In 1289, "Kuo Shou-Ching applied to build Yu Chhuan Shui to dredge the waterway and the canal of irrigation.

From Lian Song, comp., *The History of Yuan* (Shanghai: Chinese Book Bureau, 1976), 64:1589–90.

In the same year (1289), the Emperor (Hu Pi-lieh) approved it.

From Ruifang Liu, *The History of Chinese Emperors* ([Beijing ?]: National Defense University Publishing House, n.d.), 614.

5
Protective Dikes and Land Reclamation

The Netherlands

DID YOU KNOW . . . ?

➤ In the past some 1,243 miles (2,000 kilometers) of dikes and 218 miles (350 kilometers) of dunes protected the Netherlands.

➤ With modern technology the Netherlands shortened the flood-defense line from 1,056 miles (1,700 kilometers) to 196 miles (350 kilometers) by dikes, dams, and storm-surge barriers.

➤ More than nine million people are protected by the coastal defense, which safeguards 65 percent of the land surface.

➤ For more than 1,000 years, adult males living near the rivers and coast have been subject to instant call-up for service in the dike army.

➤ The Netherlands was frequently referred to as "a sinking country bordering the rising sea."

➤ Rembrandt was born in a windmill.

➤ Rotterdam was long the world's busiest seaport, and is now the second-busiest in the world.

One consequence of building protective dikes and dams was the founding and growth of population centers. Thus, centuries ago, Amsterdam—named for the dam built in the Amstel River—was the richest city on earth. Similarly, the dam in the River Rotte led to the settlement and development of Rotterdam, the world's second-busiest seaport. With a population of over 16 million, the Netherlands is the most densely populated country in Europe, with 472 inhabitants per square kilometer. It is also among the most prosperous: the 2003 average

The Maeslant storm surge barrier in the Rotterdam Waterway. Courtesy of Shutterstock.

per-capita income was $26,310. An enviable standard of living has been sustained despite a permanent and unavoidable state of war against the sea and intermittent overflows of the rivers Meuse, Rhine, and Scheldt. The thousand-year-old dike army can be regarded as an authentic progenitor of the concept of "an army enlisted against nature" (Winslow and Davidson, 193) first proposed in 1906 by American philosopher William James. The dike army remains an essential part of the day-to-day maintenance of the dikes by conscripts who are appointed by the water board.

HISTORY

In the first century A.D., Pliny the Elder saw and described the peoples who, for half a millennium, had inhabited much of the present terrain of the Netherlands. One hundred years before Pliny visited this country, inland farmers had begun to construct dams in the tidal creeks to protect their land against high water. To drain the area behind the dams, they used hollowed tree trunks as sluices, each with a valve that closed at higher tide levels. This sluice/valve principle worked so well that it is still in use today.

Pliny may have observed the Caninefates or Batavians who settled along the western coast; the Frisians lived in the northern portion along the Rhine and Meuse Rivers. He noticed that, to protect their dwellings from being flooded, homes were built on mounds (*terps*) that would stay dry even at high tide;

some *terps* were refuges for cattle during periods of inundation. In 15 B.C., the area was invaded and taken over by the Roman general Drusus, brother of the eventual emperor, Tiberius. Drusus is credited with building and popularizing dikes; the Romans introduced roads, canals, and efficient agricultural estates. But in A.D. 30, they withdrew from the Frisian marshlands. To fight the Frisians, Drusus connected the Rhine with the Ijssel by the Drusus Canal.

The population in the Netherlands grew 10-fold from 800 to 1250, just as the population was increasing all over Europe due to accelerated migration from neighboring regions, mild winters, long growing seasons, and new agricultural techniques that enhanced the ability to grow and store grain. The introduction of domestic cats in the early Middle Ages helped control the rodent population, which had long ravaged stored grains. The expanding market for wheat and rye led the inhabitants of the Netherlands to dig up and sell the peat so commonplace in the region and to convert the marshes into cultivable farmlands. Previously, much of the land had been 6.5 to 9.8 feet (two to three meters) above mean sea level, but the extensive draining of the marshes lowered the groundwater table and led to land subsidence. This in turn made it imperative to build and maintain a series of linked major dikes. People continued to dig up and sell peat beyond the protective perimeter of the dikes, and this so weakened the entire system that a number of major breaches occurred. On one night, December 14, 1287, 50,000 people were said to have drowned when the North Sea broke through between Stavoren and Eems. Finally, in 1515, Emperor Charles V forbade the digging of peat in threatened areas.

About A.D. 1000, people had begun to cultivate the peat and clay areas of the Netherlands by draining the land for agriculture. Drainage lowered the level of the land that then had to be protected by dikes. By A.D. 1100, West-Friesland had built an enclosure dike *(omringdijk)*.

It was at this time that the Frisians successfully established "the dike peace," a strongly enforced consensus that whenever a dike was endangered, family feuds and civil disturbances must cease forthwith so that all available manpower could be mobilized to reinforce the dikes. Anyone who failed to respond was forced to renounce his property forever. Neighboring farmers selected a willing and able successor to maintain that portion of the dike.

Because the collapse of a dike could mean instant disaster for an entire community, in the early Middle Ages local communities began to shorten the flood defense by damming the watercourses that intersected the peat and clay areas; local dike systems were thereby interconnected. As a result of these river dams, new cities were born, for example, Amsterdam (on the River Amstel) and Rotterdam (on the River Rotte).

Local communities came together to form water boards, which had, by common consent, wide powers of inspection, discipline, and governance. Examples of agreed stipulations of these water boards are reproduced here. The aristocracy, which held large landholdings, quickly understood the advantages of the water boards, which could be depended on to call out the local dike army whenever needed (see Document I for the establishing of several local dike armies). The water board, a democratic organization, took root and remains even today a basic element of the governance and defense of the Netherlands. The water boards slowly received formal charters from the rulers of the various districts. In effect, these local committees were an efficient instrument of self-taxation: each village contributed to the upkeep of the dikes.

Cultivation was not limited to the coastal zone; clay regions along the branches of the Rhine and the Meuse also were prepared for cultivation, and the lowered land was accorded protection by dikes. Because the inhabitants of the region between the Waal and Meuse rivers could not agree about drainage and dikes, Reinald II, the ruler of Gelderland, intervened. He issued the Dike Letter of January 12, 1321, presented here as Document II. By this decree, he created a water board and prescribed a detailed drainage system, with rules, still in force today, for financing the necessary works and their maintenance. The last watercourse stipulated by Reinald II was finally built in the 1950s.

In 1282 an unknown carpenter built the Netherlands' first water-pumping windmill—probably an Arab invention observed by a crusader who later returned home and reported on its design and use. Nearly three centuries later, a windmill with a rotatable top was developed (the rotatable feature was of key importance because winds in northwest Europe came from different directions, so the windmill could always take advantage of the prevailing wind).

These windmills helped create artificial drainage in embanked areas called *polders*. The windmill brought the drainage water from the *polders* to the watercourse from which it was discharged through a sluice. The watercourse now became the intermediate storage (*boezem*). This stepwise drainage system (*polder–boezem–sea/river*) is in common use in the Netherlands. With the improved windmills it became possible to drain large lakes; a series of windmills could remove all the water from even the deepest lakes. By judicious use of sluices, unwanted water was dispatched into specially built drainage canals.

In 1568 the Netherlands began its struggle for independence from Spain. Whenever enemy troops approached, the Dutch did not hesitate to destroy the dikes, yet the defenders remained safe behind flooded land that was rendered impassable for foot soldiers and impenetrable for ships, as the water was shallow. The 80-year war ended with independence for the seven northern provinces.

The seven provinces lacked a common river policy, which sometimes resulted in mismanagement by individuals and even the provinces themselves. The need for central coordination led in 1798 to the establishment of a state water authority, the renowned Rijkswaterstaat (see Document III). This state agency planned and supervised large-scale construction of a flood-protection infrastructure consisting of dikes along the shores and main rivers as well as large land-reclamation projects.

A notable milestone in the project was the reclamation of the Haarlemmermeer (Lake of Haarlem). This lake had steadily increased in size over the centuries, threatening and sometimes flooding the nearby cities of Amsterdam and

Leiden. In the seventeenth century, plans were developed to drain the lake and reclaim the land. Based on previous experience, it was calculated that 170 windmills would be needed to reclaim the Haarlemmermeer, but the project was deemed too expensive. In 1836, storm surges once again flooded parts of Amsterdam and Leiden, finally forcing the central government to confront the danger and reclaim the lake. After long deliberations, in 1852 the lake was drained using three steam-driven pumps. Today Amsterdam's Schiphol Airport is situated on the site of the former lake. The name *Schiphol,* meaning "ships' hell," was used because of the many ships wrecked at this spot.

Another large-scale project was the reclamation of the Zuider Sea. About 1900, the western part of the Netherlands was almost overwhelmed by a major storm-surge disaster. In 1916 a further storm-surge disaster became the final impetus for closing the Zuider Sea. A 20-mile-long (32-kilometer) dike equipped with sluices to drain the excess water was needed. In order to close the final gaps in the dike, the engineers considered the use of caissons, but this idea was abandoned because the final closure could be accomplished by using impermeable boulder clay. In addition to protection against storm surges, the main dike created Lake Ijssel, the largest Dutch reservoir, assuring the northern provinces of an ever-normal freshwater supply even during dry periods. At the same time, new *polders* were built to enlarge the agricultural area and thereby feed the expanding population.

The next step was reclamation of the island of Walcheren. In the fall of 1944, the dikes of this island in the mouth of the Western Scheldt were bombarded by the Allies to open a route for ships to reach Antwerp. Because the tidal difference was greater there than at the main dike, boulder clay was insufficient to stop water currents. Contractors reluctantly had to apply new techniques involving the use of caissons in order to close the gap. The experience of the Walcheren land reclamation proved vital following the storm-surge disaster of 1953. Large gaps in the dikes were closed by caissons

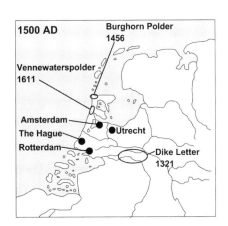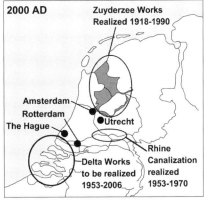

The shape of the Netherlands in 1500 and 2000. Courtesy of Pieter Huisman.

that had been originally intended for the artificial harbors ("mulberries") erected offshore in Normandy in 1944.

CULTURAL CONTEXT

That Rembrandt van Rijn, the painter, was born in a windmill reminds us that many Dutch families made their homes in windmills. The number of windmills in the Netherlands suggests the ubiquitous influence of the battle to control water in the cultural life of the country. In the past thousand years, 130 major floods have been recorded and remembered. As Erwin van den Brink remarked, "The Dutch acquired their hydraulic expertise partly in response to disaster."

In the past 2,000 years, the sea has deprived the Netherlands of 1.4 million acres, but the Dutch have reclaimed and safeguarded 1.8 million acres. In the absence of dunes and dikes, 65 percent of the nation's land area would be flooded during storm surges and high river levels. The rivers are a constant concern in a region where annual precipitation exceeds evaporation.

Beyond this crucial use, windmills, with their rotatable tops, were used for many other purposes, such as grinding corn, sawing trees for timber, and lifting the timber and other heavy items that were used in the shipbuilding industry from the fifteenth to the seventeenth centuries.

For the Dutch, science and technology are essential for survival. The loss of life to the sea has been comparable to that incurred by wars. Based on achievements in shipbuilding, navigation, and port development, the Netherlands not only experienced periods of naval predominance but became the owner of overseas territories and formed adventurous and successful trading companies.

As a result of world-renowned painters, such as Rembrandt, Vincent van Gogh, Johannes Vermeer, and Frans Hals, who lived and worked in the Netherlands, combined with the Dutch penchant for global trade and travel, the Dutch have become accustomed to artwork in their homes. That deep appreciation of art and a sense of adventure were influential dynamics in the Dutch character. Small wonder that Peter the Great, as a young and entrepreneurial czar of Russia, made the long trip to the Netherlands to apprentice in its shipyards and acquire the knowledge that made St. Petersburg possible.

PLANNING

There have been several distinct phases in planning the engineering protection of the Netherlands:

1. In ancient times, farmers built dams in tidal creeks to save their meadows from flooding.

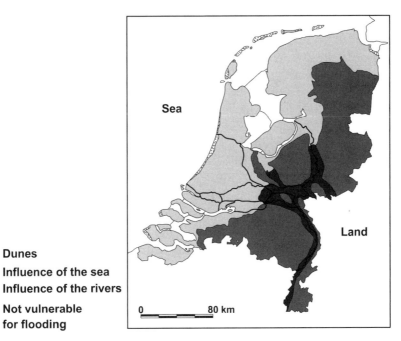

Dunes
Influence of the sea
Influence of the rivers
Not vulnerable for flooding

Sea

Land

0   80 km

Vunerability for flooding in the Netherlands. Courtesy of Pieter Huisman.

2. About A.D. 1000, when marshy peatland was systematically drained to prepare more ground for agriculture, dikes became a preferred structure for protecting the *polders* from overflows of water from the North Sea and the rivers, mainly the Meuse, the Rhine, and the Scheldt.
3. About A.D. 1300, modular sluices became commonplace, channeling excess water through dams and dikes into the rivers and the sea.
4. The invention of improved windmills with rotatable tops ensured efficient artificial drainage of the *polders*, thus making it feasible to drain large, deep lakes.
5. Fossil-fuel-driven engines in the twentieth century made it feasible to close off inland bays and estuaries and improve day-to-day water management.

BUILDING

Born in 1507, Andries Vierlingh of Breda grew up to be perhaps the first scientific hydraulic engineer. In 1570 he wrote a book on the art of dike building; only rediscovered in 1895 and published in 1920, it remains correct and useful today. He was a competent administrator and governor who approached both hydraulic engineering and institutional structure in a scientific way. In addition to his judgment of technical measures, tools, and interventions, he produced keen analyses of proposals for improving the water-management structure and practices.

Jan Adriaensz Leeghwater (1575–1650) was born in De Rijp, a village in the midst of the pools and lakes of the north Netherlands. In his early career, he was a carpenter and millwright; then he became an inventor, engineer, and constructor renowned as a hydraulic engineer and dike builder. In the course of a great career, he drained 27 lakes. His advice on reclamation was avidly sought in France, England, Denmark, and Prussia. So well established was the technology of diking that as the economy boomed, diking companies were formed and were financed by bonds offered at 2.5 percent.

In 1916, a flood disaster led to the decision to close off and largely reclaim the Zuider Sea. Completed in 1932, a freshwater reservoir, Lake Ijssel, was supplied by the northern-flowing branch of the Rhine River, and much of it eventually became high-grade farmland.

In 1953, a storm surge breached dikes in more than 900 places. The government adopted the Deltaplan. Closure of the largest estuary, the Eastern Scheldt, nearly precipitated a cabinet crisis; some ministers favored closure using a dike, while others preferred to heighten the existing dikes (almost 220 miles, or 350 kilometers, in length) for environmental reasons. Eventually the government agreed to build a storm-surge barrier consisting of 66 gates that could be closed in the event of danger. Known as the Eastern Scheldt Barrier, it was completed in 1986, eight years later than planned at a cost of more than 2 billion. Since then, the gates have had to be lowered twice a year on average to protect the inland area from flooding during stormy weather. Normally, the gates are open, which means that tidal saltwater continues to flow into the Eastern Scheldt, thus protecting the marine environment.

The requirement that dikes in the tidal region of Rotterdam meet the Delta standard was originally based on conditions in the 1950s. However, the enlargement of the Rotterdam harbors in the 1960s and 1970s, and the deepening of their entrances, changed the hydraulic conditions, as investigations showed in 1987. The current dikes along the tidal deltas were simply too low to meet the Delta standard. Instead of heightening the dikes—which would have been difficult for the cities and taken more than 30 years to accomplish—it was decided to build a storm-surge barrier in the Rotterdam Waterway. This resulted in two floating gates that can be sunk to close the 1,100-foot-wide (360-meter) waterway within one hour. This barrier includes an ingenious 30-foot (10-meter) ball-and-socket joint. One set of foundation and gate is 1,000 feet (330 meters) long, roughly the width of the base of the Eiffel Tower in Paris. The weight of one set is two times the weight of the Eiffel Tower.

The hydrodynamic security of the Netherlands has been a continuous and intricate process, both top-down and bottom-up. The two chambers of the States-General, the supreme legislative authority, continue to oversee the budget for water management, which requires one percent of the national income in a typical year.

It has become customary to deal with water issues using a comprehensive and anticipatory systems approach. Outside consultants have been employed; for instance, Delft Hydraulics and the RAND Corporation produced a policy analysis of water management for the Netherlands used in various national policy documents on water. The Water Management Act of 1989 specified the modalities and updating of management plans. Legislation was meticulous: planning was not equated with the assessment of alternative scenarios; it was carefully defined as a method of distributing and coordinating commitments so that policies, once established, were carried out.

Information exchange with Belgium and Germany was, and still is, mandatory, written into the prevailing legislation. Despite the utmost care and foresight, exogenous variables—for instance, storm surges or river floods of unprecedented intensity—could, as in the past, render the situation more precarious than anticipated.

IMPORTANCE IN HISTORY

The Netherlands has one undeniable distinction: it constitutes the only country in the world whose key areas were reclaimed from the sea, not from neighbors. The Dutch have a saying: God created the world, but the Dutch made Holland. Familiarity with the sea and a prowess for shipbuilding, navigation, and port development enabled the Dutch to prevail in bitter contests with the much larger powers of Spain and England. But the major threat to the viability of the country has come not from the changes and chances of international disputes, but from the unrelenting pressures of the unpredictable North Sea and the latent kinetic behavior of nearby rivers—the Meuse, the Rhine, and the Scheldt.

For the people living in the low-lying coastal provinces—totaling at least 60 percent of the country's population—water will always be a threat. All the engineering masterpieces that protect them—the dikes, dams, storm-surge barriers, and pumping stations—have to be maintained. Considerable effort goes into conserving the beaches and dunes by replenishing and nourishing the

The shape of the Netherlands around 0, 800, 1500, 1900, and 2000 A.D. respectively. Courtesy of Pieter Huisman.

sand, which gives the coast its natural protection. The steady rise in sea level will certainly lead to higher costs. Nature and the environment have repeatedly influenced water-management policies, and will play an even greater role in the future.

As steam pumps replaced windmills, the Dutch were able to demonstrate the benefits of their proficiency in reclaiming inland seas and lakes for agriculture and, more recently, for the industries and the growing urbanization of contemporary society. Success in water management was the *unum necessarium* in Dutch history. What the Netherlands accomplished—not mere survival but flourishing growth—can encourage other coastal countries, especially those bordering waters less dangerous than the formidable North Sea, to view land reclamation as a valid option to be assessed and considered seriously as an alternative to hostility between neighbors. As the world grows more crowded, with populations increasingly favoring coastal over inland locations, the reclamation of land from the sea may come to be seen as both sensible and politically wise.

FOR FURTHER REFERENCE

Books and Articles

Huisman, Pieter. *Water in the Netherlands: Managing Checks and Balances.* Utrecht: Netherlands Hydrological Society, 2004.

———. *Water Legislation in the Netherlands.* Delft, the Netherlands: Delft University Press, 2004.

Spier, Peter. *Of Dikes and Windmills.* New York: Doubleday, 1969.

Van den Brink, Erwin. "The Netherlands: A Country That Engineered Itself into Existence Is Tapping into Its Centuries-Old Expertise in Handling Water." *Technology Review* 108, no. 4 (2005), http://www.technologyreview.com/articles/05/04/issue/feature_gp_netherlands.asp

Van de Ven, G. P., ed. *Man-Made Lowlands: History of Water Management and Land Reclamation in the Netherlands.* Utrecht, the Netherlands: Stichting Matrijs, 2004.

Van Veen, J. *Dredge, Drain, Reclaim: The Art of a Nation.* The Hague: Martinus Nijhoff, 1962.

Winslow, Thacher and Davidson, Frank P., editors. *American Youth: An Enforced Reconnaissance.* Cambridge, MA: Harvard University Press, 1940.

Internet

For information about the Romans in the Netherlands, see http://www.livius.org/ga-gh/germania/woerden.html.

For more information about William James's "army against nature" (a quote from *Moral Equivalent of War*), see http://www.constitution.org/wj/meow.htm.

For information on land reclamation in the Netherlands, see http://www.knag.nl/pagesuk/geography/engels/news99engelstekst.html (in Dutch language).

For information on the *polders* and dikes, see http://geography.about.com/library/
weekly/aa033000a.htm.

For pictures of windmills, see http://www.kinderdijk.nl/.

For the home page for the Rijkswaterstaat, see http://www.rijkswaterstaat.nl/en/ (in Dutch
language).

For information on and pictures of mulberries, see http://www.valourandhorror.com/
DB/BACK/Mulberry.htm.

For information about precipitation/evaporation ratios, see the glossary of the
American Meteorological Society: http://amsglossary.allenpress.com/glossary/
browse?s=p&p=54.

For information about the Rijksmuseum, home to artwork by famous Dutch painters,
see http://www.rijksmuseum.nl/.

For information about population growth in western Europe, see http://www.tulane.
edu/~august/H303/handouts/Population.htm.

For information about the influence of cats on grain storage in Europe, see http://www.
isabellevets.co.uk/new_cat/newcat.htm.

Documents of Authorization—I

ESTABLISHING LOCAL DIKE ARMIES

The ruler of West-Friesland, bishop van Zuden, commanded in April 1319:
"Everybody shall come to work at the dike on instruction of the bailiff or dike
reeve."

*["ende alman sal ten menen werke comen op den dijc, daers hem die Baeiliu,
of die Dijcgrave vermaent"].* The letter from the Water Board of the Lekdijk
Benedendams commanded in 1405 and in 1454 for the Ijsseldijk:

"They (meaning the dike reeve and aldermen) will inform without delay
by a sworn-in messenger, the bailiffs of all parishes within the territory of the
Ijsseldyk and Leckendyk, to ring the bells in order to summon every house
working a quarter Lands *uyt* (a unit of land) or more within the territory, to send
an able man to the Leckendyk, . . . " and further

"Nobody will be excused, except bakers, bedridden, persons abroad and
clergymen. The dike reeve and bailiff may each keep one person at home as
messenger for their labour without penalty."

*["soo sullen zij (bedoeld zijn dijkgraaf en heemraden) sonder eenige dingtaal by
eenen geswooren zekeren boode een weete doen doen aen de schouten tot haaren
huyse, in alle Kerspelen, gelegen binnen den voorsz. schouwe van den Ijsseldyk
en Leckendyk, ende die Schouten sullen dan elk in haren Kerspelen terstond die
klocke slaen en tot dien klockeslage sal elck huys, in derselver schouwe gelegen,
daar men een viertel Lands uyt (een oppervlaktemaat) bruyckt, ofte daar en boven,
eenen weerachtigen maan senden opten voorsz. Leckendyk . . . " en verder "en hier
in en sal niemand verschoont wesen, uytgesondert die broodsakken, die bedde vast
leggen, die uytlandig zijn, of die des Heeren chynsen gekocht hebben; behoudelyk,
dat die Dykgrave in elcken Gerechte wel twee personen, die gesworen zeker boode*

in elcken Gerechte wel een persoon ende die Schout in synen Gerechte een persoon thuys houden mogen voor haaren arbeyd en kost, sonder eenig verbeuren; . . ."]

The Court of Holland defined in its ordinance for the Vriesendijk on 25 October 1497:

"In case of storm surge or river flood causing large pressure on the dikes, every water board is obliged to ring the bells in their village and is empowered to command every adult to immediately go to the dike. He who refuses to do so must pay a fine of 10 shilling to the dike reeve"

["Ende of eenige stormen ende opwateren quamen en eenige groote laste op die Dycken waren, soo sal een yegelyck waerschap in synen Dorpe gehouden zijn terstondt de Klocke te doen slaen, ende macht hebben te bevelen yegelyck binnen syn Jaren zijnde haestelyck op den Ddyck te komen ende dat elck die in gebreke ware, op de pene van thien schellingen 's Heeren geldts ot behoef van den Dycgrave."]

From A.A. Beekman. *Het dijk—en waterschaprecht in Nederland vòòr 1795.* (Gravenhage: Martinus Nijhoff, 1905) 371. (English translation of these Dutch excerpts by Pieter Huisman).

Documents of Authorization—II

THE DIKE LETTER OF JANUARY 12, 1321

We, Reinald, the son of the count of Guelderland, proclaim the confirmation of the force of law with regard to the following regulations.

The inhabitants of the region between the Meuse and Waal, and those living in the county and the Realm of Nimegue [modern-day Nijmegen], wish to construct dikes and watercourses to improve their land. However, people in the region cannot agree on the issues and therefore cannot carry out these plans without our intervention and assistance.

For that reason, and after consulting our advisors, We have appointed a dijkgraaf (dike reeve) and heemraden (other members of the executive of the dike board) which shall oversee and increase the prosperity of the region. Dijkgraaf and heemraden, and their successors, shall take an oath of office and swear to undertake the following.

The Water Board shall dig watercourses and construct sluices and bridges in the region. The Board shall develop these works in locations that will be the most adequate and profitable for the region and in such ways that the inhabitants downstream profit equally from these works as will the inhabitants in the upstream region.

The Board shall choose land that will profit from such watercourse, sluice, weir, or bridge. The cost of all work will be equally shared by the owners of such land.

All watercourses constructed by the board will be protected and defended by Us, the son of the Duke of Guelderland. This is valuable for property under our own jurisdiction as well as for private properties. This concerns Heuman, Balgoy, Leur, Hernen, Batenburg, Appeltem, Altforst, and Puifluik, as well as the other private properties between the Meuse and Waal which will profit from water courses to be built in the future.

In accordance with the heemraden, the dijkgraaf shall be given the right to claim, by standing execution, all charges of watercourses, sluices, weirs, and bridges from those who are indicated as taxpayers.

All people possessing land between the Meuse and Waal shall stand ready to pay for the cost and maintenance of the watercourses, sluices, weirs, and bridges according to the ratio of their land ownership.

The dijkgraaf will inspect the dikes between the Meuse and the Waal as well as throughout the Realm with the same frequency in every village as done in our name. The Heemraden may reinforce or demolish a dike if the interest of the regions requires this.

The first inspection is every March, on St. Gertude's Day. The heemraden can order other inspections. If the heemraden declare a dike unfit, the dijkgraaf will fine the tardy tenant four shillings and six pence. (The Grand French tournois equals sixteen pence.) The fine can also be paid in kind equal to the same value. However, people who oppose members of the board will be fined six pounds. Half of the fine will go to the dijkgraaf; the heemraden will receive the other half.

During the last inspection the dijkgraaf may not pass a dike stretch that has been declared unfit by the heemraden. In such case, he will invite contractors to make the necessary repairs to meet his standards. When the dike stretch meets such standards, he can approve the dike. The dijkgraaf may double the cost charged to the tardy tenant to meet the standards. The tardy tenant must also pay the hostel stay for the dijkgraaf and heemraden. The dijkgraaf may order reinforcement of the dike when this measure is necessary in the opinion of the heemraden.

Similar to the dikes, the dijkgraaf will also inspect the watercourses, sluices, weirs, and bridges. The heemraden will define the inspection date during and after their realization. The dijkgraaf is free to ask more than two heemraden to accompany him on the inspection. The inspection has to be timely announced in all villages.

For the land area transferred to Us by the dijkgraaf because of dikes, sluices, watercourses, weirs, and bridges that have fallen into disrepair, we do accept the consequences. We will pay all expenses necessary to meet the requirements of the heemraden which were not met by the responsible person. If We cannot repay the cost, he may claim it from our possessions in the Meuse and Waal region. As the dijkgraaf may pay the claim from our goods, he also may claim the payment from the goods of others.

Two watercourses have to be realized, one from the Maasdijk in Dreumel to the Teerse Broek and one from Appeltern to the Bergharense Meer, when the heemraden think these works are necessary.

The drainage has to be equal for the entire region as well as the discharge to the rivers, therefore every watercourse has to be equipped with three weirs.

The most downstream weir in the watercourse near Dreumel will be situated between Dreumel and Wamel at a former inner embankment (sydwinde). This inner embankment connects the Meuse between Appeltern and Maasbommel. To be approved by the heemraden, the second inner embankment has to follow the trace Holtmeer, Harense Geest to Waalwaard downstream (the latter mainly property of the duke). The third inner embankment in this watercourse starts at the windmill of Haren, Harense Geest and Waalwaard upstream.

The most downstream weir in the watercourse beginning at Tuut in Appeltern will be situated above Tuut in the jurisdiction of Appeltern. The other is planned between Haren en Hernen and the third upstream of Wesel.

All inner embankments, including weirs, will operate according to the following conditions. If the most downstream weir has to be opened due to high water levels upstream, the second and third weir may be opened. As the most downstream weir is closed, every upstream weir has to close and remained closed. As the downstream sluice evacuates water, all sluices upstream may discharge. The heemraden are obliged to control the discharging process to verify the equal right of downstream and upstream dischargers.

Cost and maintenance of these inner embankments and weirs have to be paid by the downstream proprietors of land according to their land surface ownership ratio.

Every area profiting from the aforementioned works has to contribute to cost and maintenance. The dijkgraaf may behave in these areas as in his own. We will give him protection, support, and compensation when he suffers damage by exercising his duty. We will appoint substitutes for the dijkgraaf and heemraden and increase or reduce the number of heemraden according to the region's interest. We will take care to appoint heemraden who represent all interests (including their own) of the region in order to promote the watercourses in such a way that land owners upstream and downstream are equally served.

Owners of land in Nimegue and surroundings are also obliged to contribute according to their land surface ownership ratio.

We promise to renew these regulations when people want a modification.

Realizing these regulations, the water board also has to verify that the downstream area is equally served as the upstream part of the board. The intention shall be further improvement of the water management conditions in this region.

We promise that our successors will maintain these regulations and conditions.

Signed 12th January 1321

From H. van Heinigen, *Tussen Maas en Waal.* (Zutphen: De Walburg Pers, 1972) 440–42.

Documents of Authorization—III

CREATION OF THE RIJKSWATERSTAAT

24th May 1798
ARA
Entry number: 2.01.01.04
Inventory number: 489
1798, 24th May; No. 38.

Reading the Letter of the Minister of Home Affairs of 23rd and sending the Plan to control the Waterstaat [Water Management] of this Republic to the Administration, which is inserted beneath.

(fiat insertio)

The Letter also contains:

1. Although the Plan for the Waterstaat has not yet been confirmed because of the many affairs daily arriving, the Minister has deemed it necessary to present the citizens F.J. Leemans and J. Sabrier, Fabriek bij Delfland as members for the Inland Waterstaat.

   Further as bookkeeper for the finances, the citizen Benjamin de Ruijter; as clerk at the Office, the citizen F.A. de Hartog; and as Administrative Assistant to the President of the General Waterstaat, the citizen F.W. Conrad.
2. As since 22nd of January 1798 no payments for the Waterstaat could be made to pay the contracted and realized works, the Minister had to ask this Administration an amount of 300.000 guilders.

After discussion, it has been decided:

First.

To confirm the Plan to control the Waterstaat of this Republic as submitted by the Minister.

Second.

To appoint according to the mentioned Minister, the citizens F.J. Leemans and J. Sabrier, Fabriek bij Delfland as members for the Inland Waterstaat.

Further as bookkeeper for the finances, the citizen Benjamin de Ruijter; as clerk at the Office, the citizen F.A. de Hartog; and as Administrative Assistant to the President of the General Waterstaat, the citizen F.W. Conrad.

Third.

To request the Representative Body by Letter to dispose an amount of 300.000 guilders to this Administration in order to pay the contracted and realized works for the Waterstaat of this Republic.

The Secretary has to send the Letter.

An extract of this decision as well as a copy of the above mentioned Plan will be sent to the Minister to inform him.

6
The Founding of Baghdad

Iraq

DID YOU KNOW . . . ?

➤ The construction of Baghdad began in A.D. 762.
➤ Its original size was 10,000 feet (3,000 meters) in diameter, with four gates.
➤ The reigning emperor, al-Mansur, designed the city himself.
➤ The city was circular, enclosed by three concentric walls, and was sometimes called "the Round City."
➤ Baghdad was named "the City of Peace" by its founder.
➤ At its peak, it was the cultural capital of Islam.

One of the world's great architectural marvels, Baghdad has its roots in the spirit of peace. When it was built by Caliph al-Mansur in the Islamic calendar year 145 (A.D. 762), the caliph named the new capital of Persia *Madinat as-Salam*, "the City of Peace." The name *Baghdad* is derived from an ancient Babylonian site known as *Bagh-da-do: bagh* was an idol or god; *dad* means *gift* or *given*. So the city's name originally meant "gift from God." According to scholar Guy Le Strange, the correct translation of Baghdad is "the city founded by God" (LeStrange, 11).

Baghdad's unique history was further revealed in 1870 with the discovery of the Michaux Stone, found by a French physician who lived in Baghdad.

An Iraqi man walks past the statue of Muslim caliph Abu Jaafar al-Mansur, situated in a square in Baghdad, 1998. Courtesy of AP / Wide World Photos.

Carvings on the stone tell of a twelfth-century B.C. Babylonian city known as *Bak-da-du*.

HISTORY

More than five million people live in Baghdad, Iraq's capital, which straddles the Tigris River. There is evidence of settlement in the area since ancient times, but the true origins of today's metropolis date to the period following the Arabs' conquest of Mesopotamia in the 630s. Al-Mansur, who ruled from 754 to 775 as the second caliph of the Abbasid dynasty, decided in 762 to move his capital from Damascus to a site on the western bank of the Tigris River opposite an ancient Persian village called Baghdad. He mobilized architects, engineers, and surveyors, along with thousands of workers, and they proceeded to build—in accordance, it is said, with a design conceived by al-Mansur himself—a circular city enclosed by three concentric walls. The result, a remarkable early example of city planning, became known under such names as the Round City, the City of Peace, and the City of al-Mansur.

Historical sources supply various measurements for the size of al-Mansur's city. Its likely diameter was 10,000 feet (3,000 meters). The caliph's palace and the chief mosque stood in the center, surrounded by army quarters; inside the outermost wall were residents' homes. Bazaars and merchants' houses sprang up outside the walls. The city had four gates, equally spaced around the periphery; they were named for the parts of the empire to which they led. The Khurasan (or Kurasan) Gate on the northeast also provided access to a bridge of boats leading to the Tigris's eastern bank.

The city grew quickly under the fifth caliph, the storied Harun al-Rashid (ruler from 786 to 809) of The *Arabian Nights*. At its peak, it was the cultural capital of Islam and a major trade nexus, and was reputed to be the wealthiest city in the world. In medieval times, Baghdad was regarded as a splendid place of parks, gardens, broad streets, and marble palaces that were a pleasure to its large and prosperous population. By the tenth century, the Abbasids' power was waning, and the caliphate's center, temporarily located for a few decades in Samarra, had moved to the east bank of the Tigris. Baghdad sank into prolonged eclipse as the following centuries brought incursions by the Turks and the Mongols. On occasion portions of the metropolis were laid to waste, and the city was relegated to the status of a provincial capital. In the late nineteenth century, as a lengthy period of Ottoman rule drew to a close, relatively progressive governors fostered stirrings of modernization. Baghdad was designated the capital of the new state of Iraq after World War I, and in the second half of the twentieth century the city experienced considerable growth, largely thanks to revenue garnered by Iraq from its abundant petroleum reserves. A spurt of rebuilding followed destruction suffered during the 1991 Persian Gulf War. The 2003 U.S.-British invasion of Iraq resulted in renewed devastation, raising the possibility of another period of major reconstruction.

Two texts reproduced here illuminate the founding of Baghdad by al-Mansur. The first, recounting a mythical prediction often attributed to the caliph himself, relates a vision of a marvelous future for the city given its eminently central location. The prediction largely came to pass, although Baghdad's period of glory turned out to be short-lived. The second text is a critical summary by K.A.C. Creswell, a prominent scholar of early Islamic architectural history, giving a salient account of al-Mansur's construction of the city.

A third piece, from Jacob Lassner, presents a text from the historian Shaykh al-Kahtib about the building of Baghdad. Al-Kahtib tells of refinements made to the city's design at the suggestion of a visiting Byzantine emperor.

"Would you like a tour?" asked the caliph, proud to show off his architectural accomplishments to the visiting emperor from Byzantium. After the tour, al-Mansur inquired, "What did you think?" and the answer was surprising. The emperor said that while he certainly found the city impressive, there were three problems: (1) drinking water was in short supply, would soon become scarce, and would thus limit growth; (2) there were no gardens, and the eye wanted gardens; and (3) security would be a problem, because the marketplace was inside the city walls.

The medieval reports tell us, "Al-Mansur stiffened . . . and said, 'As for your statement concerning the water, we have calculated the amount of water necessary to moisten our lips. With regard to the second shortcoming, indeed we were not created for frivolity and play; and as for your remarks concerning my secrets, I hold no secrets from my subjects.'" But immediately upon leaving the emperor, the caliph told his staff to dig two canals leading from the Tigris to increase the amount of water, to landscape al-Abbasiyah and make it a beautiful garden, and to construct al-Karkh, the new marketplace outside the city walls (Lassner, 61).

CULTURAL CONTEXT

Al-Karkh, the marketplace, is one of the most interesting pieces of the building of Baghdad. It was true, as the Byzantine emperor had observed, that the first market was inside the city walls. This was often the case, because having supplies nearby was important to those who lived within the city. Indeed, we might not have today the spectacular Taj Mahal of India if it were not for a similar kind of royal marketplace, for that is where the future Shah Jahan spied a beautiful young girl who was to become his wife, Mumtaz Mahal.

But if a market within the city walls allowed for flirtation, it also allowed for treachery. As the Byzantine emperor had warned, spies could come right into the market and quickly learn secrets, assassins could travel in and out disguised as merchants, and merchants were not above selling access and information, just as they sold other services, for the right price. Therefore, using his own money, the caliph constructed al-Karkh, thus moving commerce outside the palace walls and making the caliphate more secure.

Security was the primary reason for building Baghdad. When the previous dynasty had been overthrown by al-Mansur's brother, and when al-Mansur succeeded to the throne, he declared the first order of business would be to move the capital from the old caliphate capital, Damascus, to a more secure and strategic location.

Some legends say that the caliph rode on horseback to explore the realm. Some say he looked at a map of the empire and pointed to its very center. Both are likely, because the area around Baghdad had served as a strategic base since the time of the ancient Babylonians. But when al-Mansur rode to the banks of the Tigris and saw the fast-moving water, the warrior in him spotted the perfect key to security—a river that was difficult to cross. He may have been so preoccupied with security that he forgot about the importance of drinking water for his people, as the Byzantine emperor later pointed out, much to al-Mansur's chagrin.

The founding of Baghdad represented the transition from a transient Bedouin lifestyle to a way of living that was more anchored. Creating a home for a nation is a momentous time in a civilization, and so it was with Baghdad. While cities typically feature a sedentary culture, it is that relative tranquility that makes possible the massive projects needed for constructing a permanent settlement. Once a sedentary culture has been established, there is a shift in power. The scholar Ibn Khaldun notes that among large numbers of people who must cooperate, there must be a restraining influence, that is, a religious or royal authority. When all these factors come together, a new dynasty is born.

But the Caliph left neither dynasty nor destiny to chance. It was common knowledge at the time that there had been a popular legend among Christian monks that a magnificent and important new city would someday be built on the Tigris near the Sarat Canal by someone named Miklas. The caliph heard this and claimed (was it just cleverly made up on the spot, or was it true?) that Miklas was the very same nickname his childhood nurse called him. The caliph clearly saw himself as entitled to power.

PLANNING

Once al-Mansur had found the place to build, he did not act; instead he planned. As recounted by Lassner in Document III when al-Mansur decided to build the city, he summoned engineers, architects, and experts in measuring, surveying, and the division of plots. Then he traced the city plan (some say he drew it in the sand) making the city round. They saw that there was no other round city known in all the regions of the world. The foundations were laid at a time chosen for al-Mansur by Nawbakht, the astrologer, in the year 145 (A.D. 762).

At the time, the plan was for a round city. Why? Because al-Mansur demanded round so that all his subjects would be equally close to him; with squares and rectangles there could be outer and inner quarters. But architectural critics, and historians such as Fouad Kazaanxhi, suggest that while this may have been one of the first round Islamic cities, the planners of Baghdad may have been acting in a tradition that had its origins in the architecture of Mesopotamia, where Hatra City had round walls as early as the eighth century. In both periods, the underlying purpose of the design was security.

BUILDING

Round and well fashioned, meticulously built and refined, the great city of Baghdad was much admired. Bricks were used in the construction, and they were cemented with wet clay. To make the bricks, engineers had to dig wells and build pipelines to bring water from the canals into the city. The city had four gates, and al-Mansur could look out the gates and see who was approaching. Some accounts say the caliph took materials—perhaps even the old gates themselves—from Persia's former capital in Damascus and brought them to the new city as a signal that civilization would continue despite the change of location.

IMPORTANCE IN HISTORY

Baghdad has captured the imaginations of many. It inspired Germany's Kaiser Wilhelm II to propose a Berlin-to-Baghdad railway. In 2004, the Organisation

A wooden railroad bridge spans the Euphrates River as part of the Berlin-to-Baghdad railway. Courtesy of the Library of Congress.

for Economic Co-operation and Development expressed interest in a technology-oriented, multi-industry transport corridor from Oslo to Beijing, with pipelines, communication lines, railways, and roads. One could even think of this as a technological Silk Road. Could such a concept be helpful to those considering future visions of Baghdad? Could there be a Bosnia-to-Baghdad technology corridor?

When al-Mansur came to the ancient area that ultimately became Baghdad, he brought a new vision and a high aim. In constructing this city of peace, did al-Mansur set a tone that can carry far into the future?

FOR FURTHER REFERENCE

Books and Articles

Kazaanxhi, Fouad. "A History of Baghdad." *Baghdad Bulletin*, July 16, 2003. Available online at http://www.baghdadbulletin.com/pageArticle.php?article_id=55. An article covering the redevelopment of Iraq.

Khaldûn, Ibn. *The Muqaddimah: An Introduction to History.* Translated by Franz Rosenthal. Edited by N. J. Dawood. Princeton, NJ: Bollingen Series, Princeton University Press, 1967.

Lassner, Jacob. *The Topography of Baghdad in the Early Middle Ages: Text and Studies.* Detroit, MI: Wayne State University Press, 1970.

Le Strange, Guy. *Baghdad during the Abbasid Caliphate from Contemporary Arabic and Persian Sources*. Oxford: Clarendon Press, 1940.

Internet

For background on the Abbasids, see http://www.saudiaramcoworld.com/issue/196209/building.of.baghdad.htm.

For course syllabi from leading professors worldwide on Islamic art, architecture, and culture, see http://archnet.org/courses/.

While Kaiser Wilhelm's proposal for a Berlin-to-Baghdad railway link never became reality, for information on rail connections from Europe to the Middle East, see http://www.seat61.com/Syria.htm.

Documents of Authorization—I

Legend says that the founder of the city, Caliph al-Mansur, made the following prediction:

"This is indeed the city that I am to found, where I am to live, and where my descendants will reign afterward. Princes had lost all track of it, before and since Islam, so that the previsions and orders of God could be accomplished through my efforts; thus traditions are confirmed, signs and prognostics become clear. Assuredly, this island, bounded on the east by the Tigris and on the west by the Euphrates, will prove to be the crossroads of the universe. Ships on the Tigris, coming from Wasit, Obolla, Ahwaz, Fars, Oman, Yamama, Bahrein, and neighboring countries will land and drop anchor there. It is there that merchandise will arrive by way of the Tigris from Mosul, Azerbaijan, and Armenia, that too will be the destination of products transported by ship on the Euphrates from Raqqa, Syria, the borderlands of Asia Minor, Egypt, and the Maghred. This city will also be on the route of the peoples of Jebel, Ispahan, and of the provinces of Khurasan. I shall erect this capital and live in it all my life. It will be the residence of my descendants; it will certainly be the most prosperous city in the world."

From Gaston Wiet, *Baghdad: Metropolis of the Abbasid Caliphate* (Norman: University of Oklahoma Press, 1971), 10–11.

Documents of Authorization—II

THE FOUNDATION OF THE CITY. Having decided on the site, 5 Jumada I, 145 (1 Aug. 762), al-Mansur wrote to every city to send engineers and people acquainted with building, surveying, and mensuration. Engineers, architects, and land surveyors from Syria, Mosul, al-Jabal (Western Persia), Kufa, Wasit, and Basra were gathered together, and not until thousands of men had been assembled did the work begin. According to Ya'qubi, the number of workmen were fixed at 100,000. He says that the architects were Abdallah ibn Muhriz,

Hajjaj ibn Yusuf, Imran ibn al-Waddah, and Shihab ibn Kathir. The second of these, according to al-Khatib, was charged with the marking out of the plan. A fifth architect, Bishr ibn Maimun, is mentioned by Baladhuri; he says that he built the arcades *(taqat)* at the Bab ash Sham. Four chief overseers were appointed, one of these being Abu Hanifa, founder of the oldest of the four orthodox schools of Muslim theology. He was superintendent of the brick-making and, being an intelligent man, discarded the slow method of counting the bricks, and measured the stacks instead with a graduated rod, computing the number from the measurements obtained.

Tabari says that the plan of the city was first traced on the ground with lines of ashes *(ramad)* for al-Mansur wished to see its actual form; he then began to enter from each gate and pass into its *fasils,* and *taqat,* and *rahabas.* On the lines so traced were placed seeds of cotton, saturated in naphtha, which were set on fire, after which they dug the foundations exactly on the lines traced. The plan, which we are told had been conceived by al-Mansur himself, was circular, with four equidistant gateways named after the city or province towards which they opened: the Kufa Gate (SW), the Basra Gate (SE), the Khurasan Gate (NE), and the Syrian Gate (NW).

The foundations were laid at a moment chosen by the astrologer Naubakht, and the Arabic authors are almost unanimous as to the date: 145 H. (762–3).

In the year 146 H. (763–4), al-Mansur transferred the public treasuries, repositories, and registers from Kufa to Baghdad and established himself there with his army.

From *"The Foundation of the City"* from *"Early Muslim Architecture: Early 'Abbasids, Umayyads of Cordova, Aghlabids, Tulunids and Samanids, A.D. 751–905, Part Two"* by Cresswell, Keppel (1932–1940). By permission of Oxford University Press.

Documents of Authorization—III

THE BUILDING OF MADINAT AS-SALAM

The Shaykh, al-Kahtib

I received an account that when he decided to build the city, al-Mansur summoned the engineers, architects, and experts in measuring, surveying, and the division of plots. He showed them the plan which he had in mind, and then brought in ordinary laborers, and skilled personnel including carpenters, diggers, blacksmiths, and others. The Caliph allotted them their salaries, and then wrote to every town asking them to send inhabitants with some knowledge of the building trade. But he did not begin construction until the number of craftsmen and skilled laborers in his presence reached many thousands. Then he traced the city plan, making the city round. They say that no other round city is known in all the regions of the world. The foundations were laid at a time chosen for al-Mansur by Nawbakht, the astrologer.

From Jacob Lassner, *The Topography of Baghdad in the Early Middle Ages: Text and Studies* (Detroit, MI: Wayne State University Press, 1970), 45.

Muhammad b. Khalaf

They say, when al-Mansur built his city he constructed four gates, so that if one came from al-Hijaz, he entered by way of the Kufah Gate; if he came from al-Maghrib, he entered by way of the Damascus gate; if he came from al-Ahwaz, al-Basrah, Wasit, al-Yamamah and al-Bahrayn he entered through the Basrah Gate; and if he arrived from the east he entered through the Khurasan Gate. The reference to the Khurasan Gate was omitted from the text, and was not mentioned by Muhammad b. Jafar on the authority of as-Sakuni: but we corrected the text from a different account. He, that is to say, al-Mansur, placed every gate opposite the palace. He capped each gate with a dome, and erected eighteen towers between each gate, with the exception of the wall between the Basrah and Kulfah Gates where he added an additional tower. He fixed the length between the Khurasan Gate and the Kufah Gate at 800 cubits, and from the Damascus Gate to the Basrah Gate at 600 cubits. There were five iron gates between the main entrance to the city and the gate which led to the courtyard.

Waki related in the account which I have on his authority, that Abu Ja-far built the city in a circular form because a circular city has advantages over the square city, in that if the monarch were to be in the center of the square city, some parts would be closer to him than others, while, regardless of the division, the sections of the Round City are equidistant from him when he is in the center. Al-Mansur then built four main gates, dug moats, and erected two walls and two fasils. Between each main gate were two fasils, and the inner wall was higher than the outer wall. He commanded that no one be allowed to dwell at the foot of the higher inner walls or build any dwelling there; but ordered construction along the wall in the second fasil because it was better for the fortification of the wall. Then he built the palace and mosque.

From Jacob Lassner, *The Topography of Baghdad in the Early Middle Ages: Text and Studies* (Detroit, MI: Wayne State University Press, 1970), 51–52.

Waki related in the account, which I have received on his authority, that the city was round, and encompassed by round walls. The diameter measured from the Khurasan Gate to the Kufah gate was 1,200 cubits; and the diameter measured from the Basrah gate to the Damascus Gate was 1,200 cubits. The height of the interior wall, which was the wall of the city, was thirty-five cubits. On the wall were towers, each rising to a height of five cubits above it, and battlements. The thickness of the wall at its base was approximately twenty cubits. Then came the fasil, 60 cubits wide between the inner and outer walls, followed by the outer wall which protected the fasil, and beyond which the moat was situated. The city had four gates: east, west, south and north, and each gate in turn was made up of two gates, one in front of the other, separated by a corridor and a court opening unto the fasil which turned between the two walls. The first gate was that of the fasil, the second that of the city. When one entered the Khurasan Gate, one first turned to the left in a vaulted corridor constructed of burnt brick cemented by gypsum. The corridor was twenty cubits wide by thirty cubits long. The entrance was in the width, and the exit, which was in the length, led to a court sixty cubits by forty cubits. This court extended to the second gate, and was walled on both sides from the first to the second wall. In back of the court was the second gate which was the gate of the city: flanking to the right and left of the court were two doors leading to the two fasils. The one to the right led to the fasil of the Damascus Gate; the left door led to the fasil of the Basrah Gate, which then turned from the Basrah Gate to the Kufah Gate. The fasil which led to the Damascus Gate turned to the Kufah Gate in exactly the same fashion, since the four gateways were identical in regard to gates, fasils, courts and arcades. The second gate was the gate of the city and was protected by the large wall which we have described. The main gate gave access to a vaulted passage, constructed of burnt brick, cemented by gypsum, which was twenty cubits long and twelve cubits wide. This was true as well for the other of the four gates. Above the vaulted passage of each gate was an audience room with a staircase against the wall by which means one ascended to it. Crowning this audience room was a great dome which reached a height of fifty cubits. Each dome was surmounted by a figure different from those of the other domes, which turned with the wind. This dome served as the audience room of al-Mansur when he desired to look at the river or see whoever approached from the direction of Khurasan. The dome of the Damascus Gate served as the audience room when he desired to look at the suburbs and the countryside surrounding them. When he wished to look at al-Karkh and whoever might approach from that direction the dome of the Basrah Gate served as his audience room; and the dome of the Kufah Gate served this function when al-Mansur desired to view the gardens and the estates. The gateways in both the city's walls were protected by a heavy iron double door of large dimensions.

From Jacob Lassner, *The Topography of Baghdad in the Early Middle Ages: Text and Studies* (Detroit, MI: Wayne State University Press, 1970), 53–54.

THE MARKETS

Muhammad b. Khalaf:

Al-Harigh b. Abi Usamah related the following account to me: When Abu Ja'far al-Mansur completed Madinat as-Salam, establishing the markets in the arcades on each side of the city, a delegation from the Byzantine emperor visited him. At his command, they were taken on a tour of the city, and then invited to an audience with him at which time the caliph asked the Patrikios, "What do you think of the city?" He answered, "I found it perfect but for one shortcoming." "What is that?" asked the Caliph. He answered, "Unknown to you, your enemies can penetrate the city anytime they wish. Furthermore, you are unable to conceal vital information about yourself from being spread to the various regions." "How?" asked the Caliph. "The markets are in the city," said the Patrikios.

"As no one can be denied access to them, the enemy can enter under the guise of someone who wishes to carry on trade. And the merchants, in turn, can travel everywhere passing on information about you." It has been suggested that, at that time, al-Mansur ordered moving the markets out of the city to al-Karkh, and ordered the development of the area between the Sarat and the 'Isa Canal. The task was entrusted to Muhammad b. Hubaysh al-Katib. Al-Mansur called for a wide garment, and traced the plan of the markets on it, arranging each type of market in its proper place.

From Jacob Lassner, *The Topography of Baghdad in the Early Middle Ages: Text and Studies* (Detroit, MI: Wayne State University Press, 1970), 61.

Charlemagne opens a palace school, headed by Alcuin, 791. Courtesy of the Library of Congress.

7

Charlemagne's Works

Western Europe

DID YOU KNOW . . . ?

➤ Charlemagne linked Europe through a network of canals and bridges.
➤ He was the progenitor of the European Union.
➤ He mandated education for all clerics in his empire, laying the groundwork for monastic and university learning.
➤ He conceived the innovation of a relocatable bridge, forerunner of the World War II Bailey Bridge.

It took 1,200 years for Charlemagne's design for a canal linking the Main and Danube Rivers to become a reality, but the great medieval general had the right idea. Macroengineering projects of comparable duration include the Suez Canal, whose precursor was built, it is said, by the Egyptian pharaoh Sesostris in 1900 B.C., and the Grand Canal of China, begun around 600 B.C. and still being expanded today.

HISTORY

Charlemagne (literally, "Charles the Great") is an epochal figure in European history. Born around 742, he became king of the Franks in 768 and proceeded to extend his rule over much of central and western Europe. He laid the foundation of the Holy Roman Empire, and in 800 was crowned emperor of the Romans by Pope Leo III.

Some of Charlemagne's projects were directly related to military endeavors. In 789, he commanded two bridges to be erected across the Elbe River, one with a fortress at each end. In 792 a pontoon bridge suitable for crossing the Danube was completed. In 808 a project was initiated to build a rampart protecting his northern frontier, and the following year he decided to erect and garrison a major fortress for protection against the Danes.

Charlemagne's creative energy continued to find outlets in both military and social spheres: he launched major construction projects for churches, palaces, roads, bridges, fortifications, and canals. Original records of his instructions initiating these works are virtually nonexistent, but the projects are described in historical sources written during and after his reign. Reproduced here are extracts that exemplify the range of Charlemagne's construction and technological interests. These sources include the important chronicles known as the *Royal Frankish Annals*, which were presumably compiled during his reign, and the *Revised Royal Frankish Annals*, dating from the early years of the rule of Charlemagne's heir, Louis the Pious. Another major source for the history of the Carolingian period is the *Capitularies*, which are records of administrative decrees or ordinances.

Charlemagne's concern for the efficient functioning and regulation of his kingdom's economy is reflected in a capitulary banning unjustified tolls and in *Annals* entries describing his efforts in 793 to construct a canal linking the Rednitz and Altmühl Rivers, and thereby the Rhine (via the Main) and the Danube. This latter project proved to be beyond the capabilities of the technology of his day. Charlemagne's visionary plan was not implemented until the construction of the Ludwigskanal, or Ludwig–Danube–Main Canal, in the nineteenth century.

If, as many scholars believe, the *Annals* are an official or semiofficial account compiled in Charlemagne's court, then one may speculate that the charming entry from 807 about a wondrous water clock from Persia—each hour marked by the ringing of a bell and the emergence of a miniature horseman from a little window—reflected an interest in clever technology not only on the part of the chronicler but also the sovereign himself.

Charlemagne is remembered not just for military campaigns but for achievements such as the codification of both secular and church law and a system to administer his realm through a network of dukes, counts, and margraves, overseen by officials called *missi dominici* ("envoys of the lord"). His reign was marked by a quickening of cultural life that became known as the Carolingian Renaissance, and after his death in 814, he took on the aura of a legendary leader.

CULTURAL CONTEXT

Charlemagne was aware of the power and influence that would accrue to whoever connected the Black Sea with the North Sea. He sought a canal that would link the two seas via the Main and Danube Rivers. Charlemagne tried numerous times, with various plans, to implement this vision, but technology was not yet sufficiently developed to conquer geography.

It was often on the "other side" of the river that diverse groups developed into separate tribes. When geography separated people, those on either side formed their own languages, customs, and ruling administrations, and became ripe targets for Charlemagne's acquisition. Like the soldiers of the Roman army before them, Charlemagne's troops were not just warriors but also engineers and builders. They had to devise ways of crossing the rivers that separated them from their military targets.

A prodigious acquirer, Charlemagne's activities could be considered aggressive. However, his conquests also brought education, fair laws, and improved infrastructure. For example, when he built a bridge (described in the capitulary from 805), Charlemagne suggested tolls. But he was benevolent and thoughtful in levying taxes, charging only merchants or those doing business via the waterways and bridges. He ruled that "tolls shall not be exacted from those who have no intention of conducting trade and are carrying their property from one of their residences to another, or to the palace or to the army" (Capitulary, 805).

PLANNING

Charlemagne did not plan so much as improvise needed structures. If a river stood in the way of his conquering army, he devised a suitable solution on the spot. For example, in 789 Charlemagne decided to put an end to continual challenges from the Slavic residents of the German seacoast. So he crossed the Rhine at Cologne and advanced through Saxony to the Elbe River. Pondering how to move his troops across the river, he realized the solution was to protect both sides of the waterway, for defending just one side would leave the troops

A depiction of digging the Fossa Carolina by Lorenz Fries, 1546. Courtesy of the Library of Congress.

vulnerable to counterattack. By building two bridges across the river, he secured each one with a fort and garrison. Although the structures were simple, built from wood and earth, they were sufficient to do the job. Charlemagne crossed the river, launched a fierce attack, and soon obtained the surrender of the Slavic chiefs. That done, he marched back across the river on one of the bridges he had just built.

For Charlemagne, there appeared to be no long-range planning, no feasibility studies, no ambition to build a monument or lasting edifice. It was pure and simple efficient military improvisation. In the *Annals* of 792 presented here, we read about a pontoon bridge constructed and joined together with ropes and anchors so that it could be put together and taken apart again. A temporary bridge had much to recommend it. First, it could be taken down once the army had used it, thus not leaving a route for escape or retribution. Second, the army did not have to stop and build a bridge every time they wanted to cross somewhere. A portable bridge was a brilliant innovation.

Was this the first documentation of such an engineering feat? Certainly it was not the last. The idea resurfaced in the form of the famous Bailey Bridge. Fast-forward to World War II and a civil servant named Donald Bailey, an employee in the British War Office and a fellow who loved to tinker. He made things at home, sketched ideas, and doodled; his specialty was model bridges. We do not know if Donald Bailey was a history buff, so he may not have read Charlemagne's *Annals* for the year 792. But Bailey did devise the same idea— a temporary bridge that could be carried with the troops and used, then disassembled and moved again to a new location.

BUILDING

In one instance in which Charlemagne was a true planner and builder, having the right idea, he failed to successfully execute the plan. In the *Annals* of 793, it is recounted,

> Certain men, who claimed to have thought of the idea, persuaded the king that one could easily travel by ship from the Danube to the Rhine if a navigable canal could be dug between the Rednitz and Altmühl Rivers, since one of these flowed into the Danube and the other into the Main. Thus, the king went there immediately with all his retinue. He brought together a great number of men and spent the whole autumn engaged in the project. A canal two miles in length and 300 feet wide was built. Yet it was all in vain. Because of the incessant rain and the sogginess of the ground, which was naturally waterlogged, whatever work they accomplished did not remain standing for long. Instead, the dampness caused all of the earth that the workmen dug up by day to slip down and settle back into place during the night.

Building the canal proved to be impossible because of the sedimentary river with soft banks. Later a technical term was developed for this problem: *lateral erosion*.

As Charlemagne discovered, even though a river may look navigable, sediment is continually pushed up onto the banks by displacement caused by the weight of heavy passing vessels. The Romans encountered a similar problem when constructing aqueducts, but of course their channels were much narrower, so the banks could be reinforced with wood and other materials. Given the width and power of the river, Charlemagne had no such option. Charlemagne's Fossa Carolina, in the region now known as Karlsgraben, was a failure.

It took another 1,200 years to implement the design first sponsored by Charlemagne. An exhibit on hydraulic engineering at the Deutsches Museum reveals that the Main–Danube canal at Karlsgraben was replaced by Louis I in 1845. He eventually built what came to be called the Ludwig–Danube–Main Canal, or the Ludwigskanal. Once again, however, technology was a problem. This time it was new technology in the form of railroads. Still, the canal was used, because during World War I there was much need for rail transport, and the Ludwigskanal became an overflow channel. Additional construction occurred between 1845 and 1945. Further construction of the Main–Danube canal began in 1959, with the goal of making the waterway suitable for ships over 1,250 tons.

Final success was achieved on September 25, 1992. On that date, the Rhine and Danube Rivers were linked by a 171-kilometer passage formed by the completion of the Main–Danube Canal. The resulting water route is 3,500 kilometers long. The canal brought with it financial success. The waterway enables 18 million tons of cargo per year to be transported, and millions of euros in revenue are collected.

Another unimagined derivation from Charlemagne's initiative was the development of hydroelectric energy. The Rhein-Main-Donau AG, founded in 1921, has built 57 power stations with a total energy output of 500 megawatts. The realized revenues will pay back construction loans throughout the license tenure, which runs until 2050.

A final outcome is a total waterway network for Germany under laws that launched Project 17 of the Ministry of Transport. The Elbe–Havel Canal will link to the Mittelland Canal between Berlin and Hanover, with the center of the project located in Magdeburg, thereby completing an east–west link.

IMPORTANCE IN HISTORY

Charlemagne was not just a builder of bridges and canals; he built a united Europe. As he conquered tribes and subdued warring factions, Charlemagne encouraged a new European order, one that ultimately became today's European Union.

But it takes more to form a union than merely conquering or offering tax relief. History should thank Charlemagne for other kinds of learning. At the age of 39, he recalled that while on his travels, he had met a scholar in England who impressed him. Alcuin by name, he was a learned cleric and person of

A tourist stands next to a section of a Bailey Bridge in Normandy. Courtesy of John Flaherty.

mild, quiet temperament. Would he come to the court of the Franks and tutor its leader, Charlemagne? He agreed. Charlemagne attacked his studies the same way he waged war—he was up at dawn reading St. Augustine and Jerome. After five years, the importance of education was so clear to him that he issued an order that all the clergy were to be educated. So began the great clerical tradition that would preserve learning during the Dark Ages and give rise to European and worldwide universities.

The tradition has now come full circle. In 2000, Germany established a scholarship in honor of Charlemagne to encourage American students to participate in an international atmosphere of learning. The vision is "to train students for a world in which national boundaries no longer limit the range of knowledge or entrepreneurial spirit."

FOR FURTHER REFERENCE

Books and Articles

Barbero, Alessandro. *Charlemagne: Father of a Continent*. Translated by Allan Cameron. Berkeley: University of California Press, 2004.

Butt, John J. *Daily Life in the Age of Charlemagne*. Westport, CT: Greenwood Press, 2002.

Internet

For an overview of river and waterways engineering, and a synopsis of the Main–Danube Canal, see http://www.deutsches-museum.de/ausstell/dauer/wasserle_wass2.htm.

For information on the Bailey Bridge, see http://www.fact-index.com/b/ba/bailey_ bridge.html.

For more about the Rhine River, see http://www.public.asu/edu/~goutam/gcu325/ rhine.htm.

For the Charlemagne scholarship, see http://www.rwth-aachen.de/zentral/charlemagne_ scholarship.htm.

For J. H. Lienard's audio program on Alcuin and Charlemagne, including quotes and a bibliography of interest to those exploring the scholarly side of Charlemagne, see http://www.uh.edu/engines/epi797.htm.

Documents of Authorization

Editors' note: There are three types of documents presented here: annals, revised annals, and capitularies. Annals were written during Charlemagne's lifetime; revised annals were written after Charlemagne's death and therefore contain much more information, especially concerning difficulties or failures encountered. To illustrate this point, the reader can compare Annal 793 with Revised Annal 793. Capitularies are regulations said to have been written or authored by Charlemagne for the governance of his realm.

A.D. *789 (Revised Royal Frankish Annals)*

There is a certain tribe of Slavs dwelling along the sea-coast in Germany who are called *Welatabi* in their own tongue and *Wilzi* by the Franks. They had always been enemies of the Franks and they continually harassed the neighboring tribes—who were either subjects or allies of the Franks—and pressed and provoked them in battle. The king, concluding that he could no longer tolerate their effrontery, decided to wage war against them. Gathering together a great army, he crossed the Rhine at Cologne. From there he marched through Saxony, and, when he had reached the river Elbe, he set up camp on the bank. He then built two bridges across the river; at each end of one of these he built a fort, and secured it with a garrison, made out of wood and earth. He then crossed the river at this point with his army and, entering the land of the Wilzi, he ordered that everything should be laid waste by fire and sword. Although they were fierce warriors and trusted in their numerical strength, the tribesmen could not endure the king's attack for long. For that reason, as soon as the invading forces arrived at the town of Dragawit (for he was by far preeminent among their chiefs because of his noble lineage and the respect afforded to his age), he left the town at once along with all of his men and went to the king. He handed over the hostages that were demanded of him and swore an oath to remain loyal to the king and to the Franks. The other chiefs of the Slavs followed his example and all submitted themselves to the king's rule. After the king had subjugated the Slavs and taken the hostages from them he had demanded to be given, he returned to the Elbe along the same route by which he had come and led his army back across the bridge. After he had dealt with the affairs of the Saxons as far as time would allow, he returned into Francia and celebrated Christmas and Easter in the city of Worms.

A.D. 792 *(Royal Frankish Annals)*

In this same year no military expedition was undertaken. On the river a pontoon bridge was constructed and joined together with ropes and anchors, so that is could be put together and taken apart again. The king celebrated Christmas and Easter at Regensburg, but it is not entirely clear if the pontoon bridge was constructed there or somewhere else.

A.D. 793 *(Revised Royal Frankish Annals)*

Certain men, who claimed to have thought of the idea, persuaded the king that one could easily travel by ship from the Danube to the Rhine if a navigable canal could be dug between the Rednitz and Altmühl rivers, since one of these flowed into the Danube and the other into the Main. Thus, the king went there immediately with all his retinue. He brought together a great number of men and spent the whole autumn engaged in this project. A canal two miles in length and 300 feet wide was built between the aforementioned rivers. Yet it was all in vain. Because of the incessant rain and the sogginess of ground, which was naturally waterlogged, whatever work they accomplished did not remain standing for long. Instead, the dampness caused all of the earth that the workmen dug up by day to slip down and settle back into place during the night.

A.D. 805 *(Capitulary)*

Concerning tolls, it is our will that those that are fair and have been collected in the past should be exacted from those doing business at bridges, at ferries, and at markets; new or unjust tolls are not to be exacted, in cases where ropes are stretched (across rivers), or when ships pass beneath bridges, and in similar situations in which no assistance is provided to the traveler. Likewise, tolls shall not be exacted from those who have no intention of conducting trade and are carrying their property from one of their residences to another, or to the palace or the army. If there is any occasion for doubt, however, let the matter be brought up at the next assembly that we hold with our officials.

A.D. 807 *(Revised Frankish Annals)*

Radbert, an emissary of the emperor, died during his journey back from the East. An ambassador from the Persian king named Abdullah, along with some monks from Jerusalem who were carrying out a diplomatic mission for the patriarch Thomas . . . came to the emperor bearing gifts sent to him by the Persian king, namely a pavilion and multicolored canopies for its foyer of great size and remarkable beauty. Everything—the canopies as well as their cords—was made of linen dyed in different colors. The king had also sent many valuable silk cloaks, along with perfumes, ointments, and balsam. There was also a marvelously fashioned clock made out of brass, on which the courses of the twelve hours moved according to a water-clock, with an equal number of little bronze balls, which fell down when the clock struck the hour and caused the cymbal placed under them to ring. In addition, there were twelve mounted figures that issued from windows upon the hour and caused the windows to close again by the force of their movement. There were other things on the clock as well, though it would take too long to mention them all now. Among these gifts there were also two large candelabra made out of brass. All of these things were brought to the emperor at his palace in Aachen. The emperor kept the ambassador and the monks with him for a while and then sent them into Italy, where he ordered them to wait until it was time to sail.

Translated from the original Latin by Justin Lake, Harvard University, 2004.

The opening of the New London Bridge, witnessed by William IV, king of England, 1831. Courtesy of the Library of Congress.

8
London Bridge

DID YOU KNOW . . . ?

➤ London Bridge was the first major stone arch bridge in Britain; it was built from 1176 to 1209.

➤ It had 19 arches, each spanning 24 feet (7 meters).

➤ The deck of the bridge held the first shopping mall in England.

➤ Workers' compensation was given to bridge workers.

➤ Nonesuch House was an early prefab building.

➤ Engineer John Rennie's rendition of the bridge was moved piecemeal to a permanent home in Lake Havasu, Arizona.

Erecting the old bridge across the Thames in London was the most formidable enterprise of its kind undertaken in England during the Middle Ages."
> —Samuel Smiles, *Lives of the Engineers*, 253

London Bridge is falling down,
falling down, falling down. . . .

> —*Old English nursery rhyme*

Both statements are true. It is certain that London Bridge was an important engineering achievement in medieval times, connecting the ever-expanding environs of London. It is also true that over the years it required major repairs. In fact, the body of bridge-engineering knowledge available today can be attributed, in part, to the building and rebuilding of London Bridge.

As for the nursery rhyme, the original reads, in part,

> London bridge is broken down,
> Dance over my Lady Lee,
>
> London bridge is broken down,
> With a gay ladye . . .
>
> We'll build it up with gravel and stone,
> Dance over my Lady Lee . . .
>
> Gravel and stone will be washed away
> Dance over my Lady Lee . . .
>
> We'll build it up with iron and steel,
> Dance over my Lady Lee . . .

There is more to this nursery rhyme than meets the eye. The song's original lyrics mention "my Lady Lee and a mysterious gay ladye." These two are none other than the sister of the poet Thomas Wyatt, who after her marriage became Lady Lee, and her childhood friend Anne Boleyn. The story of Anne Boleyn, the "gay ladye" who was hated by the common folk after she replaced the beloved Katherine of Aragon to become the second wife of Henry VIII, is known by many. When the daughter of Anne Boleyn and Henry VIII, Elizabeth I, took the throne, the official position on Anne Boleyn had improved. But it was a posthumous change in status, for Anne Boleyn was hanged for adultery, with her friend and lady-in-waiting Lady Lee standing nearby.

Besides telling an allegorical story of daily life in the British court, the nursery rhyme also refers to the engineering process of strengthening London Bridge. At first, it tells of being built up "with gravel and stone," but the rhyme cautions that these can be "washed away." Various reinforcement methods are mentioned: iron, steel, silver, and gold. Such methods bring the evolution of London Bridge up to the present—it can be compared to the Millennium Bridge of 2000.

HISTORY

The name *London Bridge* has been given to many bridges connecting what is now London (on the north side of the Thames River) and Southwark (on the south side). Several wooden spans were built in the Roman and medieval eras, only to be destroyed by flood or fire, perhaps even by rampaging Vikings. But the most celebrated London Bridge—the first major stone arch bridge built in Britain—was erected in the late twelfth and early thirteenth centuries.

The church played a key role in erecting the bridge, for it had long encouraged Londoners to give alms to fund bridges, which enabled parishioners in outlying areas to cross the river and attend services. Peter, the chaplain of St. Mary Cole Church and head of the Fraternity of the Brethren of London Bridge, also known as Peter de Colechurch, was largely responsible for construction of the last wooden bridge. In 1176 King Henry II entrusted Peter with building a new stone bridge. The construction site was slightly west of the timber span in use at the time. As the years passed, Peter fell into ill health, and work slowed. Henry's son John assumed the throne in 1199 and, displeased with the slow pace of construction, put a new man in charge, the clerk Isenbert, "Master of the Schools of Xainctes." King John's letter to the mayor and citizens of London about this matter is reproduced here in translation from the Latin original. Little is known about Isenbert, but progress on the bridge quickened, and it was completed by others in 1209. Peter died in 1205.

The original design called for a wooden drawbridge with 19 arches, each spanning about 24 feet (7 meters). The supporting piers, resting on stone foundations, were to be 20 feet (6 meters) wide, each to be surrounded by a protective "starling"—a structure formed by piles and filled with loose stone. During piledriving, obstructions were found that made it necessary to vary the widths of the 19 arches from 15 feet (4.6 meters) to 34 feet (10.4 meters). Because the starlings were so large, the passageways for the flowing water became even narrower, about one-fourth the original width. During tidal change, the flow became extremely fast—dangerous for boat traffic but exciting for thrill seekers who liked to "shoot the bridge" in small boats.

The Chapel of St. Thomas of Canterbury stood on the middle pier, and at each end of the bridge there were fortified gates on the tops of which were displayed the heads of executed traitors. The upper surface of the bridge was just like a street. Shops, with living quarters on the upper floors, stood on both sides of the roadway. These shops generated substantial revenue from taxes and rent. (In his letter King John explicitly ordered that "the rents and profits" from these establishments erected by Isenbert be allotted to bridge maintenance and repair.)

The bridge's centuries-long history was shadowed by misfortune, but though it was often maintained in less-than-perfect condition, the bridge was never allowed to fail. It remained the city's only span across the Thames until 1750, when Westminster Bridge was opened. A catastrophic fire had destroyed the buildings on London Bridge in 1212, three years after it was finished; the death toll may have reached 3,000. But the structures were rebuilt, leaving a mere 12 feet (3.7 meters) for the roadway. Winter ice in 1282 caused five arches to collapse. They were restored. By 1358, 138 premises were registered. Walkways and additional rooms were extended between the buildings. Water mills were installed in the 1580s.

Following the completion of Westminster Bridge, it was decided to repair London Bridge. By 1762 the buildings atop the span were gone, the roadway was broadened to 46 feet (14 meters), and the two middle arches were replaced by a single large arch. But the elimination of the middle supporting pier led to erosion of the riverbed and increased maintenance costs for the remaining piers. A stone replacement bridge, designed by engineer John Rennie, was built some 60 yards (55 meters) upstream. The old London Bridge was demolished in 1832. Less than one and a half centuries later, Rennie's bridge was itself removed; its facing stone was transported to Lake Havasu, Arizona, where it was mounted on a core of reinforced concrete to form a five-span version of London Bridge and become a major tourist attraction. Today's London Bridge over the Thames, constructed between 1968 and 1972, is a two-pier cantilever structure of prestressed concrete.

CULTURAL BACKGROUND

The Romans built a bridge over the Thames, possibly the first London Bridge, but it was made of wood and vulnerable to fire. The Saxons built a bridge that was equally susceptible to both fire and flood. So the issue during medieval times was how to build a bridge that would last. Of course, stone was the answer.

London Bridge illustrates the critical role of the church in medieval engineering. By teaching the Lincolnshire peasants to drain swamps, St. Botolph succeeded in founding the town of Boston ("Botolph's Town") in northeastern England.

The Cistercians were for centuries leaders of agricultural engineering in western Europe, designing reliable water-supply systems and storage methods for hundreds of monastic centers.

PLANNING

In the construction of London Bridge, we have an early instance of changing general contractors during a macro project. The first plan for the bridge was conceived by Peter de Colechurch, the engineer who built the previous wooden bridge. When he died he was laid to rest in a chapel on the bridge. However exalted his interment, Colechurch had never been able to get the bridge built to last. A new plan was needed.

Exactly the right person was found. King John extended his search to France, where engineering was enjoying a renaissance, and he hired Isenbert, the French engineer and cleric who had prior experience building bridges in France. Because those bridges were located in an area with many islands, bridges were an engineering specialty of Isenbert's region. While there are no records yet found of an academy of engineering with a major research program in stone building, it is reasonable to infer that a region where rocks are found and where bridges were frequently built might also harbor experts who were skilled in stone bridge building.

As the document presented here shows, King John had to sell the idea to the lord mayor of London. Acceptance of Isenbert required assurances that this engineer had the right credentials. The authorization describes Isenbert as someone who "in a short time hath wrought in regard to the Bridges of Xainctes and Rochelle, by the great care and pains of our faithful, learned and worthy Clerk, Isenbert, Master of the Schools of Xainctes."

Besides expertise, there was the matter of how to pay for the new structure. Isenbert proposed a plan that may have been a forerunner of the first shopping mall in England. King John continues, in his letter of persuasion, "Wherefore, without prejudice to our right, or that of the City of London, we will and grant, that the rents and profits of the several houses which the said master of Schools shall cause to be erected on the Bridge aforesaid, be for ever appropriated to repair, maintain and uphold the same." Thus was sealed the real estate deal in which London Bridge would rent retail space to merchants and the rent would be used for upkeep of the structure. Such an arrangement was a far better option for the citizens of London, who had been paying tribute on all goods containing wool, skins, and leather. Now, at least, Londoners could enjoy the bridge without excessive taxes.

A final clause in King John's letter to the lord mayor provided a kind of insurance or worker's compensation plan for the building crew. "Should any injury be offered to the said Isenbert, or to the persons employed by him, which we do not believe there will, see that the same be redressed as soon as it comes to your knowledge." While we do not know what the redress and reimbursement was, we do know that many people were injured and that approximately 150 workers lost their lives during the building of the bridge.

BUILDING

Construction originally begun by Peter de Colechurch was resumed at a quickened pace by Isenbert and his crew. The bridge's piers, protected by starlings, were planted deep in the riverbed to withstand the fierce rush of water. The piers were key to the successful building of the bridge. According to Samuel Smiles, an authority on English engineering, "comparatively few of the older bridges failed from the unskillful construction of their arches, but many were undermined and carried away by floods where the piers were insecure" (Smiles, 255).

However the foundations were anchored (and sources are too meager to verify), some engineering historians have supposed that the entire course of the Thames River must have been diverted so that the riverbed could be dry while the foundations were driven in and fortified. We know that the starlings were made of elm-tree beams pushed down into the riverbed as close together as possible, then fortified by piles of stone rubble. Across the elm and stone structures were laid oak beams, and then a second layer of beams, and then around that more stones and rubble. Over time, the width of the piers was expanded to make the foundation more secure, resulting in a dangerous increase in the rapids flowing through the piers. It is said that the rapids became so vigorous that their sound could be heard from far away.

For a complete view of London Bridge as it looked in 1540, the reader can visit an interactive BBC Web site to see the structure and aspects of the bridge,

A modern day shot of the London Bridge. Getty Images / PhotoDisc.

the chapel, the great stone gate providing defense and fortification, the shops, and the drawbridge gate where severed heads once stared out in ignominy (see "For Further Reference").

When the drawbridge gate was removed in 1577, one of the first prefabricated buildings of medieval times replaced it—the so-called Nonesuch House, which was constructed in sections, shipped from the Netherlands, and assembled on-site. With the arrival of the Nonesuch House, the severed heads were moved to the Great Stone Gate, and the practice of displaying the executed continued until 1678. The gallery of rogues was such a feature that as late as 1598 a manuscript written by a German traveler, Hentzner, mentions 30 visages staring over the bridge from the poles on which they were impaled.

By the mid-1700s the bridge had changed considerably. The shops were gone. With the newly acquired space, a larger middle arch was created, and one of the foundation piers was removed due to increased support from the arch. Removing a pier made it possible for more watercraft to pass under the bridge.

IMPORTANCE IN HISTORY

A traveler to Lake Havasu, Arizona, can tour the 1831 London Bridge in a resort setting. Yet surely the importance of the remarkable old London Bridge, which so changed medieval Britain, extends well beyond that of a tourist attraction. Like many structures created as a result of great engineering innovations, London Bridge united a growing city while providing a transitway for commerce and the exchange of ideas. Cities on the water—whether sea, river, or lake—are a common theme in the story of civilization. When the aquatic resource is a river, one can be sure that the building of bridges will soon follow. What is important about London Bridge is its innovative construction in durable and fireproof stone, and the advances in engineering that evolved from this innovation.

Zooming forward to the present, one must mention the Millennium Bridge, built in London to mark the turn of the century in 2000. A few days after the bridge opened in June 2000, it was closed because of severe swaying caused by the unforeseen challenges of resonant frequency. When pedestrians crossed the newly opened bridge, the resulting vibrations caused the bridge to sway more than eight inches from side to side. For this same reason, soldiers are taught to march out of step when crossing bridges. The troubles that accompanied the construction of that bridge make one appreciate even more how well Isenbert built the original London Bridge, which lasted—intact—for six centuries.

FOR FURTHER REFERENCE

Books and Articles

Jackson, Peter. *London Bridge*. London: Cassell, 1971.
Smiles, Samuel. *Lives of the Engineers with an Account of Their Principal Works Comprising Also a History of Inland Communication in Britain*. 3 vols. London: John Murray, 1861. London Bridge is discussed in volume 1.

Internet

For a view of London Bridge in 1540, see http://www.bbc.co.uk/history/3d/bridge.shtml.
For the full lyrics of the old nursery rhyme and the legend behind the seemingly innocent children's song so familiar to many, see http://www.rhymes.org.uk/london-bridge-is-broken-down.htm.

Music

Williams, Ralph Vaughan. "As I Walked over London Bridge." Study in English Folk Song, No. 6. CD In: Bassoon Bon-bons. White Line, Cat. No. 2052, Track #16. Daniel Smith, bassoon and Roger Vignoles, piano. CD, recorded 1993.
———. "A London Symphony" No. 2. The London Symphony. Conducted by Bryden Thomson. Chandos 8629, 1988.

Art

Cooke, Edward William. *Old and New London Bridge: A Selection of Drawings Reproduced from Originals in the Possession of the Guildhall Art Library*. London: Topographical Society, 1970.

Documents of Authorization

John, by the Grace of God King of England, etc. to his faithful and beloved the Mayor and Citizens of London, greeting. Considering how the Lord in a short time hath wrought in regard to the Bridges of Xainctes and Rochelle, by the great care and pains of our faithful, learned and worthy Clerk, Isenbert, Master of the Schools of Xainctes: We therefore by the advice of our Reverend Father in Christ, Hubert (Walter), Archbishop of Canterbury, and that of others, have desired, directed and enjoined him to use his best endeavour in building your bridge, for your benefit, and that of the public: For we trust in the Lord, that this Bridge, so requisite for you, and all who shall pass the same, will through his industry and the divine blessing, soon be finished. Wherefore, without prejudice to our right, or that of the City of London, we will and grant, that the rents and profits of the several houses which the said master of Schools shall cause to be erected on the Bridge aforesaid, be for ever appropriated to repair, maintain and uphold the same.

And seeing that the requisite work of the bridge cannot be accomplished without your aid, and that of others, we charge and exhort you, kindly to receive and honour the above-named Isenbert, and those employed by him, who will perform everything to your advantage and credit, according to his directions, you affording him your joint advice and assistance in the premises. For whatever good office or honour you shall do to him, you ought to esteem the same as done to Us. . . . But, should any injury be offered to the said Isenbert, or to the persons employed by him, which we do not believe there will, see that the same be redressed as soon as it comes to your knowledge.

Witness myself, at Molinel in the Province of Bourbon, France,—the eighteenth day of April, 1209.

From C. W. Shepherd, *A Thousand Years of London Bridge* (London: John Baker and New York: Hastings House, 1971).

9
The New River

England

DID YOU KNOW . . . ?

➤ The New River was constructed between 1609 and 1613.
➤ It begins 20 miles (30 kilometers) north of London.
➤ 157 wooden bridges cross the river along its length.
➤ Its original capacity was 20 million gallons (80 million liters) per day; today it is more than double that.
➤ The people of London use approximately 40 gallons (150 liters) of water per day.
➤ The river is an early example of a public utility launched through mixed public/private finance.
➤ It is also an early application of the principle of an "ever-normal water supply."

The city of London remains a metropolis today because of a piece of jewelry. King James I sought to advance favor with his wife, the queen, by buying for her a pendant with a single beautiful diamond.

Was it a diamond that changed the realm? Or the relationship of trust between a husband and a jeweler? Perhaps a bit of both. The king trusted in England's future, and in Hugh Myddleton, whom he allowed, in 1612, to cut a deep trench through the lawn of the royal country estate, Theobald's Park, to bring freshwater from the countryside into the teeming and thirsty city of London. And who was Hugh Myddleton? The king's jeweler!

Without Hugh Myddleton's vision and tenacity, London would likely not be what it is today. As is often the case with great macroengineers—like Frontinus,

The first View of the New River—from London.

After manifold windings and tunnelings from its source, the New River passes beneath the arch and forms a basin within a large walled enclosure, from whence diverging main pipes convey the water to all parts of London. Courtesy of the Library of Congress.

who masterminded the Roman aqueducts—Myddleton was totally committed to his project. In the words of the historian Samuel Smiles: "That power of grappling with difficulties which emboldened him to undertake this great work was more like that of a Roman emperor than a private London citizen" (Smiles, 1:109).

HISTORY

The New River is not a river, but it definitely was new at one time. It was the result of an early-seventeenth-century project to provide the growing city of London with a new source of freshwater, brought from Hertfordshire via a hand-dug channel of dimensions unprecedented at the time in England. It remains in service today, albeit somewhat revamped.

Four centuries ago, London's population was approximately 500,000. Not only was this far more than the city's water sources at the time could support, but pollution was degrading the quality of major wells. An obvious solution was to bring water to the city from other sites. Edmund Colthurst pushed for authorization to use springs in Hertfordshire, and in 1604 King James I gave Colthurst full power to make the water available within seven years.

The following year the city of London obtained an act of Parliament that officially authorized the work. Perhaps indicative of a growing concern for property rights in a country where power no longer flowed solely from the monarch, the measure (reproduced here as Document I) dealt in some detail with various issues involved in constructing "the Trench"—issues such as supervisory authority, compensation to property owners, damage caused by "Breaches" in the trench, transport across the "New Cutte or River," and responsibility for future maintenance. For instance, it quickly became apparent that at least in some places an open trench would not be the best means of conveying the water. The following year Parliament passed a supplementary act (reproduced here as Document II), which explicitly permitted, where it seemed "more meete," the construction of "a Trunke or Vault of Bricke or stone . . . to be layed in the Earth or upon Arches."

However, work on the project did not immediately commence. One reason for delay was probably the need to obtain permissions from property owners; another was apprehension among city authorities engendered by the daunting anticipated costs of the project. Ultimately, businessman, engineer, and goldsmith Hugh Myddelton stepped forward and offered to shepherd the scheme to completion at his own expense. Work began in 1609, but Myddelton ran out of money midway through. The project continued thanks to timely financial assistance from King James in return for a share of the profits (see Related Cultural Document) and was completed in 1613.

Parliament authorized drawing water from Hertfordshire's Chadwell and Amwell Springs, among others in the area. In time, water was also diverted from the nearby River Lee. The starting point was situated about 20 miles (30 kilometers) north of London, but a circuitous route was required to ensure a downhill path for the flow of water; 157 wooden bridges were built across valleys and other low-lying patches. A gradient of about two inches per mile (three centimeters per kilometer) was maintained, and the actual length of the New River was nearly twice the direct distance. The channel had a capacity of about 20 million gallons (roughly 80 million liters) a day. It ended in Islington at the New River Head; from there, the water was distributed to various areas through wooden pipes.

Over the centuries, the New River's meandering course has been considerably straightened and shortened, with some portions moved underground. Pumping stations were built. Today, the conduit's capacity has doubled, and the termination point is a reservoir at Stoke Newington. A statue of Myddelton can be found in north London's Islington Green where Essex Road and Upper Street come together.

CULTURAL BACKGROUND

With London built beside the majestic Thames, why did the city need additional supplies of freshwater? By the time of Queen Elizabeth I, the population of London was growing rapidly; Londoners were polluting the very river

Hugh Myddleton, circa 1610. © Archive Photos / Getty Images.

on which they depended by tossing debris into it and dumping refuse in ditches and streams. The Corporation of London debated the issue: would it be easier to stop people from tossing garbage into the Thames, or to find fresh drinking water from another source? London may well have been saved by the decision to rely on the innovative side of human nature rather than on its more tidy side. Plan for failure and achieve success? In 1570 an act of Parliament resolved to bring fresh drinking water to London from another source.

Meanwhile, a new trade had sprung up in London. Water bearers were taking in shillings and pence as fast as the populace made tea, bringing freshwater to the city in barrels. But the water bearers were only a temporary solution, and certainly not one available to the average citizen. Soon enough, there was a water emergency. The government became aware, the realm was motivated, the water was available; all that was needed was a plan.

PLANNING

Getting the project from idea to fresh liquid in England's famous "cuppa" proved to be challenging. There were two early ideas of a new water system. One was to begin building at the River Colne at Uxbridge and bring water to the town of Holborn in north London. But that effort failed. Then Edmund Colthurst proposed using the fresh springs near Hertford and channeling them to London, stopping the flow in Islington. Why Hertford? A simple matter of gravity. Like the Romans before them, the Elizabethans knew they could rely on the gradual downward slope of the land from the springs to the city. Once the water was near the city, cisterns or ponds could be created and then pipes could do the rest of the job. Why Islington? Like Holborn, it was uphill of the city. Colthurst had the plan, he had a route, he even had permission from the Corporation of London. But Queen Elizabeth delayed commencement of the project due to concerns about tampering with navigation on the Lee River.

Elizabeth's passing brought James I to the throne and with him the long-needed backing of a sovereign. Letters patent were issued stipulating that the new water project or channel be restricted in width to 5.9 feet (1.8 meters) and that it be completed within seven years. Delighted, Colthurst starting digging, spent £200 pounds of his own money, and then returned to the Corporation of London for more funds. The project quickly ground to a halt. After years of careful planning and even a sample section dug, the situation—and the nascent river—dried up along with the Corporation's coffers.

BUILDING

Enter—*deus ex machina*—the hero of the story, Hugh Myddleton. As the king's jeweler, he had amassed a fortune and had become a banker and member of Parliament. Here was someone the Corporation liked. He had enough money to bankroll the project and get it moving again. He was an MP. And he was a friend of the king's—or at least had enjoyed some special commerce with him. Besides, Myddleton promised to complete the entire project in just four years. In 1609, the deal was sealed.

It tells us a lot about Myddleton's character that he immediately hired Colthurst, and the two worked together cordially until the project's completion, and even beyond. Together they measured and surveyed the terrain, determined the course of the cuts to be made, and ensured that the water ended up in London. A surveyor mapped the entire course and called in hundreds of laborers to complete the job.

Problems mounted and were surmounted. At first the soil was a problem—it was too porous. The solution? Clay was brought to the site and stamped into the bottom and sides of the channel by scores of workers. Then obstacles as inconvenient as roads were met and crossed by bringing in skilled carpenters and masons to build arches and bridges. By the completion of the New River, 157 bridges had been crafted, as well as many tunnels and arches.

Myddleton followed a contour line from the level of the beginning of the New River at the Chadwell Spring to the final pond at Islington. The original stream had a fall of about two inches per mile and the city end was 82 feet above the high-water mark. If the fall of the water was too steep and rapid, stopgates were introduced and weirs broke the fall just enough to keep the flow even. To fit the channel to the terrain, it was necessary to curve the river around the sides of hills, with carpenters shoring up the banks with timber and laborers padding the bottom with clay. Once the water reached Islington, it was stored in ponds and then piped into the city, with pipes fashioned of timber or lead. While one must wonder whether lead entered the water supply (and of course it is possible), protection against this unwanted outcome came from the water flow itself. Lead leaches out more quickly in standing water.

The worker system was well delineated, with categories of workers and varying pay scales. Carpenters earned 1 shilling (s) 4 pence (p) per day, while bricklayers earned slightly more, 1s 6p. Excavators were paid by poles and rods, with wages adjusted for depth and width, so their pay ranged from 1s 2p to 3s 6p. Laborers' pockets were much lighter, with just 10p per day, and a little more if they had to work standing in water, as many did.

As the project drew near the homes of the wealthy, it slowed considerably. NIMBY (not in my backyard) was alive and well even in those times, with farmers complaining that their fields would be washed out if the river was allowed to cross their land. Displeasure also erupted from London water bearers, who viewed the New River as a threat to their livelihood. Again, Myddleton came to the rescue. Thanks to their relations, Myddleton and James I reached an agreement. That original agreement, located in the Rolls Office and dated May 2, 1612, is included at the end of this chapter. The deal was this: Myddleton and the king would own the project 50-50, the New River could go through the king's land, the king and Myddleton would share the expenses, and the two would split the profits "for ever."

There is an exception clause to the deal, early evidence in English history of the charitable or humanitarian aspect of a private enterprise. The profits were available from the whole of the New River with the "exception of a small quill or pipe of water which the said Myddleton had granted, at the time of his agreement with the City, to the poor people inhabiting St. John-street and Aldersgate-street, which exception His Majesty allows" (Smiles,1:117). In later times, it became common to reserve a portion of the profits for charity; later, in some countries, tax deductions were systematically allowed for charitable contributions.

As the New River neared the city, doubts turned to cheers. The project was completed early—one of the few of this magnitude to boast such an outcome. At the conclusion, a knighthood was bestowed on Hugh Myddleton.

This spring near Ware served as the original source of the New River. Courtesy of Shutterstock.

IMPORTANCE IN HISTORY

"It was the greatest enterprise of the time yet to be attempted in England," according to Smiles (1:120). But it is not merely an ancient work or a medieval relic. The New River still supplies water to London.

The time line of the New River illustrates a veritable progression of technology. At first, gravity did the work, then people, then horses, and eventually pumps. At first the pumps were manual, then steam, and eventually electric. At first, timber was used for pipes, then lead, and even later, cast iron. At first, the water flowed from the spring, but in 1846, a cholera epidemic caused the first water-treatment process to be invented and applied in England; sand was used as the filtering material, and later, more sophisticated water-treatment plants. All were eventually replaced by the Coppermills Water Treatment Works and the Water Treatment Works at Walthamstow. Responsibility for the overall business of London's water eventually left Myddleton's control; it was taken over by the Metropolitan Water Board (1974) and thereafter by Thames Water PLC (1989).

The system continues to produce water, and London continues to need it. Today's Londoners use an average of 33 gallons (150 liters) of water daily, all of which requires treatment (ancient Roman aqueducts supplied an average of 198 gallons (750 liters) per person per day). When there is surplus water, it is brought to an aquifer 249 feet (76 meters) below London; if there is a drought, water is pumped up from the aquifer back into the New River.

Water was plentiful; so was the money, as is apparent in the financial records of the New River. According to Smiles:

> After 1640, the financial success of the New River Company flowed as fast as the rapids. As London grew, so did water consumption. The capital was divided into 72 shares; half to Sir Hugh Myddleton and half to King James. Sir Hugh sold as many as 28 of his own shares. . . . At the end of the 17th century, the dividend paid was 200 pounds per share. At the end of the 18th century, the dividend was 500 pounds. At the end of the 19th century, the dividend was 800 pounds per year. . . . This is an extraordinary return because the original cost of the New River did not amount to more than 18,000 pounds. (1:129–32)

The New River is one of the earliest great projects for which we have such accurate financial data, and it may also be one of the first public utilities to be privately (and profitably!) financed.

FOR FURTHER REFERENCE

Books and Articles

Cosh, Mary. *A Historical Walk along the New River.*
Lopresti-Essex, Michael. *Exploring the New River.* London: Brewin Books, 1997.
Smiles, Samuel. "The New River Works Begun." Part 2, chaps. III and IV, in *Lives of the Engineers with an Account of Their Principal Works Comprising Also a History of Inland Communications in Britain.* London: John Murray: 1861.

Internet

For an excellent overview of the project, with helpful readings and concise material, see http://www.waterinschools.com/newriver/story.html.

For the technology and science of the New River, with suggested experiments that enable students to explore the physics and hydraulic-engineering principles, see http://www.waterinschools.com/newriver/technology.html.

For information leading to material about the aqueducts of the New River and the tunnels, see http://www.citiesofscience.co.uk/go/London/ContentPlace_2525.html.

For more on the water authority, see http://www.thames-water.com/.

For water use and treatment, see http://www.riverlee.org.uk/.

For a walking route along the New River, a path of 27 miles (43 kilometers) noted as an easy stroll with historic highlights (along with lists of several publications, including *The New River Path: A Walk Linking Hertford and Islington,* which can be ordered from Thames Water or the Ramblers Association), see http://www.ramblers.org.uk/info/paths/newriver.html.

Music

Songs of London. Capital Radio 194, HALC3 CD UK, 1979. This album, made for the benefit of London children, includes classics such as Joni Mitchell's "Chelsea Morning" and Rick Kemp of Steeleye Span's rendition of a nineteenth-century song contrasting London dandies and the poor from the album *Rocket Cottage.*

Documents of Authorization—I

A. STATUTORY FOUNDATIONS 1605

A.1 3 Jas. I c. 18 (1605)

An Acte for the bringing in of a fresh Streame of running water to the North parts of the Citie of London.

For that it is found very convenient and necessary to have a fresh Streame of running water to bee brought to the North parts of the City of London, from the Springs of Chadwell and Amwell, and other springs in the Countie of Hartford not farre distant from the same, which upon view if found very fesible, and like to be profitable to many: It is therefore enacted by the Kings most Excellent Majestie, and by the Lords Spiritual and Temporal, and Commons in this present Parliament assembled, and by the Authoritie of the same, that

> 1 It shall be lawfull to the Lord Mayor, Comminaltie and Citizens of the Citie of London, and their Successours, at any time or times hereafter to beginne and continue the laying out of such convenient Limits of ground for the making of the Trench of the sayd River at the breadth of tenne foot and not above, as to them and their Deputies and workemen, with the allowance of the Commissioners hereafter mentioned or any seven of

them shall bee seene convenient and meete for the same, And in that place that they shall find to be most apt and meet for that purpose, to have and take for the purpose abovesaid the use and libertie of such and so much ground, as shall contain Ten foot in breadth, and not above, during and by all the length, as the sayd new Channell, Cut, or River shall passe for the conveying of the new said water from the sayde Springs to the Citie of London, leaving the inheritance of the New Cut in the owners thereof.

2 And that the sayd Mayor, Comminaltie, and Citizens of London, and their Successours for ever, for the consideration hereafter expressed, shall have liberty to digge the same ground to be imployed for the sayd River or New Cut, not exceeding tenne foot in breadth alongst all the sayde whole length of the sayd River or New Cut, and from time to time for ever to maintaine and preserve the same, and to lay the earth there digged or to be digged on either side of the same River or New Cutte in such places, as shall be thought meete for that purpose, and to have free Passage to and from the sayd New Cut or River with men, horses, cart and carriages at all times convenient for making of the same New Cut or Trench and for the preserving of the same, and the Banckes thereof, from time to time for ever, to the intent that no part of the sayde Streame be at any time after the making of the New Cut, without the consent of the Mayor, Comminaltie, and Citizens of London, turned or conveyed out of the same New Cutte or Watercourse.

3 In consideration whereof, The Mayor, Comminaltie, and Citizens of London, and their Successours shall make such satisfaction or composition to and with the Lords, Owners, and Occupiers of the same grounds, through which the newe Cut or River shall be made, and with all such person and persons as shall sustaine any Damage, Losse or hinderance in their Mils standing upon any of the Rivers or Streames, from which the water shall be taken through the sayd New Cut or River, as shall be to the contentment of the Lords, Owners and Occupiers of the sayd grounds and Mils, and in default of their agreement by mutual assent such satisfaction or recompense, as shall bee limited and appoynted by the Commissioners to bee assigned for that purpose, according to the intent of this Statute, by the Lord Chancellour or Lord Keeper of the great Seale of England for the time being, by Commission under the great Seale of England, or by any nine of them, whereof foure of them to bee Citizens of the City of London.

4 And for the better effecting of the premisses, and for the due rating of the value of the things to be compounded for by the true intent of this Statute (if the parties shall not agree,) Be it enacted by the Authoritie of this present Parliament, that at the request and charges of the Mayor, Comminaltie and Citizens of London, Commission or Commissions, under the great Seale of England, shall be granted to such persons, as the Lord Chancellour or Lord Keeper of the great Seale of England for the time being, shall

nominate and appoint, wereof foure shall bee of the Countie of Middlesex, foure of the Countie of Essex, and foure of the Countie of Hartford, and foure of the City of London, and every of them having lands and Tenements of the cleare yeerely value of Fortie pounds at the least: which sixteene or any nine of them, whereof two to bee of the Citie of London, shall have power to order and set downe what Rate or Rates, summe or summes of money shall be payed by the Mayor, Comminaltie and Citizens of London to the Lords, Owners, and Occupiers of the grounds, and Soyle and Milles, for which Composition is to be made by the intent of this Acte, if the parties cannot of themselves agree, and in what manner the same shall be payed, and that for the recovery of such money, as shall be so ordered and set downe by the said Commissioners, or any nine of them, whereof two to be of the Citie of London, the partie or parties to whom the same money shall bee due to be payd by the true intent of the said order, shall or may recover the same against the sayd Mayor and Comminaltie of London by Action of Debt in any of his Maiesties Courts at Westminster, wherein no Essoigne, Protection, or wager of Law shall be allowed.

5   Provided alwayes, and bee it enacted, that if in the New Cutte there happen any Breaches, Inundation, or Hurts, the Mayor, Comminaltie and Citizens of London, shall from time to time stoppe the Breaches at their owne charges, and sufficiently maintaine them from time to time, and make sufficient recompense to the Partie grieved for the Damage susteined by the same Breaches rising by their default, to bee recovered by Action of the Case grounded upon this Statute.

6   And bee it further enacted by the Authoritie aforesayd, that the Mayor, Comminaltie, and Citizens of London, and their Successours for ever, shal make and maintaine at their costes and charges from time to time convenient Bridges and wayes for the passage of the Kings Subiects, and their Cattell and carriages over or through the sayd New Cutte or River, in places meete and convenient.

7   And further be it enacted, that untill or before a full agreement with the Lords, Owners and Occupiers of the premisses be had, or that such Order or meanes be devised and agreed upon by the Commissioners by the L. Chancellour or L. Keeper of the great Seale of England to bee nominated and appointed, or the more part of them, as shall seeme meete for the due effecting of the premisses, and the sayd Commission returned into the High Court of Chancerie, it shall not bee lawfull to the sayd L. Mayor, Comminaltie, and Citizens to put the said digging, trenching or New cutting of the said new River from the said Springs in Execution, nor to cut or take in any ground for Passage of water from the sayd Springs or any of them by force of this Statute, Anything in this Acte to the contrary not withstanding.

8   And bee it further enacted by the Authoritie aforesayd, that after such time as there shall bee a new Cut, Streame, or River brought from the Springs

aforesaid, or any of them, to the City of London, that for the better maintenance and preservation of the sayd River or new Cut, and of all the water therein running, to be brought to the Citie of London; the same shall be subject to the Commission of Sewers, and to the Lawes and Statutes made for Sewers, as fully to all intents and purposes, as if the same River or new Cut had bene expressly mentioned in the sayd Statutes of Sewers to bee under the Survey of the sayd Commissioners.

9   Provided neverthelesse, and bee it enacted, That all such things as shall be done, at any time hereafter, for the Scouring, cleansing, amending, and conservation of the sayd new River or Cut, shall be at the onely costs and charges of the Mayor, Citizens, and Communaltie of the Citie of London; And that all Fines and amerciaments which shall bee imposed by vertue of the sayd Commission of Sewers, for any wilful annoyances and offences which shall be at any time hereafter committed to the hurt or prejudice of the sayd new River or Cut, or any thing thereunto appertaining, shall bee to the onely use, benefit, and behoofe of the sayd Mayor, Communaltie, and Citizens of London, and of their Successours for ever.

Documents of Authorization—II

A.2 STATUTORY FOUNDATIONS 1606

4 Jas. I c. 12 (1606)

An Act for explanation of the Statute made in the third yeere of the reigne of King James, Intitulted

An Act for the bringing of a fresh Streame of running Water to the north parts of the Citie of London.

Whereas of late in the Parliament holden at Westminster, in the third yeere of the Reigne of our Soveraigne Lord King James, An Acte was made for the bringing of a fresh Streame of running water to the North parts of the City of London, as by the same Act appeareth, Now for that sithence the making of the law, upon view of the grounds through which the Waters are to passe, by men of skill, and upon advised consideration of the premises, it is thought more convenient and lesse damage to the ground. That the same running water bee brought and conveyed in and through a Trunke or vault of Bricke or stone inclosed, and in some places where need is, raysed upon Arches, then in an open Trench or Sewer, which maner of conveyaunce of the same water in a Trunke or vault of bricke or stone, is doubtful whether by the wordes of the former Law it may be lawfully effected by the Lord Maior and Communaltie and Citizens of the Citie of London, albeit they doe duely performe every part, clause, matter and thing in the said Statute conteined, which on their part are by the true intent of that law to be performed:

1   For cleering of which doubt and plaine declaring of the true meaning of the sayd Lawe, Bee it enacted by our Soveraigne Lord the King, and by the Lords Spirituall and Temporall, and Commons of this present Parliament assembled, and by the Authoritie of the same, that at any time or times after the laying out of such convenient limits of ground for the making of the Trenche, or Conveyaunce of Water to the North parts of the sayd Citie of London, at the breadth of tenne foote, and not above, as to the Maior, and Communaltie, and Citizens of the Citie of London, and their Deputies, and Workemen, with the allowance of the Commissioners in the sayd former Acte mentioned, or any seven of them shall be seene convenient and meete for the same, and in that place that they shall find to be most apt and meete for that purpose, according to the true intent of the said Statute, that the Maior and Communaltie, and Citizens of the sayd Citie of London, and their Successours, Deputies and Workemen, for the consideration of the sayd former Acte expressed, shall have liberty not onely to digge the same ground to be employed for the said River, or new Cut, as in the sayd former Acte is expressed, but also in the same place, where they shall think more meete for the sayd new Cut, or Passage of Water, to frame, erect, and make a Trunke or Vault of Bricke or stone for the Passage of the sayd Water to the North parts of the sayd Citie of London, not exceeding tenne foote in breadth, in such maner and forme to be layed in the Earth or upon Arches, as to the Maior and Communaltie and Citizens of London shall seeme meete: And from time to time for ever to maintaine and preserve the same Trunke or Vault of Bricke or Stone, and for that purpose to have like liberty and free passage to and from the said Trunke or Vault of Bricke or Stone, for making erecting, maintaining, and preserving thereof from time to time for ever, as they had or might have had by the intent of the sayd former Acte, to, and from the sayd new Cut or River, With men, horses, cart and carriages, at all times convenient, and in places convenient, for the making of the said new Cut or Trench, and for the preserving of the same from time to time for ever: Any thing in the said former Statute, or in any other Law or Statute to the contrary thereof in any wise notwithstanding.

From Bernard Rudden, *The New River: A Legal History* (Oxford: Clarendon Press, 1985).

Related Cultural Documents

AGREEMENT BETWEEN KING JAMES I AND HUGH MYDDELTON

Myddelton, having had dealings with His Majesty as a jeweler, seized the opportunity of making known his need of immediate help, otherwise the project must fall through. Several interviews took place between them at Theobalds and on the ground; and the result was that James determined to support the engineer with his effective help as King, and also with the help of the State purse, to enable the work to be carried out.

An agreement was accordingly entered into between the King and Myddelton, the original of which is deposited in the Rolls-office, and is a highly interesting document. It is contained on seven skins, and is very lengthy; but the following abstract will sufficiently show the nature of the arrangement between the parties.

The Grant, as it is described is under the Great Seal, and dated the 2nd of May, 1612. It is based upon certain articles of agreement, made between King James I and Hugh Myddelton, "citizen and goldsmith of London," on the 5th of November preceding.

After reciting that Hugh Myddelton had "begun his new cutt or river," and that it promised to be of great convenience and profit to the several districts through which it passed, and more particularly to the city of London, and His Majesty being desirous of seeing perfected so advantageous an undertaking, stipulates, in the first place, to discharge a moiety of all necessary expenses for bringing the stream of water within "one mile of the city";

secondly, to pay a moiety of the disbursements "already made" by Hugh Myddelton, upon the latter surrendering an account, and swearing to the truth of the same;

thirdly, provision is made for the appointment of an "expenditor," who is to be made "acquainted with and privy to" each item of expenditure, for which purpose he is to keep proper books of account, to be duly subscribed by himself and Myddelton—the King agreeing, upon the oath of the last mentioned, to discharge a moiety of such account or accounts within twenty-one days next ensuring;

fourthly, His Majesty grants an exclusive right to Hugh Myddelton and his assigns to bring water from the springs of Hertfordshire, and the King further stipulates for himself and successors to assent to any Act or Acts of Parliament which may be necessary for confirming and enlarging the powers, &c., originally granted to the Mayor and citizens of London for bringing water to their city, and by them assigned to Myddelton;

fifthly, the King grants to Hugh Myddelton, a right of way, &c., through his manors, parks, lands, and premises, through which it may be necessary to carry the New River, without charge for loss or damage to the same;

sixthly, His Majesty contracts to provide for a moiety of the expenses to be incurred for cisterns and ponds for holding the water, as well as for pipes for distributing the same into small houses, &c.

In consideration of these aids and concessions on the part of His Majesty, Myddelton assigns to him a moiety of the interest in, and profits to arise from, the New River, "for ever," with the exception of a small quill or pipe of water which the said Myddelton had granted, at the time of his agreement with the City, to the poor people in habiting St. John-street and Aldersgate-street,—which exception His Majesty allows.

From Samuel Smiles, *Lives of the Engineers with an Account of their Principal Works Comprising Also a History of Inland Communication in Britain* (London: John Murray, 1861).

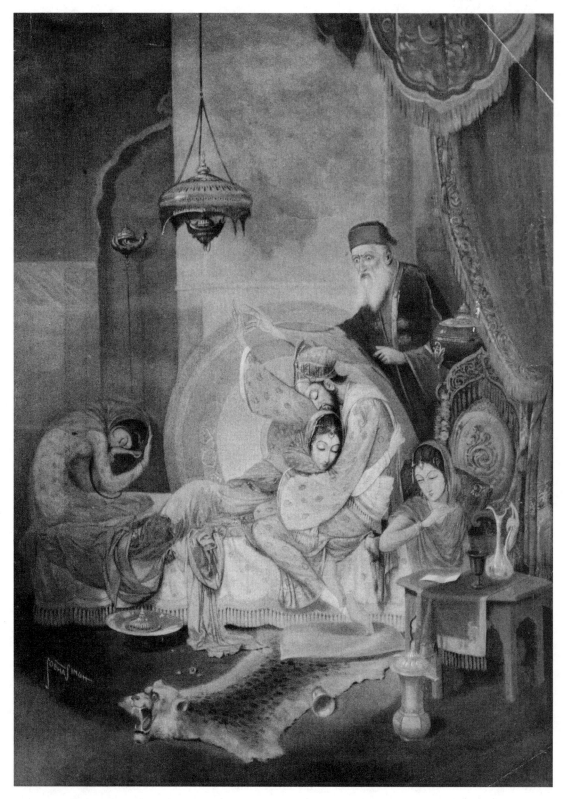

Emperor Shah Jahan embraces his dying wife, the empress Mumtaz Mahal, 1629. © Hulton Archive / Getty Images.

10
The Taj Mahal

India

DID YOU KNOW . . . ?

➤ The Taj Mahal was constructed from 1632 to 1643.
➤ It measures 1,000 by 1,900 feet (305 by 580 meters).
➤ 20,000 workmen labored to build the monument.
➤ The center dome is 240 feet (73 meters) high.
➤ 43 different kinds of jewels were used for adornment.
➤ It utilizes a unique system of landscape watering.
➤ The shah's wife accompanied him to battle, and died giving birth on the battlefield.
➤ The shah's grief was so great that it turned his hair gray in one week and almost caused the loss of his eyesight.
➤ It is the world's most famous monument to love.

Like London Bridge in the children's song, the Taj Mahal may be falling down. In October 2004, during a festival celebrating the 350th anniversary of the monument, officials noticed that one of the four minarets of the Taj was leaning, which could mean the whole structure is in trouble. Probable cause? Building decisions made in 1632. The Taj Mahal was constructed next to the Yamuna River, with its foundation on a wooden platform sunk deep into the earth below the water level of the river. Engineers used the force of the river to maintain constant pressure against the platform, thereby counterbalancing the giant heavy structure. This balance would be compromised if the river changed appreciably.

According to the *New Zealand Herald* (October 22, 2004), one reason for the Taj's tilt was the drying up of the riverbed adjacent to the monument. According to Ram Nath, former head of history at Rajasthan University: "Originally the pressure of the river water flowing by the Taj kept the building erect. But over the years, the Yamuna has dried up and the building has no support."

HISTORY

The Taj Mahal is a white marble mausoleum, considered by many to be one of the most beautiful buildings in the world—and certainly the most romantic. The fifth emperor of India's Muslim Moghul dynasty, Shah Jahan, ordered construction of the Taj Mahal as a memorial to his beloved wife, Arjumand Banu Begam (also known as Mumtaz Mahal), who died in childbirth in 1631. The name *Taj Mahal* was derived from her name.

Construction of this immense project, which involved thousands of laborers and craftsmen (over 20,000 according to some accounts) began the following year. The central mausoleum was finished around 1643, while the associated outbuildings and gardens were completed a decade later.

The Taj Mahal complex is the most renowned example of the Moghul style of architecture, which blends Indian and Persian elements. The complex is an intricately structured marvel of symmetry and balance. Its patterns are extensively based on the number four, which has particular symbolic significance in Islam. Each of the building's four facades is identical, dominated by a central arch 108 feet (33 meters) high. The structure's central dome rises more than 240 feet (73 meters) and is surrounded by four smaller domes. The mausoleum rests on a square marble platform, or plinth, with minarets at the corners. At each side is a red sandstone structure, one a mosque and the other, its mirror image, the *jawab* ("answer"), which created visual balance and served as a resting place for pilgrims. The *jawab* could not be used for prayer since, in contrast to the mosque, it faced away from Mecca.

The complex, which measures about 1,000 feet (305 meters) wide and 1,900 feet (580 meters) long, features a garden 1,000 feet by 1,000 feet, centered on an elevated reflecting pool. Gardens, as early as the times of Zoroaster, were ardently encouraged as symbols of paradise. Marble canals divided the Taj garden into four equal squares. Fruit trees (symbols of life) were planted in the gardens; the canals were lined with cypress trees (symbols of death). Inside the mausoleum, which lies at the rear of the gardens and is embellished with calligraphy and floral designs, the jewel-studded cenotaph of Mumtaz Mahal is situated at the center of an octagonal hall; at its side is the slightly larger cenotaph of Shah Jahan. (Cenotaphs are merely monuments; the couple's actual remains are buried in tombs in the crypt below.)

The *farman*, or command (reproduced here with its various endorsements and certifications), is one of the few pieces of existing documentary evidence pertaining to the construction of the Taj Mahal. It confirms that Raja Jai Singh

The Taj Mahal. Courtesy of Corbis.

has been granted possession of four properties in compensation for the donation of his estate near Agra as the site of Mumtaz Mahal's mausoleum.

CULTURAL BACKGROUND

When the Moghuls took control of India in 1526, they introduced many new ideas. Some historians consider Akbar (1543–1605) to be one of the most influential rulers because of his efforts to promote appreciation and blending of Muslim and Hindu ideals. For example, Akbar hosted debates in his court where Hindu and Muslim leaders exchanged ideas, resulting in a broadening of the culture. Akbar included non-Muslims in his strategy advisory team, and even dropped a hated tax that penalized Hindus. In 1581, Akbar proposed a new "Divine Faith," which he hoped would bring a peaceful union among the different religions, and while it was never adopted officially, it had wide-ranging influence.

The Moghuls brought a new kind of art to India. In Muslim culture, it is not permitted to depict the human visage in any form—unlike Hindu art, which displays a pantheon of deities. In response, a new art form, a graceful and elaborate calligraphy often in the shapes of plants and flowers can be seen in the marble filigree of the Taj Mahal.

The Taj Mahal is one of only a few architectural monuments that was built to honor a woman. The Moghuls encouraged recognition and elevation of women, who played strong roles in their society. Aristocratic Muslim women were taught to read and write, some worked for pay, and many owned land. There was a tradition of women joining their men during military campaigns, a practice that leads to the story of the Taj Mahal itself and the love story it symbolizes.

Emperor Jahangir, Moghul ruler of India in the sixteenth century, had a son named Prince Khurram who, although not the eldest (he was the fifth), was the brightest and his father's choice to be his successor. Prince Khurram's romantic destiny was sealed when he was just 15, and he went to the bazaar located within the royal enclosure. There he saw a young lady so beautiful, so delightful, that his heart was won instantly. The sentiment was returned, in a glance that was to last through the adventures of a lifetime and the remembrance lasting far longer. They fell in love, quickly became engaged, and were married within five years.

Lucky in love and life, Prince Khurram, although the fifth son, was renamed Shah Jahan, "King of the World," by his father even before he was elevated to the actual position. Never before was such a title given to one whose succession was not yet resolved. But Shah Jahan was not one to leave matters to question—in love or power. He battled enemies as well as his brothers until the former were subdued and the latter had mysteriously "disappeared." In 1628, when Shah Jahan became head of state with his beautiful young wife by his side, he gave her a new title, Mumtaz Mahal, meaning "Chosen One of the Palace," and he promised to build a splendid white marble residence for her. Little did he know he would soon enough build her a white marble mausoleum.

Their marriage produced 13 children, and the king and queen remained as deeply in love as when, as teenagers, they had first glanced into each other's eyes in the royal marketplace. So devoted were the couple that Mumtaz accompanied her husband into battle (not unheard of for Moghul women) in 1631, even though she was nine months pregnant. However, birth on the battlefield was arduous and unsafe, and Mumtaz died. Legend says that her beloved Shah Jahan promised then that he would build a great tomb to enshrine her forever.

It is said that he cried so much the first week after her death that he ruined his eyesight and had to wear glasses ever after, and also that his hair turned gray in the first week following Mumtaz's death. The only comfort he found was in planning the Taj Mahal. It was not until construction commenced that the shah's heart began to find a release from the sorrow that consumed him. Great love was transformed into deep grief, which then was transfigured through art.

PLANNING

Most people imagine that the Taj Mahal was built on land owned by the shah, but that was not the case. In reality, the chosen site was located next to the Yamuna River on land that had first to be acquired. It is this acquisition that is documented in the *farman* featured in this chapter. The document is an excellent example of the kind of behind-the-scenes work often required to build something of enormous scale. It is also noteworthy that the signers—"the endorsement in the handwriting of Jumlat al-Mulki Madar al-Mahami," "the endorsement in the handwriting of . . . Mir Jumla," "the endorsement in the handwriting of . . . Makramat Khan," and the entry of the agreement in the news register—are the medieval Indian equivalents of a lawyer, a notary public, and the registrar of deeds.

Why was this particular site so desirable? One answer relates specifically to the structural difficulties facing the Taj today. We know that the monument was placed on a wooden platform sunk deep beneath the banks of the Yamuna, and that it depended on the force of the river to keep the structure in balance. This would explain why Raja Jai Singh's land was so necessary.

Once the legalities were resolved, then planning could begin. Is it known who really designed and built the Taj Mahal? Certainly Shah Jahan put up the money. But was there an architect? Some believe Geronimo Veroneo from Italy was the creator. There are other rumors about Austin de Bordeaux, a silversmith from France. But the more probable chief architect was someone more local, although artisans from around the world were hired.

One factor that had to be scrupulously observed in building the Taj Mahal was the Islamic law, which mandates that a tomb cannot be altered or changed in any way once it has become the resting place of a body. No additions and certainly no subtractions can be made. Thus, according to Muslim law, the Taj Mahal had to be built as one single project. But this was impossible. The way to circumvent this stricture was to *plan* all at once. This meant the tomb, the dome, the two mosques on either side of the main dome, the outbuildings, and the reflecting pools all had to be in the plan. Then building could proceed in sections.

BUILDING

It took 20,000 people to build the Taj Mahal. Both local laborers and distinguished visiting experts contributed. When a great dome designer was needed, the Turkish architect Ismail Khan was hired. For mosaics in fanciful shapes of vines and flowers, Chinranjilal from Delhi was hired. Calligraphy, so central to Muslim art, was contributed by Amanat Khan from the city of Shiraz, and confirmed by his own signature on the Taj's gateway.

The work team's organization is well documented. There was a leadership cadre of 37 who headed sections of carvers, inlayers, calligraphers, and stone-cutters. There were specialists known for their exquisite turrets—an important feature in Moslem architecture, for it is from such minarets that the sacred Call to Prayer emanates.

If there were 20,000 workers, one can only imagine the construction materials needed and the means required to deliver the heavy marble that had been mined. It took a fleet of 1000 elephants for transport of building materials.

But this was no ordinary marble structure and no ordinary tomb, and so gems were now inlaid. Shah Jahan's enormous wealth and equally enormous grief were both well represented in the magnitude of jewels encrusted in the marble. This was an era when the land in India yielded huge diamonds and rubies, sometimes filling giant trunks in the treasuries of the nobility. But local materials were not enough. From China came jade and crystal. Burma supplied amber, while turquoise came from Tibet, and lapis lazuli from Afghanistan. In fact, there were 43 different kinds of jewels in the building. Between the materials procured and incorporated into the building and the workers who were called together to build it, the Taj Mahal was a true multinational project.

One of the most charming and interesting aspects of building the Taj Mahal was its water gardens and fountains—a characteristically Persian means of cooling. In a desert culture or warm region of India like Agra, the use of water is an art. The first Moghul emperor, Baber, introduced the water garden to India. Because a garden in a desert setting must be human-made, some foliage combined with ample amounts of marble graced by water are the key elements.

So expert was the art of the water garden that a way was devised to maintain the steady flow of water so vital in a fountain. Fountain pipes were attached to large copper pots under each pipe. Water went first into the pot and then into the fountain conduit pipe, so the pipe was constantly full. For less pressure-sensitive water needs, the builders switched from copper pipes to ones made of earthenware. A recent excavation revealed an earthenware pipe nine inches in diameter and buried five feet below a walkway. The quadrants of the garden near the center get the most water, which keeps flowers and low plants moist; farther out, where the water supply is naturally diminished, a gentle trickle waters the roots of trees. Certainly a master craftsman was at work, because the pipes were sunk to such a depth that no repair work was expected or needed.

What did all this cost? No expense was spared in the quality of marble; in the use of gems; for artists, calligraphers, and artisans; or for the water systems. In Document III, there is record of cost estimates for the tomb: "The master artisans and the architects of discerning eye have estimated its cost would be 20 lakhs of rupees." And later: "The far-seeing engineering and art-creating architects estimated the cost of this building would be 40 lakhs of rupees." (*Lakh* is a Hindi word derived from the Sanskrit *laksha*, perhaps relating to Lakshmi, the goddess of plenty and wealth. The term means a "the sum of 100,000" or sometimes, less exactly, "a very large number.")

IMPORTANCE IN HISTORY

Does the Taj Mahal hold an important position in history? It is important for its aesthetics, psychology, sociology, architecture, and recognition of the role of women. Besides being a wonder of architecture, the Taj is a unique recognition of that essentially human sentiment, respect for the dead and, by implication, respect for and acknowledgment of the past.

While scholars know of many threnodic monuments, and while history or travel buffs will favor some over others, virtually everyone knows of the Taj Mahal. Its preservation is important both to India and to the world.

FOR FURTHER REFERENCE

Books and Articles

Horn, Paul. *Inside Paul Horn: The Spiritual Odyssey of a Universal Traveler.* New York: HarperCollins, 1990..

"No Official Word Yet on Taj Minaret." *Times of India*, October 21, 2004. http://timesofindia.indiatimes.com/articleshow/msid-893982,prtpage-1.cms.

Orland, Brian, Amita Sinha, Terry Harkness, and Vincent J. Bellafiore. "Taj Mahal Cultural Heritage District Development Plan, Agra, India." *Landscape Architecture.* Online: Landscape Architecture News Digest, 2002 award winner. Washington, DC: American Society of Landscape Architects, 2004. To contact Bellafiore, e-mail address: vbellari@uiuc.edu

Scarre, Chris, ed. *Seventy Wonders of the Ancient World: The Great Monuments and How They Were Built.* London: Thames and Hudson, 1999.

Shors, John. *Beneath a Marble Sky: A Novel of the Taj Mahal.* Kingston, NY: McPherson, 2004.

Internet

For the building details, see http://www.tajmahalindia.net/ For details on the unique plumbing and water systems, see http://www.tajmahalindia.net/taj-mahal-garden-water-devices.html.

For a photo gallery of images, see http://www.greatbuildings.com/buildings/Taj_Mahal.html.

For background on the love story, see http://www.pbs.org/treasuresoftheworld/taj_mahal/tmain.html.

To test your knowledge of the Taj Mahal, try the crossword puzzle at http://www.pbs.org/treasuresoftheworld/puzzle/pgm3_puzzle_page.html.

Film and Television

Treasures of the World. Videocassette. Seattle: Stoner Productions, 1999. http://www.pbs.org/treasuresoftheworld/. The Web site also has a family of products, including an education module.

Music

Horn, Paul. *Inside the Taj Mahal.* Kuck Kuck Records #11062, 1991. Paul Horn, a renowned musician whose flute artistry has graced classical and jazz ensembles, traveled to India in the 1960s to study meditation. Inspired by the culture, he arranged to record this tribute to the Taj one night, seeking permission from the government of India to enter the monument when no one was there. Horn's tribute suite was played in the great dome of the Taj where the unique acoustics make the sound float.

McGurk, Tom. *Treasures of the World* soundtrack. Music from the PBS series *Treasures of the World*, with tracks for the Taj Mahal and Borobudur, available through http://www.pbs.org/treasuresoftheworld/ and http://www.flopsy.com/.

Documents of Authorization—I

ROYAL FARMAN TO RAJA JAI SINGH, DATED 26 JUMADAD II 1043, SIXTH REGNAL YEAR

His Exalted Majesty (Hazrat-I-A'la):

Be it known through this glorious farman marked by happiness, which has received the honor of issuance and the dignity of proclamation, that the mansions detailed in the endorsement, together with their dependencies, which belong to the august crown property, have been offered to that pride of peers and vassal of the monarch of Islam, Raja Jai Singh, and are hereby handed over and transferred to his ownership—in exchange for the mansion, formerly belonging to Raja Man Singh, which that pride of the grandees Willingly and voluntarily donated for the mausoleum of that Queen of the ladies of the world and Lady of the ladies of the Age, that honor of the daughters of Adam and Eve and upholder of the stature of chastity of the Time, that Rabi'a of the world and chastity of the World and Religion, that recipient of Divine Mercy and Pardon, Mumtaz Mahal Begam.

And it shall be incumbent upon all present and future governors, officials, overseers, agents and inspectors, in the implementation and execution of this august lofty order, to hand over to his possession the said mansions, and convey to that one worthy of bounty their absolute ownership. Moreover they should never and by no means bring about any obstruction or deviation, nor should they ever require a fresh farman or deed; and they should neither depart or deviate from this order, nor fail to execute it promptly.

Written on this date, the 7th of the month of Dai, Ilahi year 6, corresponding to the 28th of Jumadad II, year 1043 Hijri.

ENDORSEMENT ON THE REVERSE SIDE OF THE FARMAN

Sunday, the 28th of the month of Dai, Ilahi year 6, corresponding to the 14th of Rajab, year 1043.

The memoranda of the Pillar of the State, of the government and support of the kingdom, the trust of the great and organizer of the affairs of kingship, the Plenipotentiary of the government and pivot of important affairs, 'Allami Fahami Afzal Khan; and that asylum of ministership of and mainstay of good fortune and glory, Mir Jumla; and that asylum of ministership Makramat Khan; and the holder of the secretariat, the least of the servants, Mir Muhammad:

The ever-obeyed farman, as effulgent as the sun and exalted as the sky, was issued:

The mansions, together with their dependencies, belonging to
The august crown estate, in exchange for the mansion belonging
To Raja Jai Singh, which that Pillar of the State, for the sake of the
Illumined Tomb, willingly and voluntarily donated as a gift, have
Hereby been granted to us by the said Raja and settled on him in
Full ownership.

And by way of attestation, this note has been put into writing. And the endorsement in the handwriting of Jumlat al-Mulki Madar al-Mahami is that: "this should be entered in the news-register." Another endorsement in the handwriting of Jumlat al-Mulki is that: "The haveli of the late Shahzada Khanam which was granted to the said Raja is confirmed."

The endorsement in the handwriting of that asylum of ministership and mainstay of good fortune and glory, Mir Jumla, is that: "As specified in the memorandum of Jumlat al-Mulki Madar al-Mahami, it should be entered in the news-register." The endorsement in the handwriting of that asylum of good fortune and mainstay of glory, Makramat Khan, is that: "It should be entered in the news-register."

The endorsement on the margin is in the handwriting of the news-writer, attesting its entry in the news-register.

Another endorsement in the handwriting of the Jumlat al-Mulki Madar al Mahami, 'Allami Fahami, is that: "It should be resubmitted." The endorsement in the handwriting of that favorite of the royal court, Hakim Muhammad Sadiq Khan, is that: "It should be placed again before the august notice on Tuesday."

Another endorsement in the handwriting of that favorite of the royal course, the administrator of the foundations of the Gurgani rule and enforcer of the rules of justice, the model of the lords of high station and choice of the peers of the world, Jumlat al-Mulki Madar al-Mahami, 'Allami Fahami Afzal Khan, is that: "A farman of high dignity should be issued."

List of Properties

Four properties have been granted to the Raja:
Haveli of Raja Bhagwandas
Haveli of Madho Singh
Haveli of Rupsi Gairagi, in the locality of Atga Khan Bazar
Haveli of Chan Singh, son of Suraj Singh, in the aforementioned locality.

ATTESTATION AND SEAL

Certified as a true copy of the original:
THE SERVANT OF THE RELIGIOUS CODE OF
MOHAMMAD . . . ABU'L-BARAKAT.

From W. E. Begley and Z. A. Desai, comp. and trans., *Taj Mahal: The Illumined Tomb; An Anthology of Seventeenth-Century Mughal and European Documentary Sources, Sponsored* by Aga Khan Program for Islamic Architecture, Harvard University and the Massachusetts Institute of Technology, Cambridge, Massachusetts (Seattle: University of Washington Press, 1989). For the scholar who wishes to consult this reference in its place of origin in India, the following sources are useful: Jaipur, City Palace, Kapad Dwara Collection, K.D. No 176/R. See Descriptive List of Documents in the Kapad Dwara Collection, Jaipur, National Register of Private Records, No. 1, Park I (Delhi: National Archives of India, 1971); and G. N. Bahura and Chandramani Singh, *Catalogue of Historical Documents in Kapad Dwara, Jaipur* (Jaipur, India: Jaigarh Public Charitable Trust, 1988).

Documents of Authorization—II

ROYAL FARMAN TO RAJA JAI SINGH, DATED 28 SHARIWAR, FIFTH REGNAL YEAR

GOD IS GREAT!

SEAL: ABU'L-MUZAFFAR SHIHAB AL-DIN MUHAMMAD SAHIB-I-QIRAN-SANI, SHAH JAHAN, PADSHAH GHAZI, SON OF NUR AL-DIN JAHANGIR PADSHAH, SON OF AKBAR PADSHAH, SON OF HUMAYUN PADSHAH, SON OF BABUR PADSHAH, SON OF 'UMAR SHAIKH MIRZA, SON OF SULTAN ABU SA'ID, SON OF SULTAN MUHAMMAD MIRZA, SON OF MIRAN SHAH, SON OF AMIR TIMUR SAHIB-I-QIRAN.

TUGHRA: ROYAL FARMAN OF ABU'L-MUZAFFAR SHIHAB AL-DIN MUHAMMAD SAHIB-I-QIRAN-I-SANI, SHAH JAHAN, PADSHAH GHAZI.

To the best of equals and grandees, the pride of peers and contemporaries, worthy of attention and favors, the sincere, loyal and devoted servant Raja Jai Singh. Having been distinguished and exalted by imperial favors, he should know that we have dispatched Mulukshah to Amber to have white marble brought out from the new quarry (kan-i-jadid).

And We hereby order that, whatever the number of stone-cutters (sang-bur) and carts-on-hire (araba-i-kiraya) for loading the stone that may be required by the aforesaid Mulukshah, the Raja should make them available to him; and the wages of the stone-cutters and rent-money of the carts, he will provide with funds from the royal treasurer (tahwildar). It is imperative that the pride of peers and contemporaries should assist Mulukshah in all ways in this regard; and he should consider this a matter of utmost importance, and not deviate from this order.

Written on the date of 28 Shariwar, Ilahi year five (5 Rabi'I 1042/20 September 1632).

From W. E. Begley and Z. A. Desai, comp. and trans., *Taj Mahal: The Illumined Tomb; An Anthology of Seventeenth-Century Mughal and European Documentary Sources,* Sponsored by Aga Khan Program for Islamic Architecture, Harvard University and the Massachusetts Institute of Technology, Cambridge, Massachusetts (Seattle: University of Washington Press, 1989).

Documents of Authorization—III

BODY OF MUMTAZ MAHAL TAKEN TO AKBARABAD FOR BURIAL, AND RETURN OF PRINCE SHAH SHUJA' AND OTHERS TO BURHANPUR

5.1. Qazwini

As there was on the southern side of Akbarabad, adjoining the city, on the bank of the river Jumna, a tract of land (zamini), which formerly (sabiqa) was the house (khana) of Raja Man Singh, but at this time was in the possession of his grandson Raja Jai Singh, and which from the point of view of eminence and pleasantness appeared to be worthy of the burial of that one whose residence is Paradise, it was selected for this purpose. And the Raja, as a token of his sincerity and devotion, donated the said land (zamin) and considered this to be the source of happiness. However, His Majesty, in exchange (iwad) for that, granted to the Raja a lofty house (khana-i-'ala) which belonged to the crown estate. And even though the Raja's consent was obvious, the Emperor obtained permission for the repose of that companion of the Houris of Paradise.

And it was decided that her auspicious body should be buried in that heart-pleasing land (zamin); but until its arrival at the Abode of the Caliphate, everywhere on the roads, there should be distributed food and drink and innumerable coins should be given in alms to the poor and the deserving. And the prince of the people of this world and his companions, having carried the blessed dead body to the Abode of Caliphate, entrusted it to that holy earth. And having halted in the seat of the kingdom for three or four days, they returned to the exalted court.

And in compliance with the order, which is obeyed by the World, the overseers (mutassaddiyan) of the affairs of the Abode of Caliphate hurriedly covered the top (bala) of that grave (turbat), having the signs of divine mercy, so that it remained hidden from the public gaze (nazar).

Foundation of the Tomb

And in that bountiful and heart-pleasing land (zamin) a Paradise-like tomb (rauza) was laid out (tarah afgandand) and a dome (gumbad) of exalted foundation (buniyan) and a magnificent edifice ('imarat) was founded (bina nihadand)—the like and peer of which the eye of Age has not seen beneath these nine vaults of heaven, and the resemblance and equal of which the ear of Time has not heard of in any of the past ages.

And the master artisans (ustadan-I-sanatgar) and the architects (mi'maran) of discerning [lit., mature] eye have estimated its cost would be 20 lakhs of rupees. And after its completion, it will be the masterpiece (karmama) of the days to come, and that which adds to the astonishment of humanity at large.

As there was a tract of land (zamini) of great eminence and pleasantness towards the south of that large city, on which there was before this the mansion (manzil) of Raja Man Singh, and which now belonged to his grandson Raja Jai Singh, it was selected for the burial place (madfan) of that tenant of Paradise. Even though Raja Jai Singh considered the acquisition (husul) of this to be his good fortune and a great success, by way of utmost care, which is absolutely necessary in all important things, particularly in religious matters, a lofty mansion from the crown estates (khalisa-sharifa) was granted to him in exchange ('iwad).

And plans were laid out (tarah afganand) for a magnificent building ('imarat-i-alishan) and a dome (gumbadi) of lofty foundation (rafi-'buniyan), which for height (dar bulandi) will, until the Day of Resurrection, remain a memorial to the sky-high aspiration of His Majesty the Second Sahib Qiran and which for strength (dar ustwari) will display the firmness of the intentions of its builder. And the far-seeing engineers (muhandisan) and art-creating architects (mi'maran-i-sanat-afrin) estimated the cost of this building ('imarat) would be 40 lakhs of rupees.

Subsequently, in that heaven-like tract of land (sarzamin), the heavenly plinth (asman asas) was laid for a mausoleum (rauza) of lofty foundation ('ala-bunyan), which, in strength and loftiness and high dignity and magnificence of rank, is the honor of the terrestrial world, which is completely of white marble slabs, and which has arranged round it a pleasing garden having the marks of Paradise. On one side of it, a lofty mosque was built and on the other side, a replica thereof, a guest house (mihman khana) of lofty expanse: and on its sides (atrafash), there were constructed (bunyad nazirafta) rooms (hujaraha) and portals (aiwanha), and before its gate, several newly fashioned (nau-a'in) plazas and joy-increasing sarais (sara), which have no like and equal on the surface of the earth in spaciousness of area and novelty of design.

In the space of twenty years that building ('imarat), the foundation (buniyadash) of which is the eighth layer of the world and whose cap is the tenth roof of the sky, was completed at a cost of 50 lakhs of rupees; and through its extreme loftiness of dignity and rank and excellence of decoration and ornament, it has become the honor of the ancient roof of the azure sky.

11
The Canal des Deux Mers

France

DID YOU KNOW . . . ?

➤ Nine million cubic yards (seven million cubic meters) of dirt was dug to build the canal, enough to build a square-based pyramid higher than the Eiffel Tower.

➤ More than 10,000 workers were involved in its construction.

➤ The canal is 149 miles (240 kilometers) long from its beginning at Toulouse to its terminus at the Etang de Thau.

➤ It averages over 30 feet (10 meters) wide and up to 6 feet (2 meters) deep.

➤ The canal cost 17 million livres, 4 million of which came from Pierre-Paul Riquet, the designer and builder.

➤ Parts of the canal are built *over* rivers; it includes more than 100 locks.

➤ Its construction involved one of the earliest uses of explosives in subterranean construction.

➤ It is regarded as the greatest macro project in Europe between Roman times and the nineteenth century.

> The Canal des Deux Mers is comparable to the very largest works attempted by the Romans.
>
> —*Denis Diderot*

It had long been a dream to link the Atlantic Ocean and the Mediterranean Sea. Charlemagne wanted to do it; before him, the Roman emperor Augustus. Francois I (1494–1547) proposed it. King Henry IV (1553–1610) authorized

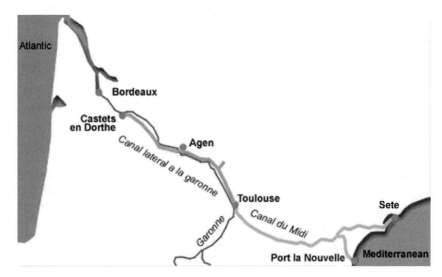

A map showing the Canal des Deux Mers.

a feasibility study by a special commission headed by the archbishop of Toulouse; the report was favorable to the project. Such a canal would obviate a long sailing trip around the coast and the associated hazard of pirates in the Straits of Gibraltar. The clever traveler could avoid taxes charged by the Spaniards for passage. If the French built such a canal, it would not only strengthen the country's strategic position in world trade, it would also weaken their competitors, especially the Spanish, and improve the security of France's commerce.

HISTORY

In its current form, the Canal des Deux Mers ("Canal of Two Seas") consists mainly of the Canal du Midi and the Canal Latéral à la Garonne, which together form France's longest navigable artificial waterway. Its key link is the Canal du Midi, also called the Canal du Languedoc. (Midi is a general name for southern France; the region of Languedoc encompasses local capitals such as Toulouse and Montpelier.) The construction of the Canal du Midi in the latter half of the seventeenth century was a signal event in the history of civil engineering, ranked by some as the greatest project in Europe in the period between the Roman Empire and the nineteenth century.

The chief stumbling block to the completion of the canal was the daunting uplands of the route between Toulouse and the Mediterranean. This problem was solved by the Baron de Bonrepos, Pierre-Paul Riquet, with a complicated design that was a colossal undertaking for the technology of the day. The project employed more than 10,000 workers and required excavation of an estimated nine million cubic yards (seven million cubic meters) of earth. Starting from Toulouse, the route took 32 miles (51.5 kilometers) to rise 200 feet (70 meters) to the highest segment, a stretch of about 3 miles (5 kilometers). It then descended 620 feet (189 meters) over a distance of 114 miles (183.5 kilometers). More than 100 locks were built, a quarter in the uphill segment. A rocky hill near Béziers was a particularly challenging obstacle. Riquet blasted a tunnel—measuring 515 feet (157 meters) long, 22 feet (6.7 meters) wide, and 27 feet (8 meters) high—through it with black powder. It was one of the earliest uses of explosives in subterranean construction. The problem of supplying water to the summit segment year-round was resolved by an intricate water-storage reservoir and flow-diversion network. Three major aqueducts were built to lift the canal over rivers.

Riquet, born in 1609, was a tax collector by profession. Self-trained as an engineer, he drew up a design for the canal, which he presented to King Louis XIV's finance minister, Jean-Baptiste Colbert, in 1662. With Colbert's help, Riquet obtained loans to help finance the difficult project, along with the necessary approval by the king, which was conditioned on Riquet's committing his own considerable wealth to the endeavor. Reproduced here is Louis's edict, issued in October 1666. Riquet did not live to see his plans fully implemented; he died in 1680 while work was underway on the harbor of Cette (now Sète) on the Mediterranean. The waterway, averaging over 30 feet (9 meters) wide and up to 6 feet (2 meters) deep, was officially opened in May of the following year by Colbert, although it was not fully completed until more than a decade later.

The Canal Latéral, the other part of today's Canal des Deux Mers, was built along the Garonne in the nineteenth century to improve passage from Toulouse to the Atlantic. Unable to compete with alternative means of transporting cargo, the Canal des Deux Mers is no longer used for heavy freight, but it remains a popular tourist destination. In December 1996, UNESCO named it a World Cultural Heritage site. In UNESCO's estimation, the canal served as a model for the flowering of technology that culminated in the Industrial Revolution and the modern technological age. In UNESCO's words, "The care that its creator, Pierre-Paul Riquet, took in the design and the way it blends with its surroundings turned a technical achievement into a work of art" (http://whc.unesco.org/sites/770.htm).

CULTURAL CONTEXT

If building a canal was so obvious, why hadn't it been done before? One reason—typical of macro projects—is that they may take several generations to ripen. Another reason is the need for a true visionary, one who has a passionate dream and will put full effort into making it happen. Sometimes the visionary is so committed that he or she even funds the project initially or through hard times. Such was the case with the Canal des Deux Mers. The cost of the canal was estimated at 17 million livres, and 4 million came from Riquet, the canal's champion.

Riquet was funding his childhood dream. When he was a boy, his father took him to a business meeting of the Council of the Counts of Languedoc.

Louis XIV of France. Courtesy of the Library of Congress.

Expecting to be bored, instead the child became fascinated when he heard a presentation for a canal that would link the Atlantic and the Mediterranean. This childhood dream never lost its hold even for the 40 years when he was controller of the salt tax. Riquet retired at 58, wealthy, influential, and ready to make his dream come true.

The time and the technology were also ripe. Louis XIV of France had formed an alliance with the middle class in France, making the construction industry his right arm. It was Colbert, the minister of commerce, whom Riquet approached about the idea for the canal, and it was Colbert who arranged everything so the king could sign the decree for the concession. It was Colbert who developed the infrastructure of France. But the canal was the most ambitious of the projects undertaken in that period.

PLANNING

When Riquet went to see Colbert about the canal, Colbert demanded that the entrepreneur/engineer prove that water would be available through the full length of the proposed canal, even uphill when necessary. Riquet agreed to construct a trial canal in miniature along the entire route, complete with actual working locks. Instead of starting at both ends and coming to the middle, he stationed groups of workers every few kilometers, so the time needed to complete the entire trial canal was just the time required to complete one section. The whole prototype was completed in one summer, water flowed and was available even at the high points, and full-scale construction was approved.

Quite possibly the miniature test canal (the *ruisseau d'essai*) was one of the first engineering test models in history. In other canals before this, we find instances of sections built as demonstrations, but the Canal des Deux Mers may be the first to have used a complete miniaturized working model to test the entire process.

Riquet's negotiating skills were legendary. It is said that when he was planning the canal route, he visited small towns along the way. If he did not get a contribution from the town, they were taken off the route and another village was selected. Due to Riquet's financial contributions and planning success, it was possible for the king, when announcing the Canal de Languedoc, to proclaim that "we have to find the funding without charging our subjects of our Province of Languedoc and of Buyenne with any new taxation which they would be obliged to contribute, because they will receive the first and most considerable advantages."

The founding document presented here includes detailed provisions about ownership, how land would be obtained, and reimbursement procedures "following the estimates which were made by the Experts who were named by Commissioners whom We deputized." No doubt there were many informal dealings behind the scenes of these negotiations.

Louis XIV had considerable experience in construction, for he included in the proclamation specific provisions for administering the new work as "an independent Fief." With the financial entity established legally, the King also provided details about the construction of commercial entities—"houses, buildings, stores, and warehouses, mills." There was even discussion of fishing and hunting rights.

The most important aspect of the king's declaration, which makes the Canal des Deux Mers such a milestone in history, is that it established a strategic route for the purpose of peace. In the document, Louis states, "We were persuaded that this would be a great work of peace," and he also mentions "all the nations of the world as well as our own subjects." Later, the Suez Canal also upheld access to all nations in time of war and peace. The strategic issue of who controls access to a macro work, such as a cut-through from the Mediterranean to the Atlantic, remains critical to this day. Macro works that open access to water systems, energy, transport, and other essentials all share this strategic purpose.

BUILDING

The canal was 149 feet (240 kilometers) long from its beginning at Toulouse to the Etang de Thau. The initial route was established by Riquet's plan, and the first section from Toulouse to Trebes was financed by Riquet personally. The second section stretched from Trebes to the Etang de Thau and was financed by the government. It was the third section that was undoubtedly most dear to Riquet's heart, for it involved a new port that would become a great center of wealth. Awakening the quaint and quiet fishing town of Sète, Riquet turned it into a boomtown when it became the Mediterranean terminus of the canal.

Riquet proposed tunnels to avoid the massive amounts of digging that would have been required to conquer the many hills along the canal's route. One such tunnel is the Malpas Tunnel.

Malpas Tunnel. Courtesy of Shutterstock.

Another feature of the canal is the round lock at Agde, an ingenious invention equivalent to a water rotary, with three gates rather than the customary two.

The canal has 63 locks, 33 of which are manually operated, and some sections have as many as 8. The locks have a distinctive oval shape. Like lighthouse keepers who stand ready to help vessels on the water, lock keepers still operate the locks.

During one modernization of the canal, the Fonseranne site was improved by the addition of an inclined plane lock or *pente d'eau*, a mechanical water slope completed in 1983. It consists of a concrete ramp in which wedge-shaped pools of water containing boats are pushed up or down the incline by a pair of engines attached to a mobile gate.

One hundred and thirty bridges were built over the canal's course. The first bridges were humpbacked, and even today their shape stands out, easily distinguishing them from bridges built later.

Building materials changed too, although not over time, as one would expect. Instead, the composition of the bridge-construction materials can be tracked from west to east. On the Atlantic end of the canal, volcanic rocks were readily available, and they are most prevalent in the construction. Limestone is more available in the middle of the country, while at the east end of the canal, brick was the material of choice. The St. Nazarre bridge is an example of a structure that combined both volcanic and limestone building materials.

There were difficulties during construction. Some were solved with simple, local solutions such as outlets and wasteweirs. But when flooding and accumulated silt threatened the canal, Riquet opted to build the canal on an aqueduct that straddled the river. This was his unique innovation and was the first such solution in European canal building. The Repudre aqueduct is the only one actually built by Riquet; 49 more were constructed by Vauban, a military engineer.

In the nineteenth century, there was an effort to modernize the canal so steamships could pass through, but the proposal was unsuccessful. As a result, today the canal remains a delightful and quaint route, free of advertising and replete with sumptuous views of the famous trees that line the banks.

IMPORTANCE IN HISTORY

The Canal de Deux Mers is one of the most successful examples of the complete historic preservation of a macroengineering project.

It is also one of the earliest examples of what is now described as a *mixed economy finance*: in effect, Louis XIV (the government) was a partner with Riquet, a private citizen. In this sense, the project financing was analogous to the financing of the New River in England.

For the 200 years of its heyday the canal was prosperous, until railroads offered a faster means of moving cargo. In a curious turnabout, in 1858 the management of the canal was taken over by the Compagnie des Chemins de Fer (the French railroad) for a 40-year period, during which time the canal fell into disrepair. Use declined, and the last cargo floated through the canal in the 1970s.

Tourism soon emerged as a new, equally lucrative use of the canal. It is noteworthy that in an English-language search on the Internet, there are over 10,000 entries offering canal travel. Imagine a commodious barge floating down the ancient route, passengers sipping a drink and enjoying the shade of the tree-lined passage.

On December 7, 1996, UNESCO declared the Canal des Deux Mers a World Heritage site; it was the 22nd site in France to be so designated, one of 582 sites worldwide, and after this recognition, tourism increased 20 percent. Since 1997, the canal has been on the list of classified sites for the French Environmental Ministry.

FOR FURTHER REFERENCE

Books and Articles

Andreossy, Comte Antoine-Francois. *Histoire de Canal de Midi ou Canal de Languedoc consideré sous les rapports de'invention, d'art, d'administration, d'irrigation et dans ses relations avec les étangs de l'intérieur des terres qui l'avoisinnet.* Paris: Impr. De Crapelet, 1804.

Davidson, Frank P. and John S. Cox. *Macro: A Clear Vision of How Science and Technology will Shape our Future.* New York: William Morrow and Co., 1983.

de Roquette-Buisson, Odile, and Christian Sarramon. *The Canal du Midi.* London: Thames & Hudson, 1983. Includes excellent photographs.

Gerard, P. *Le Voyage de Thomas Jefferson sur le Canal du Midi.* Portet-sur-Garonne: Loubatières, 1995.

Petroski, Henry. *Pushing the Limits: New Adventures in Engineering.* New York: Knopf, 2004.

Internet

For a general description of the canal, see http://www.canalmidi.com/.

For engineering and the technical aspects of the building process, with illustrations of the tunnels and locks, see http://www.canalmidi.com/anglais/ouvraggb.html.

For background on the concept of the canal through history and the leadership of Pierre-Paul Riquet, see http://www.canalmidi.com/anglais/historgb.html.

For information on UNESCO and its designation of the Canal as a World Heritage Site, see: http://whc.unesco.org/sites/770.htm.

Other Sources

American Society of Civil Engineers (ASCE) did an exhibit in 2002 named "Centuries of Civil Engineering" featuring canals, water supplies, monuments, bridges, lighthouses, viaducts, and rare books. This exhibit includes the Canal des Deux Mers. Interested readers may contact the Linda Hall Library in Kansas City, Missouri (Linda Hall Library, 5109 Cherry Street, Kansas City, MO 64110-2498, Tel: 816-363-4600).

Documents of Authorization

Edict of October 1666 for the Canal of Languedoc:

Louis, by the grace of God, King of France and of Navarre, in the presence of all and to give salutations, states the proposition that we make a joining of the Atlantic Ocean and the Mediterranean Sea, on the coast of our Province of Languedoc. There has appeared an extraordinary passage for centuries past, that the most courageous Princes and the nations who have left the most beautiful monuments to our posterity with their indefatigable labors, had never conceived of this possibility. Nevertheless, as the project's lofty designs challenge us to be worthy of them with our greatest courage, and the designs were reviewed with prudence and were carried out with success, the representatives of the enterprise were of the highest quality, and the infinite advantages that represent to us the power of the commerce of exchange, the junction of the two seas, We were persuaded that this would be a great work of peace, most worthy of our application and our care, capable of perpetuating over the centuries in the memory of Its Author, and well marking the grandeur, abundance and happiness of our reign. Indeed, We have known that the communication of the two seas would give to all the nations of the world, as well as our own subjects, the capability to complete in a few days an assured navigation by the means of the Canal to traverse the lands of our jurisdiction, and a few new ones, which we could not undertake until today, to pass from Detroite to Gibraltar without very great defenses, and a lot of time, along with the hazard of pirates, and the odds of seasickness, and then even so nevertheless with no certain view of success.

We have, after a very exact and strong discussion of the proposition that we have undertaken, concerning the rationale for the construction of the Canal, which will create the joining of the two seas, deputized a Commission drawn from the groups of the people of the three states of the aforementioned province of Languedoc, conjoining with the Commissions of the said Province of Languedoc, and Presiding heads from the aforementioned States who transported themselves from these places with intelligent persons who are necessary for the construction of the aforesaid Canal, to give Us their opinion concerning the possibility of the Enterprise that has been executed by the said Commissioners with great circumspection and knowledge. They have given Us their advice for the possibility of the execution of these aforementioned propositions, and for the form and manner in which the said construction of the aforementioned Canal can be done. But to act with the most surety, in a work so important, We have resolved to do a test, and to that goal to take the form of trying a little canal trench, and to conduct this in the same place where the construction of the big Canal is projected so that it can be so skillfully done and well executed, by the application of Monsieur Riquet, that we have promised all our subjects with certitude a very strong and happy success.

But as a work of this importance cannot be done without a very considerable expenditure, We have done a review with our Council of the diverse propositions for which we have to find the funding without charging our subjects of our Province of Languedoc and of Guyenne with any new taxation which they would be obliged to contribute, because they will receive the first and most considerable advantages, and we have settled upon the most supportable and most benign that will be necessary to get the job done.

To these causes and other considerations that have moved us, and on the advice of our Council, and of our certain science, with full power and Royal authority, We have said and ordained, and by these documents signed with our own hand, we say and ordain, we wish and it pleases Us, that we can now without hesitation proceed with the construction of the Canal of navigation and commerce of the aforementioned Ocean and the Mediterranean, following and confirming the estimates made by the Chevalier de Clerville, and we have made the determination, here attached under the counter-seal of our Chancellor, that the Entrepreneur may take all the land and inherited properties necessary for the construction of the said Canal, together with the small channels of derivation, warehouses of reserves, border banks, filters and locks, and for these said lands and inherited properties We will pay the particular Proprietors and property owners, following the estimates which were made by the Experts who were named by Commissioners whom We deputised, and also the Nobles and the Justices of the underground water springs which are situated on the aforementioned lands and inherited properties. The rights of justice and movement are indemnified by Us, and other rights remaining that have been apportioned for those said lands and inherited properties, and also all rents following the same estimates that were done by the Experts and people with this knowledge, who will accomplish this for the aforesaid lands and inherited properties carried in perpetuity as the districts of their Fiefs and Jurisdictions that compose a Fief. And so to that effect We have created and set up, and by these aforesaid determinations, we will create and will set up an independent Fief, with all the Justice—high, middle, low, and mixed—the said Canal of communication of the Seas, the channels, the stores and warehouses, the border banks to slacken or loosen to a fixed measurement on each side, the towpaths, the locks and dikes which are from the river Garonne, to the foundation that empties into the Mediterranean, including in this the Canal of derivation, from the Black Mountain to the rocks of Naurouse, without reversing or barring or excepting, assigning back to the aforesaid Fief and those immediate subordinates of our Crown under the faith and homage of Louis D'or, who has paid for each alteration in his own handwriting to Treasurer of our Lord of the Sénéchaussée de Carcassone; with the power of the Lord who is in possession of the aforesaid Fief, to build and construct for the said canals a chateau and buildings necessary for their lodging for all, with crenellations for battlements and a sufficient number of windmills to mill the wheat; and also order to be constructed on the banks of the said Canal the houses and stores to serve as lodging for those who work as employees for the waterway and for the bonded warehouses and for the security of the merchandise and commodities, to the exclusion of all others, and for the places that are judged to be appropriate without encumbering navigation or causing damage to the said works. With these houses, buildings, stores, and warehouses, mills, the aforementioned Proprietor, his heirs, successors or any claimants will enjoy a priority immutable and noble, together with the aforesaid canals, stores of reserve, and their perimeter areas, free and clear of all taxes and fees ordinary, extraordinary, municipal, and for the billeting of troops. And also the said Proprietor will have the right of hunting and fishing in this said Fief to the exclusion of all others, preventing all our subjects in those conditions which they deem from constructing buildings or stores near the banks of the said canal, nor hunting or fishing in the Fief, with the penalty of 500 pounds as a fine for each violation. And equally the said Proprietor, to the exclusion of all others, may establish on the said Canal or places which are judged necessary, the boats for transport of merchandise and for passage of people.

From M. De La Lande, *Des Canaux De Navigation, et Specialement du Canal de Languedoc* (Chez la Veue Desaint).

12
The Founding of
St. Petersburg

Russia

DID YOU KNOW . . . ?

➤ Construction began in 1703 and was completed in 1713.
➤ 100,000 laborers died of starvation and the harsh climate while working to build the city.
➤ In 1725, a standing army of 200,000 was housed in the city.
➤ The city was originally built of wood, and later reconstructed entirely in stone.
➤ It comprised 100 islands linked by 600 bridges.
➤ It was the first principal port and commercial center of Russia.
➤ The early price of admission to the city was one large stone.
➤ The imperial government was the sole source of chessboards.
➤ It experienced dramatic population growth in a short period of time.
➤ It was the site of the first Russian ballet.

The price of admission to St. Petersburg was once a stone. When Czar Peter the Great decided to build a new capital for Russia, he first used wood as the building material. But his goal was to fashion a city of lasting grandeur, and stone was preferred for endurance. Therefore the czar handed down two imperial decrees: first, he forbade building with stone anywhere in Russia except in St. Petersburg, so that all stonemasons had to come to the new city if they wished to pursue their craft; second, he put out the whimsical but direct order that no one could enter the city unless the person brought a stone for admittance.

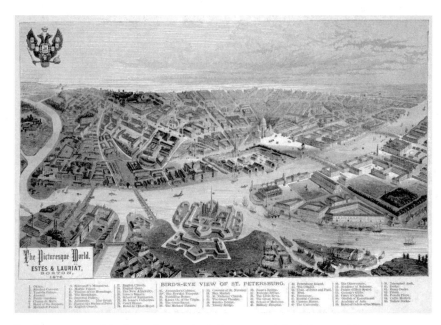

Bird's-eye view of St. Petersburg, 1876. Courtesy of the Library of Congress.

HISTORY

For two centuries, St. Petersburg was the capital of Russia and the country's foremost example of imperial splendor. In 1918, Moscow became the seat of government. Today, although no longer the capital, St. Petersburg remains Russia's largest seaport and one of the most beautiful cities in the world. A monument to the power of human enterprise to reshape the environment, St. Petersburg—a political, cultural, and industrial center—was built on a desolate swamp.

When, in 1703, Czar Peter the Great ordered that a city be established at the mouth of the Neva River on the Gulf of Finland (an arm of the Baltic Sea), he was motivated not only by a desire to develop Russian society, but by geopolitical concerns as well. He wanted to assert Russia's military and political hold on the region and give Russia a seaport that would facilitate ties with western Europe.

In the fifteenth century the area came under the rule of the Russian state centered at Moscow. Early in the seventeenth century, however, Sweden, then the major power in northern Europe, took possession of the Neva region, blocking Russia from direct maritime trade and communication with western Europe. In the ensuing Great Northern War of 1700–1721, which pitted Sweden against Russia and its allies, Russia regained the territory. Russian forces broke through to the sea in May 1703, when the small Swedish fortress of Kantsy (Nyenskans), several miles upstream from the Neva's mouth, capitulated on May 11 (according to the Julian calendar then in use, corresponding to May 22 in the modern Gregorian calendar).

Immediately thereafter, Peter resolved to fortify the area. As the accompanying extract from Peter's journal indicates, the Swedish fortress (renamed Sloteburg by Peter) was deemed inadequate, and a new site was chosen. The traditional date for the founding of St. Petersburg is May 16, 1703, according to the old calendar. On that day, digging to construct the new Peter and Paul Fortress began on what Peter refers to in his journal as the island of Lust Elant.

To build the new city, legions of workers and artisans were mobilized at Peter's command from throughout Russia. Conditions at the site were extremely difficult. The marshy islands were frequently ravaged by floods, and winter temperatures were often below zero. The early construction years were especially harsh, for housing and food supplies were inadequate, as the war with Sweden dragged on and competed for resources. Construction progressed, but at an immense cost in human lives. During the first years, thousands of people—some estimates go as high as 100,000—died from the harsh climate, hunger, or disease. St. Petersburg was, as a famous saying had it, "a city built on bones." It should be added, however, that Peter founded orphanages and hospitals as a result, and even ordered that although serfdom was legal and not discouraged, no families were to be separated to further the building and agricultural activities that drew the masses of laborers.

Despite the hard conditions, primitive tools, and need to transport huge quantities of construction materials from afar, the building of the city proceeded apace, proving once again that even extraordinarily daunting projects may be feasible if run by a boss with power and competence. The workers cut timber, hauled stone, drove piles, constructed buildings, dug canals for drainage, and erected hundreds of bridges. Upon completion, St. Petersburg spread over more than 100 islands and was linked by more than 600 bridges.

Peter exercised his powers of persuasion to induce his subjects to settle in the new city. In 1712, St. Petersburg replaced Moscow as the capital; by 1726, the year after Peter's death, the new capital accounted for 90 percent of Russia's foreign trade. The young city continued to burgeon in the following decades; a number of distinguished Russian and foreign artists and architects helped make it an imperial showcase of opulent palaces, magnificent cathedrals, and government buildings that generally adhered to a set pattern and were located on a brilliant network of wide avenues and expansive squares. By the end of the eighteenth century, St. Petersburg's population had grown to well over 200,000.

CULTURAL BACKGROUND

Peter was 10 years old when he became czar. As an impressionable 15-year-old, he traveled to the Netherlands, where he worked as a laborer in a shipyard.

Gifted with manual dexterity, he loved maritime construction, and even in his elder years spent time on the wharves; some said he was the best shipbuilder in Russia. Once, when he was in England, he asked to see firsthand the practice of execution by walking the plank. The British admiral showed him how a scaffold might be fashioned, but Peter wanted to witness someone actually being so condemned. But the admiral replied that none of his men warranted such a punishment, so Peter promptly offered one of his own. The British commander refused because even Russians accompanying their imperial ruler could not be so ordered because they were under the protection of English law. This story illustrates both Peter's interest in the traditions of the sea and his rather cold command of his subjects.

Some historians claim that Peter was brilliant, open to new ideas and innovations, and farsighted in economics, government administration, and financing. But he was better with things than people. He also drew respect because he

Czar Peter the Great. Courtesy of the Library of Congress.

combined a powerful, freethinking, visionary mind with humble tastes. He built enormous structures, but lived simply (at first) in a tiny cottage. In 1695, when he commanded his army to attack the Turks, he did not lead the charge but joined the ranks as a bombardier. In England, Germany, and the Netherlands, where he traveled in 1697, he went not as a czar but as a humble workman— in the Netherlands he labored in the shipyards.

As he matured, Peter continued to think like a shipbuilder and admiral, which led him to recognize the need for a great port—ultimately the new capital, St. Petersburg. The document featured here, taken from Peter's personal diary, is a brief notation that illustrates the power of an autocrat who requires neither consent nor approval to establish and build. Characteristic of Peter, who was inspired by open water, the document mentions proximity to the sea as a requirement of the new fortification that would became St. Petersburg.

Although the Neva River is only 42 miles long, it is a most suitable river for shipping transport because of its width—at Smolnyi, the river is 2,100 feet (640 meters) across. When the river reaches the sea, it breaks into many small streams and runs around many small islands. That is why the construction of St. Petersburg was so challenging. On the one hand, it was an excellent port for a large but (at that time) landlocked country. On the other hand, the marshy terrain in such a northern location made the land hard to work; the weather was frequently an obstacle as well.

BUILDING

It took 10 years, from 1703 to 1713, for the new Russian capital to be built, followed by moving the seat of government from Moscow. In the building and construction, there were few innovations, for when the force of power is manual labor old methods suffice. Peter first built the city in wood, erecting over 2,000 structures before deciding to rebuild it in stone, thus creating the architectural monuments with which we are familiar today.

St. Petersburg is also known as the site of institutional innovations for which Peter was responsible. As a military commander, he inaugurated what some historians call the first modern standing army: by 1725 there were 200,000 soldiers on permanent duty. For a shipbuilder like Peter, it was logical to create a great navy to protect the new port and fortress. He required noblemen to register for service in either the army or the navy, with government a third choice. He brought commoners into the vast military force. In an ingenious incentive to commoners, Peter made it possible to become a member of nobility through military service rather than by birth—soldiers and sailors who reached the top ranks were made aristocrats.

Once fortified and protected, the new Russia became productive. Iron was exported, and steel mills flourished. Peter established textile mills and sawmills. Business boomed. By 1725, Russian exports doubled, and the country was no longer forced to import goods from the same European cities that Peter strove to emulate. He created incentives of power: industrial factory entrepreneurs were rewarded with their own towns whose governance was up to them and whose enterprises were tax-free.

But while factory owners enjoyed tax-free status, such was not the case for those with facial hair. Peter taxed everything he could, even beards. (He shaved his own beard when he returned from travel in Europe, and he wanted every man to follow suit. Amusingly, however, there is a famous portrait of Peter sporting a mustache.) He also capitalized on the many houses being built for workers and the middle class, requiring them to pay a tax on cellars and windows. If one desired a private bath, it was taxed. If a hat was worn in the cold Russian winter, it was taxed. If one wanted a melon as respite from hearty Russian fare, it was taxed. Even the peasants were taxed. Sending out his financial team to new towns, Peter received reports that taxing by the household did not cover all the workers, because many were crowding into one house. Peter doubled the state coffers when he devised the hated poll tax, borrowed from the French, which meant that every person paid a tax to the government. Peter even taxed slaves.

Administering all these taxes grew unmanageable. The traditional political structure, imported in its entirety from Moscow when the seat of government moved to St. Petersburg in 1703, was a tangle of old rules and bureaucratic routines. When he threw out the Duma of Boyars, the old advisory council filled with cronies, Peter established in its place eight districts of government. Always keeping one eye on power, Peter saw that even this innovation did not drive power down to the local level, so he reformed the administration again. This time he further subdivided the former 8 districts into 50, thereby pushing decision making into the countryside and creating at the top the concept of a senate.

Peter's most innovative administrative practice, one that he invoked in a naive effort to keep the government honest and provide financial incentives, was denunciation. He encouraged citizens to identify government officials who were corrupt and turn them in, and as a reward they were given possession of the corrupt person's property.

Because his own life had been deeply influenced by study and travel in Europe, education did not escape Peter's attention. One-third of the population above six years old was unable to read. In 1701, he founded a School of Mathematics. In 1716, he required all the children of landowners to attend one of three colleges: Army, Navy, or Engineering. Education was also required for the nobility, and study abroad was encouraged and even supported by the government. Recognizing the link between education and a better society, Peter devised an ordinance that no one in Russia could marry unless they had been educated, but the law did not stand. It was evidence of Peter's vision of the importance of education in the modernization of Russia.

While Peter unleashed entrepreneurial freedom in government and enterprise, thereby creating substantial revenue for the state coffers, he retained absolute control over a few of the staples of the land. Alcohol production remained strictly a government monopoly. So were coffins. And perhaps in partial explanation of why the game's greats are often Russian, the czar was the only source of chessboards.

Still, the greatest tradition of the new capital, the reason for its founding and its economic strength, would always be the port. From 1875 to 1888, a shipping canal was completed at the cost of 1,057,000 pounds sterling, in order to make the capital a seaport. The canal began at Kronstadt and led to Gutuyev Island in the harbor, and it could accommodate as many as 50 ships. With a depth of 23 feet (7 meters) and stretching 17 miles (27 kilometers), the canal is substantial. In addition, three bridges were constructed across the Neva River.

Too much attention to architecture and not enough concern for infrastructure proved to be a mistake. At one time the Neva was so pure that in 1760, it is said a Russian could dip a glass into the river and drink safely; 100 years later, not only was the water polluted, it was the source of frequent cholera outbreaks.

IMPORTANCE IN HISTORY

Berlin, London, Paris, St. Petersburg, and Vienna remain five of the largest cities in Europe. St. Petersburg, the youngest of the five, reveals dramatic population growth over a relatively short period of time:

St. Petersburg also gave Russia and the world great advances in culture and learning. The buildings that would later house the university were constructed

Table 12.1

YEAR	POPULATION
Early 1700s	<10,000
1742	150,000
1800	220,000
1830	435,000
1850	487,300
1881	928,000
1900	1,440,000
1915	2,348,000
1920	763,000
1925	1,379,000
2002	4,700,000

Note: The Russian Revolution was the cause of the dip between 1915 and 1920.

on Vasilievsky Island between 1722 and 1742 to house the government, its 10 ministries, and the Senate. St. Petersburg University, founded in 1819 (or perhaps earlier; historians differ), was the site of the development of the periodic table of the elements (Dmitry I. Mendeleev) and the radio (Alexander S. Popov, at the same time as Marconi).

No discussion of Peter the Great's contributions would be complete without mentioning the ballet. Upon his return from a tour of Europe in 1698, Peter opened the door to cultural traditions already established in Europe and encouraged them to take root in Russia. Jean-Baptiste Lande gave a recital of his ballet students that led to the founding of a Russian ballet school in 1738, which grew into the famed St. Petersburg Imperial Ballet School. By 1801, Charles Didelot's productions at the St. Petersburg Bolshoi Theater (later called the Maryinsky and still later the Kirov) began showcasing the great innovations and artistic achievements of Russian ballet. Beautiful theaters were built and visiting artists paid so handsomely that all the great dancers came to Russia. And Russians responded with passion. It is said that when Ivan Valberkh, a ballet master trained at the St. Petersburg school, became its director in 1794, he mounted a production of *Love for the Fatherland*. Upon seeing the production, streams of people left the audience and enlisted in the war against Napoleon! This same tradition later produced the masterful artistry of Rudolf Nureyev and Mikhail Baryshnikov.

St. Petersburg has produced some of the world's greatest music. Composers who lived in St. Petersburg included Glinka, Borodin, Mussorgsky, Rimsky-Korsakov, Tchaikovsky, and Shostakovich. St. Petersburg also inspired great literature, with contributions from Pushkin, Dostoyevsky, Turgenev, and Gogol.

There are many paintings depicting the construction of St. Petersburg. Of course, the idea of capturing great cities on canvas was not new, but until this time most cities were painted from memory or after completion. Peter was the first ruler to commission paintings of a city still under construction.

St. Petersburg is home to what many consider to be one of the great museums of the world. Housed in the former Winter Palace erected at the command of Catherine the Great, the Hermitage has over three million objects of art, a collection begun when she acquired 250 paintings.

FOR FURTHER REFERENCE

Books and Articles

Almedingen, E. M. *My St. Petersburg: A Reminiscence of Childhood.* New York: Norton, 1970.

Efimova, A., O. Turkina, and V. Mazin. *Layers: Contemporary Collage from St. Petersburg, Russia.* Catonsville: Fine Arts Gallery, University of Maryland, 1995.

Steward, J. S., ed. *The Collections of the Romanovs: European Art from the State Hermitage Museum.* London and New York: Merrell, 2003.

Internet

For current news and information about St. Petersburg, see http://www.sptimes.ru/.

For the population of St. Petersburg and its changes, see http://www.fact-index.com/s/sa/saint_petersburg.html.

For antique photos, see http://www.alexanderpalace.org/petersburg1900/2.html.

For Russian history, see for example, http://mars.wnec.edu/~grempel/courses/welcome.html.

For a virtual tour of the Hermitage, see http://www.hermitage.ru/.

For photos and information about ballet in Russia, see http://michaelminn.net/andros/index.html Choose section on Russian ballet.

For more on St. Petersburg University, see http://www.cityvision2000.com/city_tour/university.htm.

Film and Television

Sokurov, Alexander. *Russian Ark.* Fox Lorber, #720917538228. DVD. 2002. The Hermitage can be explored through many sources; for a unique offering, consider this work, claimed by some to be the world's first unedited feature film. Shot by Steadicam, the film follows an unnamed seeker through 33 rooms of the Hermitage, where he is greeted by famous figures in Russian history. The film has a cast of one star and 800 supporting actors. See a preview/trailer at http://www.imdb.com/title/tt0318034/maindetails.

Documents of Authorization

Upon the capture of the fortress Kantsy, the Council of War had gathered to decide whether to further strengthen that redoubt or to look for another, more convenient place for the fortress (because the above-mentioned Kantsy was too small, too far from the sea and was not conveniently protected by its natural surroundings). The Council resolved to look for a new place. Several days later, such an appropriate place was found, an island named *Lust Elant* (i.e., Pleasure Island), and on May 16, 1703 [Pentecost Sunday] the fortress was established there and was named St. Petersburg.

From E. S. Gorlov, *The Founding of St. Petersburg as Recounted in the Journal (Diary Notes) of Tsar Peter the Great* MS in Structural Engineering (May 1956), Institute of Railroad Transportation Engineering (Moscow). Pub. Art. 1777, part 1, p. 76. Translated and adapted from the archaic Russian script of the eighteenth century.

Construction of the dome of the capitol building, 1861. Courtesy of the Library of Congress.

13
The Founding of Washington, DC

United States

DID YOU KNOW . . . ?

➤ The site for Washington, DC, was selected at a dinner party hosted by Thomas Jefferson.

➤ The city was designed in 1791, mapped in 1792, and expanded for 100 years thereafter.

➤ The chief designer of Washington, DC, was hired and fired by George Washington.

➤ Washington, DC, is an example of a city as public art.

➤ Washington, DC; Canberra, Australia; Brasília, Brazil; and Abuja, Nigeria, were each purpose-created as new seats of government.

The decision about where to locate the capital of the United States came about through a deal sealed at a dinner party. Hosted by Thomas Jefferson, the convivial occasion was also meant to highlight a cooperative agreement on two of the great issues of the day: where the capital should be placed, and how to assume the debts of the states.

Jefferson's colleague, Alexander Hamilton, was keen on establishing the finances of the new republic by having the federal government assume the states' debts—a stroke of genius. But to achieve agreement at the dinner party, and owing to reluctance by the State of Virginia, the partygoers agreed that Virginia could have the equivalent of an additional $500,000 of their indebtedness assumed by the federal government. Those in the Hamilton school of thought won their point in the political debate; they acquired the finance of the country in return for an agreement that the capital would stay in Philadelphia for 10 years. Only then would it be moved to the new federal district.

Earlier, Jefferson had been meeting with Pennsylvania's Robert Morris, a strong advocate for locating the capital in Pennsylvania. There had been some preliminary suggestions that a compromise might work if Philadelphia could be given 10 years as the first center of power. To obtain agreement, however, Jefferson knew he would have to get Hamilton to work his persuasive magic with certain Pennsylvanians; James Madison would be needed to pacify certain Virginians into accepting a compromise solution.

That solution had three parts: Virginia gained monetary benefit directly, Pennsylvania gained monetary benefit for a specified but limited period, and the government would eventually be located at a new site on the Potomac River, in a location to be decided by George Washington. In July 1790 Congress enacted two separate bills: the Assumption Act and the Residence Act (the latter included here as Document I).

George Washington was not at the dinner party: he could not leave the Potomac area, because if the landowners got even a hint of the impending escalation of the value of their land, speculation could run rampant and require presidential attention.

HISTORY

Washington, DC, is, in some respects, a purpose-created city like St. Petersburg in Russia. Both were established by government mandate; the physical layout of each city's core still reflects the original master plan. But the U.S. capital differs from St. Petersburg in a number of ways. While the Russian capital showcased imperial splendor, the U.S. capital highlighted American democratic values and history. St. Petersburg's site was chosen for political, social, and military advantages; the location of the U.S. capital was chosen because of potential river-trade advantages (which never materialized), as well as a need to achieve compromise among the competing interests of individual states.

Lest the state where the capital was located acquire undue influence over the federal government, the final site for the U.S. capital was given a unique legal status. To this day Washington, District of Columbia, remains under the authority of Congress and has no voting representation.

The creation of a capital district was envisioned in Article I, Section 8, of the Constitution, drawn up in 1787: "The Congress shall have power . . . to exercise exclusive legislation in all cases whatsoever, over such district (not exceeding ten miles square) as may, by the cession of particular States, and the acceptance of Congress, become the seat of the government of the United States." Several pieces of legislation were subsequently passed by Congress concerning the establishment and administration of such a district. Among the most notable of these measures are the three presented here:

1. The Residence Act of 1790, which required that the seat of U.S. government be moved by December 1800 to a "district or territory, not exceeding ten miles square" along a certain stretch of the Potomac River near Georgetown, Maryland, with the precise location to be determined by three commissioners appointed by President George Washington. The commissioners were also charged with overseeing the construction of the new governmental buildings, in accordance with design plans to be approved by the president.
2. An amendment to the Residence Act, passed in 1791, allowing for the inclusion of Alexandria, Virginia, but retaining the original act's injunction to erect public buildings only on the Maryland side of the river.
3. The 1802 act incorporating Washington as a city, with an elected council and a mayor appointed by the president. Congress subsequently altered the city's governmental structure on several occasions.

The three commissioners appointed by President Washington to oversee design and construction decided to call the capital city *Washington* and the district the *Territory of Columbia*—presumably in honor of Christopher Columbus. The name *Territory*, however, was soon changed to *District*. Today the limits of the city of Washington coincide with the boundaries of the District of Columbia.

The French architect and engineer Pierre Charles L'Enfant was hired by President Washington in March 1791 to map out the new city. L'Enfant's plan for a tract encompassing 9.5 square miles rejected "Jefferson's preference for a small village that would gradually expand, in favor of a massive area that would gradually fill up." Construction proceeded slowly as the contentious L'Enfant became entangled in disputes. Washington fired him in 1792, and he left the city, taking his plans with him.

This could have been a disaster but for the excellent memories of two people: Benjamin Banneker and Major Andrew Ellicott. Banneker, an African American mathematician and astronomer, had been engaged by Ellicott to help survey the federal territory during the time Ellicott was working closely with L'Enfant. When the tempestuous architect was dismissed, the plans were recreated from memory, thus saving the U.S. government considerable time, effort, and expense to have someone else design the capital. Historians are still assessing who should be credited for the recreation of L'Enfant's design, but certainly both Ellicott and Banneker deserve honorable mention.

CULTURAL BACKGROUND

Many countries start from a core that includes the most populous city, such as London in England, Rome in Italy, and New York in the United States.

Sometimes there comes a point in a country's development when a decision is made to form a permanent capital at a new site entirely dedicated to the business of government. This new city generates a sense of patriotism and national unity expressed through symbolic architecture. Such capitals are, in a sense, a form of public art. Several examples are discussed in other chapters: Madinat as-Salam (City of Peace), now known as Baghdad; Brasília; and Abuja.

Five locations were vying to be chosen as the site for the country's capital: New York, Philadelphia, Baltimore, a site along the Susquehanna River in Pennsylvania, and a site along the Potomac River. New York wanted to keep the capital there, and that would have pleased Hamilton. But Philadelphia wanted it too. The South felt somewhat tossed aside because these two northern sites were being given top consideration. Locating the capital in Washington would satisfy the South, but it was necessary to appease the Pennsylvanians by locating the capital for 10 years in Philadelphia. Then it was moved to its permanent location in Washington, DC.

PLANNING

With so much competition concerning the location for the capital, it is no wonder that an outside expert was needed to design the city. Indeed, this may be one of the first times in history that an architect or engineer proposed himself for the job, for such was the case with Major Pierre L'Enfant. Although he was born in France and trained there as an engineer, L'Enfant became an ardent supporter of the cause of America's Revolutionary War, and at 22 he volunteered and served with the Corps of Engineers of the Continental Army. He met George Washington during that service and maintained this key contact.

September 11, 1789, is the date of a letter authored by L'Enfant, in which he "solicits the favor of being Employed in the Business" of designing the new capital using what he called "a plan wholly new." He wrote, "No nation perhaps had ever before the opportunity offered them of deliberately deciding on the spot where their capital city should be fixed." He even proposed the time period for building, suggesting the new capital could be ready in 1800. Such a short time frame for this endeavor may have been advantageous. (Was it a coincidence that it was the same time period supported by George Washington for temporarily locating the capital in Philadelphia—for the 10 years from 1790 to 1800?) But perhaps L'Enfant's vision was also stirring: "The plan should be drawn on such a scale as to leave room for that aggrandizement & embellishment which the increase of the wealth of the Nation will permit it to pursue at any period how ever remote" (http://www.loc.gov/exhibits/treasures/tri001.html). Grand plans for a large place were no doubt music to the ears of the new president, George Washington. The choice of architect does not seem to have been made by committee or via a request for proposals. Based on L'Enfant's vision, and perhaps colored by George Washington's awareness of L'Enfant's clear sincerity exemplified by his service in the American army, L'Enfant would play a key role in the development of a country still in its infancy.

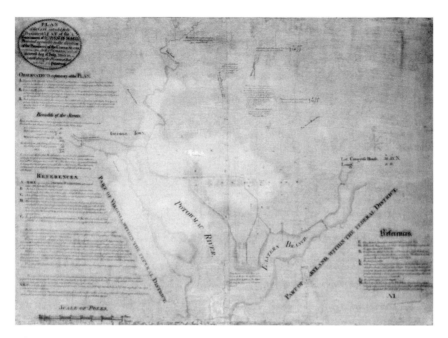

The original plan of Washingtion drawn by Pierre-Charles L'Enfant in 1791. © Corbis.

In 1791, planning began. A site survey was undertaken immediately. Based on topography, L'Enfant fashioned a baroque plan complete with large spaces that could be used for grand ceremonies. Avenues radiated out from the two most significant buildings of the new capital: the house for Congress and the house for the president. The plan provided that the avenues be wide and tree-lined—but with trees low enough to preserve views of the monuments and fountains that would follow. L'Enfant then manually shaded with a pencil and numbered 15 large open spaces, at the intersections of the avenues, to be divided among the states. Each intersection would have statues and memorials to honor special citizens. The open spaces were to be used to rest the eye and give the monuments meaning. To L'Enfant, these open spaces were as important as the buildings and the monuments.

BUILDING

With the plan completed, building commenced. As we know from Jefferson's dinner party and Washington's arrangements for a house in Philadelphia, the construction process took 10 years.

But L'Enfant's plan did not reach full expression for 100 years. In 1898, President McKinley proposed a more comprehensive park system for the city.

The first meeting of a joint committee of Congress was held in February 1900; chaired by Senator James McMillan of Michigan, the outcome became known as the McMillan Plan. The Potomac flats had just been acquired, so with the additional land, the idea of a great mall became possible, and that caught the attention of the American Institute of Architects (AIA). The country's mood was ready and oriented to grandeur and beauty, having just experienced the Columbian Exposition in Chicago in 1893.

Encouraged by the AIA and Congress, political leaders turned to the country's artistic leaders to guide them. Daniel Burnham, whose success in the Chicago Exposition brought a new era of architecture and public-space planning, joined the team, along with Frederick Law Olmsted, the great landscape architect. Sculptor Augustus St. Gaudens and architect Charles F. McKim completed the team. Burnham is famously said to have urged, "Make no little plans. . . . Make big plans; aim high in hope and work, remembering that a noble, logical diagram, once recorded, will not die," (http://www.dmreview.com/article_sub.cfm?articleid=6533) so they thought big.

However, big plans meant extending the grand vision of Burnham's "City Beautiful" movement. In 1910 President Theodore Roosevelt and Congress formed a commission to oversee the aesthetic quality of public art as it applied to parks, bridges, and paintings. Public buildings were almost an afterthought, but a stipulation was added the following year that the list of public spaces must go through design review to be sure they expressed appropriate *gravitas*. Some say that L'Enfant did such a good job with his original design that the McMillan Plan and all efforts since then have only built on the first grand vision.

IMPORTANCE IN HISTORY

While it often takes only a single passionate visionary to launch a macro project or structure, it takes many committed people to keep the vision alive. In the case of L'Enfant, the aesthetic standard he set required several presidential directives, an act of Congress, and the creation of a Commission of Fine Arts to keep the vision on track.

Washington, DC, is a remarkable case of far-sighted planning that turned out to be thoroughly realistic. Unlike many visionary designs, Washington was in fact built according to the original plan. However, it was almost not so. When L'Enfant was fired, it is widely believed that when he left Washington he took with him his detailed plans for the design of the city. L'Enfant had given some reports to the president when he presented plans in August 1791. Details of L'Enfant's design were recouped through the efforts and meticulous memory of the surveying team of Ellicott and Banneker. L'Enfant's original conceptual plan has been preserved, and George Washington showed it to Congress but retained the map himself until December 1796, when he handed it to the city commissioners of Washington. On November 11, 1918, it was given over for safekeeping to its permanent home in the U.S. Library of Congress. In 1991, the plan's 200th anniversary was honored when three authorities collaborated to produce a perfect facsimile with the help of the National Geographic Society, the National Park Service, and the U.S. Geological Survey. With the technology available today, one can now see editorial comments noted on the original in the hand of that master of the dinner party, Thomas Jefferson.

FOR FURTHER REFERENCE

Books and Articles

Allen, W. B., ed. *George Washington: A Collection*. Indianapolis: Liberty Classics, 1988.
Bedini, Silvio A. *The Life of Benjamin Banneker*. Rancho Cordova, CA: Landmark Enterprises, 1984.
Ellis, Joseph J. *His Excellency George Washington*. New York: Knopf, 2004.
Ferling, John. *A Leap in the Dark: The Struggle to Create the American Republic*. Oxford: Oxford University Press, 2003.
Stephenson, Richard M. *"A Plan Whol[l]y New"*: *Pierre Charles L'Enfant's Plan of the City of Washington*. Washington: U.S. Government Printing Office, 1993.

Internet

For an overview of the L'Enfant and McMillan Plans, see http://www.cr.nps.gov/nr/travel/wash/lenfant.htm.
For a discussion of Benjamin Banneker's contributions, see http://www.math.buffalo.edu/mad/special/banneker-benjamin.html.
For links to the letters and papers of Thomas Jefferson at the Library of Congress, see: rs6.loc.gov/ammem/collections/jefferson_papers.

Documents of Authorization—I

THE RESIDENCE ACT OF 1790

An Act for establishing the temporary and permanent seat of the Government of the United States

SECTION 1. *Be it enacted by the Senate and House of Representatives of the United States of American in Congress assembled,* That a district of territory, not exceeding ten miles square, to be located as hereafter directed on the river Potomac, at some place between the mouths of the Eastern Branch and Connogochegue; be, and the same is hereby, accepted for the permanent seat of the government of the United States. *Provided nevertheless,* That the operation of the laws of the state within such district shall not be affected by this acceptance, until the time fixed for the removal of the government thereto, and until Congress shall otherwise by law provide.

SEC. 2. *And be it further enacted,* That the President of the United States be authorized to appoint, and by supplying vacancies happening from refusals to act or other causes, to keep in appointment as long as may be necessary, three commissioners, who, or any two of whom, shall, under the direction of the President, survey, and by proper metes and bounds define and limit a district of territory, under the limitations above mentioned; and the district so defined, limited and located, shall be deemed the district accepted by this act, for the permanent seat of the government of the United States.

SEC. 3. *And be it further enacted,* That the said commissioners, or any two of them, shall have power to purchase or accept such quantity of land on the eastern side of the said river, within the said district, as the President shall deem proper for the use of the United States, and according to such plans as the President shall approve, the said commissioners, or any two of them, shall, prior to the first Monday in December, in the year one thousand eight hundred, provide suitable buildings for the accommodation of Congress, and of the President, and for the public offices of the government of the United States.

SEC. 4. *And be it further enacted,* That for defraying the expense of such purchases and buildings, the President of the United States be authorized and requested to accept grants of money.

SEC. 5. *And be it further enacted,* That prior to the first Monday in December next, all offices attached to the seat of the government of the United States, shall be removed to, and until the said first Monday in December, in the year one thousand eight hundred, shall remain at the city of Philadelphia, in the state of Pennsylvania, at which place the session of Congress next ensuing the present shall be held.

SEC. 6. *And be it further enacted,* That on the first Monday in December, in the year one thousand eight hundred, the seat of the government of the United States, shall, by virtue of this act, be transferred to the district and place aforesaid. And all offices attached to the said seat of government, shall accordingly be removed thereto by their respective holders, and shall, after the said day, cease to be exercised elsewhere; and that the necessary expense of such removal shall be defrayed out of the duties on imposts and tonnage, of which a sufficient sum is hereby appropriated. *Approved, July 16, 1790*

From "An Act for establishing a temporary and permanent seat of the Government of the United States," First Cong. Sess.II, Congressional Record, Ch. 28, page 130, 1790.

Documents of Authorization—II

AMENDMENT OF THE RESIDENCE ACT

On March 3, 1791, Congress amended the Residence Act, enlarging the boundaries of the District of Columbia and prohibiting the construction of the public buildings on the Virginia side of the Potomac River.

AN ACT to amend "An Act for establishing the temporary and permanent Seat of the Government of the United States."

Be it enacted by the Senate and House of Representatives of the United States of America in Congress assembled, That so much of the act, entitled "An act for establishing the temporary and permanent Seat of the Government of the United States," as requires that the whole of the district of territory, not exceeding ten miles square, to be located on the river Potomac, for the permanent Seat of the Government of the United States, shall be located above the mouth of the Eastern Branch, be, and is hereby repealed; and that it shall be lawful for the President to make any part of the territory below the said limit, and above the mouth of Hunting Creek, a part of the said district, so as to include a convenient part of the Eastern Branch, and of the lands lying on the lower side thereof, and also the town of Alexandria; and the territory so to be included, shall form a part of the district not exceeding ten miles square, for the permanent Seat of the Government of the United States, in like manner and to all intents and purposes, as if the same has been within the purview of the above re-erection of the public buildings otherwise than on the Maryland side of the river Potomac, as required by the aforesaid act.

Approved, March 3d, 1791.

From "An Act to amend 'An Act for establishing the temporary and permanent Seat of the Government of the United States,'" *Congressional Record,* First Cong., Sess III, Chapter 16, 17, 1791:214,215.

Documents of Authorization—III

AN ACT TO INCORPORATE THE INHABITANTS OF THE CITY OF WASHINGTON, IN THE DISTRICT OF COLUMBIA

Be it enacted by the Senate and House of Representatives of the United States of America in Congress assembled, That the inhabitants of the city of Washington be constituted a body politic and corporate, by the name of a mayor and council of the city of Washington, and by their corporate name, may sue and be sued, implead and be impleaded, grant, receive, and do all other acts as natural persons, and may purchase and hold real, personal and mixed property, or dispose of the same for the benefit of the said city; and may have and use a city seal, which may be broken or altered at pleasure; the city of Washington shall be divided into three divisions or wards, as now divided by the levy court for the country, for the purpose of assessment; but the number may be increased hereafter, as in the wisdom of the city council shall seem most conducive to the general interest and convenience.

Sec. 2. And be it further enacted, That the council of the city of Washington shall consist of twelve members, residents of the city, and upwards of twenty-five years of age, to be divided into two chambers, the first chamber to consist of seven members, and the second chamber of five members; the second chamber to be chosen from the whole number of councillors elected, by their joint ballot. The city council to be elected annually, by ballot, in a general ticket, by the free white male inhabitants of full age, who have resided twelve months in the city, and paid taxes therein the year preceding the election being held: the justices of the county of Washington, resident in the city, or any three of them, to preside as judges of election, with such associates as the council may, from time to time, appoint.

Sec. 3. And be it further enacted, That the first election of members for the city council shall be held on the first Monday in June next, and in every year afterwards, at such place in each ward as the judges of the election may prescribe.

Sec. 4. And be it further enacted, That the polls shall be kept open from eight o'clock in the morning till seven o'clock in the evening, and no longer, for the reception of ballots. On the closing of the poll, the judges shall close and seal their ballot-boxes, and meet on the day following in the presence of the marshal of the district, on the first election, and the council afterwards, when the seals shall be broken, and the votes counted: within three days after such election, they shall give notice to the persons having the greatest number of legal votes, that they are duly elected, and shall make their return to the mayor of the city.

Sec. 5. And be it further enacted, That the mayor of the city shall be appointed, annually, by the President of the United States. He must be a citizen of the United States, and a resident of the city, prior to his appointment.

Sec. 6. And be it further enacted, That the city council shall hold their sessions in the city hall, or, until such building is erected, in such place as the mayor may provide for that purpose, on the second Monday in June, in every year; but the mayor may convene them oftener, if the public good require their deliberations. Three fourths of the members of each council may be a quorum to do business, but a smaller number may adjourn from day to day: they may compel the attendance of absent members, in such manner, and under such penalties, as they may by ordinance, provide: they shall appoint their respective presidents, who shall preside during their sessions, and shall vote on all questions where there is an equal division; they shall settle their rules of proceedings, appoint their own officers, regulate their respective fees, and remove them at pleasure: they shall judge of the elections, returns and qualifications of their own members, and may, with the concurrence of three fourths of the whole, expel any member for disorderly behaviour, or mal-conduct in office, but not a second time for the same offence: they shall keep a journal of their proceedings, and enter the yeas and nays on any question, resolve or ordinance, at the request of any member, and their deliberations shall be public. The mayor shall appoint to all offices under the corporation. All ordinances or acts passed by the city council shall be sent to the mayor, for his approbation, and when approved by him, shall then be obligatory as such. But if the said mayor shall not approve of such ordinance or act, he shall return the same within five days, with his reasons in writing therefor; and if three fourths of both branches of the city council, on reconsideration thereof, approve of the same, it shall be in force in like manner as if he has approved it, unless the city council, by their adjournment, prevent its return.

Sec. 7. And be it further enacted, That the corporation aforesaid shall have full power and authority to pass all by-laws and ordinances; to prevent and remove nuisances; to prevent the introduction of contagious diseases within the city; to establish night watches or patrols, and erect lamps; to regulate the stationing, anchorage, and mooring of vessels; to provide for licensing and regulating auctions, retailers of liquors, hackney carriages, wagons, carts and drays, and pawnbrokers within the city; to restrain or prohibit gambling, and to provide for licensing, regulating or restraining theatrical or other public amusements within the city; to regulate and establish markets; to erect and repair bridges; to keep in repair all necessary streets, avenues, drains and sewers, and to pass regulations necessary for the preservation of the same, agreeably to the plan of the said city; to provide for the safe keeping of the standard of weights and measures fixed by Congress, and for the regulation of all weights and measures used in the city; to provide for the licensing and regulating the sweeping of chimneys and fixing the rates thereof; to establish and regulate fire wards and fire companies; to regulate and establish the size of bricks that are to be made and used in the city; to sink wells, and erect and repair pumps in the streets; to impose and appropriate fines, penalties and forfeitures for breach of their ordinances; to lay and collect taxes; to enact by-laws for the prevention and extinguishments of fire; and to pass all ordinances necessary to give effect and operation to all the powers vested in the corporation of the city of Washington: *Provided,* that the by-laws or ordinances of the said corporation, shall be, in no wise, obligatory upon the persons of non-residents of the said city, unless in cases of intentional violation of by-laws or ordinances previously promulgated. All the fines, penalties and forfeitures, imposed by the corporation of the city of Washington, if not exceeding twenty dollars, shall be recovered before a single magistrate, as small debts are, by law, recoverable; and if such fines, penalties and forfeitures exceed the sum of twenty dollars, the same shall be recovered by action of debt in the district court of Columbia, for the county of Washington, in the name of the corporation, and for the use of the city of Washington.

Sec. 8. *And be it further enacted,* That the person or persons appointed to collect any tax imposed in virtue of the powers granted by this act, shall have authority to collect the same by distress and sale of the goods and chattels of the person chargeable therewith: no sale shall be made unless ten days previous notice thereof be given; no law shall be passed by the city council subjecting vacant or unimproved city lots, or parts of lots, to be sold for taxes.

Sec. 9. *And be it further enacted,* That the city council shall provide for the support of the poor, infirm and diseased of the city.

Sec. 10. *Provided always, and be it further enacted,* That no tax shall be imposed by the city council on real property in the said city, at any higher rate than three quarters of one per centum on the assessment valuation of such property.

Sec. 11. *And be it further enacted,* That this act shall be in force for two years, from the passing thereof, and from thence to the end of the next session of Congress thereafter, and no longer.

APPROVED, May 3, 1802.

From "An Act to incorporate the inhabitants of the City of Washington, in the District of Columbia," *Congressional Record*, Seventh Cong., Sess. I, Ch. 53, 1802:195–197.

Related Cultural Documents—I

LETTERS FROM GEORGE WASHINGTON REGARDING MAJOR L'ENFANT

To William Deakins, Jr. and Benjamin Stoddert
Philadelphia
March 2, 1791.

Gentlemen:

Majr. L'enfant comes on to make such a survey of the grounds in your vicinity as may aid in fixing the site of the federal town and buildings. His present instructions express those alone which are within the Eastern branch, the Potowmac, the Tyber, and the road leading from George town to the ferry on the Eastern branch. he is directed to begin at the lower end and work upwards, and nothing further is communicated to him. The purpose of this letter is to desire you will not be yourselves misled by this appearance, nor be diverted from the pursuit of the objects I have recommended to you. I expect that your progress in accomplishing them will be facilitated by the presumption which will arise on seeing this operation begun at the Eastern branch, and that the proprietors nearer Georgetown who have hitherto refused to accommodate, will let themselves down to reasonable terms.

This communication will explain to you the motive to my request in a letter of the 28th. ulto. I now authorise the renewal of the negotiations with Mr. Burns agreeably to former powers, at such time and in such a manner as, in your judgments is likely to produce the desired effect. I will add however that if the lands described by the enclosed plat, within the red dotted line from A to C thence by the Tiber to D, and along the North line to A can be obtained I shall be satisfied although I had rather go to the line A B. I have referred Majr. L'enfant to the mayor of Georgetown for necessary aids and expences. Should there be any difficulties on this subject, I would hope your aid in having them surmounted, tho' I have not named you to him or any body else, that no suspicions may be excited of your acting for the public.

I am etc.

March 17 [1791]
Washington wrote to the Secretary of State:

"The P. has just recd. the enclosed. He prays Mr. Jefferson to write by tomorrows Post to Major L'Enfant agreeably to what was mentioned this morning."

A press copy of the postscript to Jefferson's letter of March 17 to Major L'Enfant reads:

"There are certainly considerable advantages on the Eastern branch: but there are very strong reasons also in favor of the position between Rock creek and Tyber independent of the face of the ground. it is the desire that the public mind should be in equilibria between these two places till the President arrives, and we shall be obliged to you to endeavor to poise their expectations."

This press copy of the postscript, together with the above notes, are in the Jefferson Papers in the Library of Congress. See http://www.geocities.com/bobarnebeck/post90letters. html. See also the Library of Congress Web site: rs6.loc.gov/ammem/collections/ jefferson_papers.

LETTER FROM GEORGE WASHINGTON TO THOMAS JEFFERSON

Mount Vernon,
March 31, 1791.

Dear Sir:

Having been so fortunate as to reconcile the contending interests of Georgetown and Carrollsburg, and to unite them in such an agreement as permits the public purposes to be carried into effect on an extensive and proper scale, I have the pleasure to transmit to you the enclosed proclamation, which, after annexing your counter signature and the seal of the United States, you will cause to be published. The terms agreed on between me, on the part of the United States, with the Land holders of Georgetown and Carrollsburg are. That all the land from Rock creek along the river to the Eastern-branch and so upwards to or above the Ferry including a breadth of about a mile and a half, the whole containing from three to five thousand acres is ceded to the public, on condition That, when the whole shall be surveyed and laid off as a city, (which Major L'Enfant is now directed to do) the present Proprietors shall retain every other lot; and, for such part of the land as may be taken for public use, for squares, walks, &ca., they shall be allowed at the rate of Twenty five pounds per acre.

The Public having the right to reserve such parts of the wood on the land as may be thought necessary to be preserved for ornament &ca. The Land holders to have the use and profits of all their ground until the city is laid off into lots, and sale is made of those lots which, by this agreement, become public property. No compensation is to be made for the ground that may be occupied as streets or alleys.

To these conditions all the principal Land holders except the purchaser of Slater's property who did not attend have subscribed, and it is not doubted that the few, who were not present, will readily assent thereto; even the obstinate Mr. Burns has come into the measure.

The enlarged plan of this agreement having done away the necessity and indeed postponed the propriety, of designating the particular spot, on which the public buildings should be placed, until an accurate survey and sub-division of the whole ground is made, I have left out that paragraph of the proclamation.

It was found, on running the lines that the comprehension of Bladensburg within the district, must have occasioned the exclusion of more important objects, and, of this I am convinced as well by my observation as Mr. Ellicott's opinion. With great regard and etc.

LETTER FROM GEORGE WASHINGTON TO MAJOR L'ENFANT

Mount Vernon
April 4, 1791.

Sir:

Although I do not conceive that you will derive any material advantage from an examination of the enclosed papers, yet, as they have been drawn by different persons, and under different circumstances, they may be compared with your own ideas of a proper plan for the Federal City (under the prospect which now presents itself to us.) For this purpose I commit them to your private inspection until my return from the tour I am abt. to make. The rough sketch by Mr. Jefferson was done under an idea that no offer, worthy of consideration, would come from the Land holders in the vicinity of Carrollsburg (from the backwardness which appeared in them); and therefore, was accommodated to the grounds about George Town. The other, is taken up upon a larger scale, without reference to any described spot.

It will be of great importance to the public interest to comprehend as much ground (to be ceded by individuals) as there is any tolerable prospect of obtaining. Although it may not be immediately wanting, it will nevertheless encrease the Revenue; and of course be beneficial hereafter, not only to the public, but to the individual proprietors; in as much, as the plan will be enlarged, and thereby freed from those blotches, which otherwise might result from not comprehending all the lands that appear well adapted to the general design; and which, in my opinion, are those between Rock Creek, the Potowmac river and the Eastern branch, and as far up the latter as the turn of the channel above Evans' point; thence including the flat back of Jenkins's height; thence to the Road leading from George Town to Bladensburgh, as far Easterly along the same as to include the branch which runs across it, somewhere near the exterior of the George Town cession; thence in a proper direction to Rock Creek at, or above the ford, according to the situation of the ground. Within these limits there may be lands belonging to persons incapacitated, though willing to convey on the terms proposed; but such had better be included, than others excluded, the proprietors of which are not only willing, but in circumstances to subscribe.

I am etc.

Related Cultural Documents—II

A LETTER FROM BENJAMIN BANNEKER TO THOMAS JEFFERSON

Maryland, Baltimore County, August 19, 1791

SIR,

I AM fully sensible of the greatness of that freedom, which I take with you on the present occasion; a liberty which seemed to me scarcely allowable, when I reflected on that distinguished and dignified station in which you stand, and the almost general prejudice and prepossession, which is so prevalent in the world against those of my complexion.

I suppose it is a truth too well attested to you, to need a proof here, that we are a race of beings, who have long labored under the abuse and censure of the world; that we have long been looked upon with an eye of contempt; and that we have long been considered rather as brutish than human, and scarcely capable of mental endowments.

Sir, I hope I may safely admit, in consequence of that report which hath reached me, that you are a man far less inflexible in sentiments of this nature, than many others; that you are measurably friendly, and well disposed towards us; and that you are willing and ready to lend your aid and assistance to our relief, from those many distresses, and numerous calamities, to which we are reduced. Now Sir, if this is founded in truth, I apprehend you will embrace every opportunity, to eradicate that train of absurd and false ideas and opinions, which so generally prevails with respect to us; and that your sentiments are concurrent with mine. . . .

Sir, I freely and cheerfully acknowledge, that I am of the African race, and in that color which is natural to them of the deepest dye; and it is under a sense of the most profound gratitude to the Supreme Ruler of the Universe, that I now confess to you, that I am not under that state of tyrannical thraldom, and inhuman captivity, to which too many of my brethren are doomed, but that I have abundantly tasted of the fruition of those blessings, which proceed from that free and unequaled liberty with which you are favored. . . .

Sir, suffer me to recall to your mind that time, in which the arms and tyranny of the British crown were exerted, with every powerful effort, in order to reduce you to a state of servitude. . . .

This, Sir, was a time when you clearly saw into the injustice of a state of slavery, and in which you had just apprehensions of the horrors of its condition. It was now that your abhorrence thereof was so excited, that you publicly held forth this true and invaluable doctrine, which is worthy to be recorded and remembered in all succeeding ages: "We hold these truths to be self-evident, that all men are created equal; that they are endowed by their Creator with certain unalienable rights, and that among these are, life, liberty, and the pursuit of happiness." Here was a time, in which your tender feelings for yourselves had engaged you thus to declare, you were then impressed with proper ideas of the great violation of liberty, and the free possession of those blessings, to which you were entitled by nature; but, Sir, how pitiable is it to reflect, that although you were so fully convinced of the benevolence of the Father of Mankind, and of his equal and impartial distribution of these rights and privileges, which he hath conferred upon them, that you should at the same time counteract his mercies, in detaining by fraud and violence so numerous a part of my brethren, under groaning captivity and cruel oppression, that you should at the same time be found guilty of that most criminal act, which you professedly detested in others, with respect to yourselves.

I suppose that your knowledge of the situation of my brethren, is too extensive to need a recital here; neither shall I presume to prescribe methods by which they may be relieved, otherwise than by recommending to you and all others, to wean yourselves from those narrow prejudices which you have imbibed with respect to them . . .

And now, Sir, I shall conclude, and subscribe myself, with the most profound respect, Your most obedient humble servant,
Benjamin Banneker

14
The Erie Canal

United States

DID YOU KNOW . . . ?

➤ Construction of the Erie Canal began in 1817 and was completed in 1825.
➤ The canal is 363 miles (584 kilometers) long.
➤ It was regularly referred to as "Clinton's ditch" and "Clinton's folly."
➤ The canal was built along a route originally followed by Native Americans.
➤ Its original measurements were 40 feet (12 meters) wide at the top, 28 feet (8.5 meters) wide at the bottom, and 4 feet (1.2 meters) deep, with 18 aqueducts and 83 locks.
➤ Its measurements today are 70 feet (21 meters) wide and 7 feet (2.1 meters) deep.
➤ Canal laborers were paid 80 cents a day for a 12-hour day.
➤ Construction costs were recouped in nine years through tolls.
➤ It provided a major route for settling and supplying the West and for transporting western produce to the East.

[George Washington] could not have imagined a world in which technology was more important than geography.
—*Joel Achenbach,* The Grand Idea, *265*

George Washington envisioned and championed a route west. He saw the bigger picture and understood the value of an internal route that would connect the populous eastern seaboard with the western territories (at that time, to the Ohio

Governor DeWitt Clinton and guests ride on the first boat on the Erie Canal. Courtesy of the Library of Congress.

Valley and eventually beyond). A similar thought had occurred to Kublai Khan in China, when he improved linkages of the internal waterways of commerce and communication and brought them right to his door in the new capital of Dada, known today as Beijing. Unfortunately, the Potomac River, Washington's favored route to the West, did not turn out to be optimal because it was difficult to navigate and did not extend far enough west. Not so the Hudson River, if it could be properly connected to the Great Lakes. Washington, DC, remained the administrative capital, but commerce flowed down the Hudson River into New York City, which became the Atlantic port of the Erie Canal.

HISTORY

The development of the United States was shaped by a number of major engineering projects. One of the earliest and most far-reaching in its effects was New York State's Erie Canal. Connecting Lake Erie with the Hudson River, the canal provided a crucial trade link between the Great Lakes region and the Atlantic coast, triggering the first substantial westward migration of American settlers and accelerating the development of the Midwest and New York City (situated at the mouth of the Hudson on the Atlantic Ocean).

The route followed by the canal had been used by Native Americans; its merits for water transportation were discussed in the eighteenth century, notably in 1724 by Cadwallader Colden, surveyor general of the province of New York. Later in the century, George Washington voiced interest in developing the route. The Western Inland Lock Navigation Company was incorporated in 1792 to construct an uninterrupted water route between the mouth of the Hudson River and Lake Ontario by connecting existing waterways. Within a few years navigation began between Schenectady and Seneca Falls, but the company had difficulty making a profit. The construction of a full-fledged Hudson–Lake Erie canal across hills, rivers, and swampland, using manual labor in a territory notably devoid of roads, in an undeveloped nation with a paucity of engineers (the country then had no school of civil engineering), was a colossal undertaking. Thomas Jefferson, for one, called the idea "little short of madness" (http://www.nycanal.com/nycanalhistory.html).

More than to anyone else, credit for the canal goes to DeWitt Clinton, a politician who served several years as first a state senator, a U.S. senator, and mayor of New York City. He was defeated by James Madison in the 1812 presidential election, and later became governor of New York State. In 1810 Clinton advocated building a canal, but a request for aid from the federal government was rejected. After the War of 1812, Clinton continued to push the project.

On April 17, 1816, the state legislature gave permission. Clinton was elected governor in the spring of 1817, putting him in a position to move the project forward, although it was regularly derided by naysayers as "Clinton's ditch" or "Clinton's folly." As the document presented here indicates, on April 15, 1817, the New York state legislature provided financing for the canal, estimated to cost $7 million.

Construction began on July 4, 1817 at Rome, New York, a starting point chosen because of convenient terrain in the area—no lock or aqueduct would be needed for 80 miles (130 kilometers). The canal measured 40 feet (12 meters) wide at the surface, 28 feet (8.5 meters) wide at the bottom, and 4 feet (1.2 meters) deep. The full canal was equipped with 18 aqueducts and 83 locks, each typically measuring 90 feet (27 meters) by 15 feet (4.6 meters). It stretched 363 miles (584 kilometers) from a point near Buffalo to the Albany area, taking advantage of the Mohawk Valley as the best route for crossing the Allegheny Mountains. It was completed in October 1825.

The Erie Canal quickly proved a tremendous success, since it halved travel time and slashed shipping costs by over 90 percent. Total collected tolls recouped the cost of construction in less than nine years, and in 1882 tolls were abolished. A decade and a half after its opening, New York City ranked as the busiest port in the United States (up from fifth).

The canal countered competition from railroads by upgrading its capacity. Between 1836 and 1862, for example, it was broadened to 70 feet (21 meters) and deepened to 7 feet (2.1 meters), and double locks were built to expedite traffic. Shortly before World War I. the locks were lengthened to accommodate

300-foot (91-meter) ships, and the channel was deepened to 12 feet (3.7 meters). Along with a few smaller canals, the Erie Canal was incorporated into the system of waterways known as the New York State Barge Canal (now called the New York State Canal System).

Today the canal channel has a width of 150 feet (46 meters) and a depth of 12 feet (3.7 meters) or 14 feet (4.3 meters), depending on the location. The route uses rivers more extensively than the original Erie Canal, with 57 locks now required if vessels wish to make the 568-foot (173-meter) rise and traverse its length from the Hudson to Lake Erie.

Commercial shipping via the canal declined in the latter part of the twentieth century owing to competition from the St. Lawrence Seaway as well as from railroads and highways, and efforts were subsequently undertaken to expand use of the canal to include recreational purposes.

CULTURAL BACKGROUND

George Washington had always been interested in land. As a young man, he was trained as a surveyor, and the idea of a practical water-based route into the westerly parts of the United States was never far removed from his thoughts. Even after his so-called retirement—long after his success as a general and his leadership of the new country for two terms as president—he never gave up the idea. Soon after he left the White House and returned to his home at Mount Vernon, he saddled up for several weeks on horseback in the West—a dangerous trip that most people would not have undertaken at any age. Washington was in his early 50s, which was considered fairly old in those days.

About the same time, Robert Fulton traveled to Washington, DC, for a meeting with President Jefferson. Fulton tried to interest Jefferson in a Hudson River canal. Fulton was a big supporter of the Canal des Deux Mers in France, and he could conceptualize a similar canal in New York. But Jefferson was emphatic—he preferred the Potomac River.

It should be noted that Jefferson had become somewhat estranged from George Washington by this time. Nevertheless, at some point the two statesmen had discussed the possibility of a Potomac canal, and this remained Jefferson's preference, perhaps for the same reasons Washington favored it: it brought resources from the West right into the nation's capital.

But whether it began on the Potomac or the Hudson, the idea of a canal that could unite the new country was extremely attractive. The sense of the country as a physically connected whole was beginning to emerge. With the governmental structure of the nation in place, it became critical to envision and develop the physical connections needed. It was not until 1956 that there would be a federal highway system that actually linked the entire nation.

But in those days there were far fewer engineers in the United States. Just as in ancient China, it was easier to connect the country using naturally occurring infrastructure, that is, rivers that could be linked by canals. Water is one of the first connectors considered when looking to forge routes of communication, commerce, and cultural exchange.

But there was one other reason why the Hudson was chosen and not the Potomac—a reason that echoes back to the Canal des Deux Mers. That canal was the result of the vision of Pierre-Paul Riquet, a finance official overseeing the salt tax. In western New York, there was also salt in abundance. And in that era before refrigeration, salt was in great demand for both flavoring and preserving meat and fish.

In the document presented here, the reader finds a fascinating and detailed discussion of salt. The first four clauses of the document discuss the canal planning. Clauses V and VI discuss the salt that is manufactured in the western part of New York State. Thus the commercial transport of salt made an immediate financial contribution to the canal's success as the conveyor of Onondaga salt.

The building of the Erie Canal unknowingly contributed to the slowly simmering divisions that were growing between the North and South, which would later erupt full-scale into the U.S. Civil War. Jeanne Krause said, "Refused by three presidents and as many Congresses, New York State built the canal itself, thereby delaying the shift of commercial and financial power [from moving] further South, but also, in a very real sense, tying together a fragmented nation" (Krause, 231).

PLANNING

In 1801, Jefferson appointed Albert Gallatin as secretary of the treasury. In 1808, Gallatin proposed that the country build a network of canals to link the nation, including the Erie Canal. In 1808, the New York state legislature authorized a formal survey of potential canal routes.

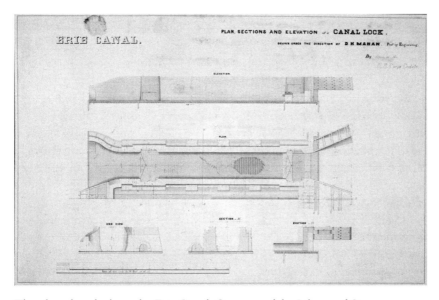

The plans for a lock on the Erie Canal. Courtesy of the Library of Congress.

In 1810, while mayor of New York City, DeWitt Clinton championed Gallatin's idea. Soon thereafter Clinton became governor of New York, and was now in a position to move the Erie Canal project forward. Clinton forged a coalition of federalists and western legislators to set the process in motion. Most of the planning for the canal seems to have taken place through political deal making, not engineering. Once permission to build was obtained, efforts commenced following the initial groundbreaking in 1817.

BUILDING

Contracts were let for many small segments, some as small as three-quarters of a mile—what was really a series of mini-canals, each one tested with water before being brought into the system. The contractors had to succeed or fail without a plan of execution to guide them; indeed, many had to invent construction techniques in order to overcome their lack of experience and training.

According to Krause, the first segment of 93 miles connected the Mohawk Valley with the western counties. This choice of the middle section for initial construction proved to be fortunate. The choice of the eastern section would have duplicated an existing and occasionally impassable route through the Mohawk Valley. The choice of the western section would have produced a canal linking wilderness with wilderness. The middle segment extended the potential for travel west from Utica, generated new trade in timber and grain, and passed through the salt flats on which Syracuse was built.

The contractors were mainly local, and they brought their own tools, hired their own laborers, and used specifications drawn up by carpenters and surveyors during construction of earlier segments. Contrary to popular belief, over 75 percent were American-born (Krause, 226). Precisely because they were local and American, they were not professionals. In fact, sources confirm that the four principal engineers who were guiding the effort had never even seen a canal. As for the laborers who built the canal, most were local folks supplemented by Irish immigrants. We know they were paid 80 cents per day for a 12-hour day.

Perhaps due to American ingenuity, perhaps to the fact that the canal was being built and tested in small segments, it opened for business in 1825—an amazingly short time. Governor Clinton celebrated the moment by pouring a symbolic keg of Lake Erie water into New York harbor. From that moment on, revenues from taxes on travel, salt, and commerce were immediate proof of the canal's potential effectiveness and guaranteed that the remaining sections would also be built.

Who were the early engineers who built the initial section? Benjamin Wright, James Geddes, Nathan Roberts, Canvass White, and John Bloomfield Jervis deserve mention, for they not only contributed to the Erie Canal but went on to build other canals and make important contributions to advances in hydraulic engineering. Canvass White manufactured the first hydraulic cement used in the United States. This was an advance that the builders of London Bridge would no doubt have liked to have had, because they had to secure the pilings of the bridge with rubble and rocks. White's innovation was cement that hardened under water. Jervis made innovative contributions to the railroads and to New York City's Croton water-supply system.

The Mohawk Valley and the canal constituted the only major east–west break in the Appalachian Mountain chain, which extended from Maine to Georgia. Mountains had kept most of the population pinned within a relatively thin strip of Atlantic coastline; mountains had kept well out of reach the abundant natural resources of the West. But now they became available with the Erie Canal linkage, and commerce flourished.

Soon, however, railroads loomed on the competitive landscape, so the canal was widened in response. Thus, just 10 years after first opening, the canal had doubled in size—in effect creating a new canal. In 1903, it was enlarged again, this time to accommodate barges, and was completed in 1918. Draft animals hauled the barges, working in 15-mile shifts. Eventually, in 1959, the canal was linked to the St. Lawrence Seaway.

IMPORTANCE IN HISTORY

The importance of the Erie Canal to the economic progress of the United States is clearly apparent in time and money costs. Prior to the canal, it took two weeks to carry a ton of cargo by road from Buffalo to New York City. After the canal opened the same cargo could be moved through in three and a half days.

The comparative cost of hauling cargo by land versus canal was equally astounding. Krause noted, "In 1812, a $400 cannon cost $2000 to move across

The Ohio Erie Canal Visitor Center, 2005. Workers are completing bicycle trails along a 110-mile stretch of the Ohio and Erie Canal, which runs from Cleveland to New Philadelphia. Courtesy of AP / Wide World Photos.

New York State" (Krause, 225). The typical cost for moving goods over land was $100 per ton; via the canal the cost was $10 per ton, and within 10 years of the canal's opening this had dropped to $4 per ton.

The Erie Canal increased commerce. In the canal's first year, 2,000 boats, 9,000 horses, and 8,000 men were involved in transporting goods on the canal. By 1845, a million tons of goods were carried on the canal each year! The operation was so successful that in 1882, tolls were abolished. From 1903 to 1918, the canal system was enlarged for use by barges, which were pulled by steam or diesel tugboats; the locks were electric. The barge canal was operational until 1994, still functioning as a viable alternative to the railroads. In 1994, New York State overhauled the system to make the canal more attractive for recreational use.

The canal was perhaps one reason why New York surpassed Boston and Philadelphia as the leading American city. Certainly the presence of the canal caused the population of New York to increase. Other cities had deep ports, but only New York had access to the West via the canal.

The Erie Canal had a major impact on the advancement of engineering education in the United States. When the canal was enlarged, the engineers and designers were trained in the newly formed engineering programs at Rensselaer Polytechnic Institute in Troy, New York, and at Union College in Schenectady.

The canal could be called the Internet of its day—a network of sorts that linked parts of the country together, sparking powerful economic change. According to Charles Stein, the parallels between the canal and the Internet are striking. Both were financed by government and developed by the private sector. In each case, what mattered was not the pathway itself but what sprung up along the path—new cities and industries on the banks of the canal, like new companies along the Internet.

FOR FURTHER REFERENCE

Books and Articles

Achenbach, Joel. *The Grand Idea: George Washington's Potomac and the Race to the West.* New York: Simon and Schuster, 2004.

Allen, W. B., ed. *George Washington: A Collection.* Indianapolis, IN: Liberty Classics, 1988.

Bernstein, Peter L. *Wedding of the Waters: The Erie Canal and the Making of a Great Nation.* New York: Norton, 2005.

Hawthorne, Nathaniel. "The Canal Boat." *New-England Magazine,* no. 9 (December 1835): 398–409. A story written by the famous American author telling of his own trip on a canal boat in 1835. The text can be found on the University of Rochester History Department canal Web site: http://www.history.rochester.edu/ canal.

Hecht, Roger W., ed. *The Erie Canal Reader, 1790–1950.* Syracuse, NY: Syracuse University Press, 2003. Contains poems, essays, and fiction by major American and British writers about the Erie Canal.

Krause, Jeanne. "The Erie Canal: Macro-engineering When the World Was Still Simple." In *How Big and Still Beautiful? Macro-Engineering Revisited,* edited by

Frank P. Davidson, C. Lawrence Meador, and Robert Salkeld. Boulder, CO: Westview Press, 1980.

Stein, Charles. "Be It Ditch or Dot.com, a World of Growth." *Boston Globe,* January 23, 2005.

Internet

For a time line, see http://www.eriecanal.org/UnionCollege/timeline.html.

For an overview, see http://www.eriecanal.org/; see also: http://www.nycanal.com/nycanalhistory.html.

For a bibliography, see http://www.eriecanal.org/books.html.

For information on the 175th anniversary of the canal, see http://www.eriecanal.org/UnionCollege/175th.html.

For information on the weigh lock, see http://www.eriecanal.org/UnionCollege/The_Weigh_Lock.html.

For more on the National Canal Museum, see http://www.canals.org/.

Film and Television

The Erie Canal: Albany to Buffalo. DVD. Cicero, NY: Media Artists, 2002. Three episodes of 25 minutes each from the PBS series *Cruising America's Waterways.*

Modern Marvels: The Erie Canal. Videocassette. New York: A&E Television, History Channel, 2000. Available at: http://www.eriecanalmuseum.org/shop.

Documents of Authorization

AN ACT RESPECTING NAVIGABLE COMMUNICATIONS BETWEEN THE GREAT WESTERN AND NORTHERN LAKES AND THE ATLANTIC OCEAN.

Passed April 15, 1817

WHEREAS navigable communications between Lakes Erie and Champlain, and the Atlantic Ocean, by means of canals connected with the Hudson River, will promote agriculture, manufacture, and commerce, mitigate the calamities of war, and enhance the blessings of peace, consolidate the union, and advance the prosperity and elevate the character of the United States: and whereas it is the incumbent duty of the people of this state to avail themselves of the means which the Almighty has placed in their hands for the production of such signal, extensive and lasting benefits to the human race: Now, therefore, in full confidence that the Congress of the United States, and the states equally interested with this state in the commencement, prosecution, and completion of those important works, will contribute their full proportion of the expense; and in order that adequate funds may be provided, and properly arranged and managed, for the prosecution and completion of all the navigable communications contemplated by this act:

I. *Be it enacted by the people of the State of New York, represented in senate and assembly,* That there shall be constituted a fund to be denominated the Canal Fund, which shall consist of all such appropriations, grants and donations, as may be made for that purpose by the Legislature of this state, by the Congress of the United States, by individual states, and by corporations, companies and individuals; which fund shall be superintended and managed by a board of commissioners, to be denominated "the Commissioners of the Canal Fund," consisting of the lieutenant governor, the comptroller, the attorney general, the surveyor general, secretary and treasurer, a majority of whom with the comptroller shall be a quorum for the transaction of business; and that it shall be the duty of the said board to receive, arrange and manage to the best advantage all things belonging to the said fund; to borrow, from time to time, monies on the credit of the people of this state, at a rate of interest not exceeding six per centum per annum, and not exceeding in any one year a sum which, together with the net income of the said fund, shall amount to four hundred thousand dollars; for which monies, so to be borrowed, the comptroller shall issue transferable certificates of stock, payable at such time or times as may be determined by the said board; out of the said fund to pay to the canal commissioners hereafter mentioned, the monies so to be borrowed and the income of the said fund, reserving at all times sufficient to pay the interest of all monies that shall have been borrowed by the said board; to recommend from time to time to the legislature, the adoption of such measures as may be thought proper by the said board for the improvement of the said fund, and to report in the Legislature at the opening of every session thereof the state of said fund, and then the comptroller and treasurer shall open separate books and keep the accounts of the said fund distinct from the other funds of the state.

II. *And be it further enacted,* That the commissioners appointed by the Act, entitled "An Act to Provide for the Improvement of the Internal Navigation of This State," passed April 17, 1816, shall continue to possess the powers thereby conferred and be denominated "the Canal Commissioners;" and they are hereby authorised and empowered, in behalf of this state, and on the credit of the fund hereto pledged, to commence making the said canal, by opening communications by canals and locks between the Mohawk and Seneca Rivers, and between Lake Champlain and the Hudson River; to receive from time to time from the Commissioners of the Canal Fund, such monies as may be necessary for and applicable to the Canal Fund, such monies as may be necessary for and applicable to the objects hereby contemplated; to cause the same to be expended in the most prudent and economical manner, in all such works as may be proper to make the said canals; and on completing any part or parts of the works or canals contemplated by this act, to establish reasonable tolls and adopt all measures necessary for the collection and payment thereof to the Commissioners of the

Canal Fund; that a majority of the said Commissioners shall be a board for the transaction of business, each of whom shall take an oath well and faithfully to execute the duties of his office, and shall report to the Legislature at each session thereof, the state of said works and expenditures, and recommend such measures as they may think advisable for the accomplishment of the objects intended by this act: and in case of any vacancy in the Office of Commissioner, during the recess of the Legislature, the person administering the government may appoint a person to fill such vacancy until the legislature shall act in the premises.

III. *And be it further enacted,* That it shall and may be lawful for the said Canal Commissioners, and each of them, by themselves and by any and every superintendent, agent and engineer employed by them, to enter upon, take possession of, and use all and singular any lands, waters and streams necessary for the prosecution of the improvements intended by this act, and to make all such canals, feeders, dikes, locks, dams and other works and devices as they may think proper for making said improvements, doing nevertheless no unnecessary damage; and that in case any lands, waters or streams, taken and appropriated for any of the purposes aforesaid, shall not be given or granted to the people of this state, it shall be the duty of the Canal Commissioners, from time to time, and as often as they think reasonable and proper, to cause application to be made to the justices of the Supreme Court, or any two of them, for the appointment of appraisers; and the said justices shall thereupon, by writing, appoint not less than three nor more than five discreet, disinterested persons as appraisers, who shall, before they enter upon the duties of their appointment, severally take and subscribe an oath, or affirmation, before some person authorised to administer oaths, faithfully and impartially to perform the trust and duties required of them by this act; which oath or affirmation shall be filed with the Secretary of the Canal Commissioners; and it shall be the duty of the said appraisers, or a majority of them, to make a just and equitable estimate and appraisal of the loss and damage, if any, over and above the benefit and advantage to the respective owners and proprietors, or parties interested in the premises, so required for the purposes aforesaid, by and in consequence of making and constructing any of the works aforesaid; and the said appraisers, or a majority of them, shall make regular entries of their determination and appraisal, with an apt and sufficient description of the several premises appropriated for the purposes aforesaid, in a book or books to be provided and kept by the Canal Commissioners, and certify and sign their names to such entries and appraisal, and in like manner certify their determination as to those several premises which will suffer no damages, or will be benefited more than injured by or in consequence of the works aforesaid; and the Canal Commissioners shall pay the damages so as to be assessed and appraised, and the fee simple of the premises, as appropriated, shall be vested in the people of this state.

IV. *And be it further enacted,* That whenever, in the opinion of the Canal Commissioners, it shall be for the interest of this state, for the prosecution of the works contemplated by this act, that all the interest and title, (if any) in law and equity, of the Western inland lock navigation company, should be vested in the people of this state, it shall be lawful for the said Canal Commissioners to pass a resolution to that effect; and that it shall then be lawful for the President of the Canal Commissioners to cause a copy of such resolution with a notice signed by himself and the Secretary of the said Commissioners, to be delivered to the President or other known officer of said company, notifying the President and directors of the said company, that an application will be made to the justices of the Supreme Court, at a term thereof to be held not less than thirty days from the time of giving such notice, for the appointment of appraisers to estimate the damages to be sustained by the said company, by investing in the people of this state, all the lands, waters, canals, locks, feeders and appurtenances thereto acquired, used and claimed by the said company, under its act of incorporation, and the several acts amending the same; and it shall be the duty of the justices aforesaid, at the term mentioned in the said notice, and on proof of the service thereof to appoint by writing, under the seal of the said court and the hands of at least three of the said justices, not less than three, nor more than five, disinterested persons, being citizens of the United States, to estimate and appraise the damages aforesaid; and it shall be the duty of said appraisers, or a majority of them, to estimate and appraise the damages aforesaid, and severally to certify the same, under oath, before an officer authorised to take the acknowledgment of deeds, to be a just, equitable and impartial appraisal, to the best of their judgement and belief, and shall thereupon deliver the same to one of the Canal Commissioners, who shall report the same to the said court; and if the said court shall be of opinion, that the said damages have been fairly and equitably assessed, the said justices, or any three of them, may certify the same on the said report; and the amount of the said damages, and the expense of the said appraisal, shall be audited by the comptroller, and paid, on his warrant, by the treasurer, out of the Canal Fund; and the people of this state shall thereupon be invested with, and the said Canal Commissioners may cause to be used, all the lands, waters, streams, canals, locks, feeders and appurtenances aforesaid, for the purposes intended by this Act.

V. *And be it further enacted,* That for the purposes contemplated by this Act, and for the payment of the interest, and final redemption of the principal of the sums to be borrowed by virtue hereof, there shall be and hereby are appropriated and pledged, a duty or tax of twelve and an half cents per bushel upon all salt to be manufactured in the western district of this state; a tax of one dollar upon each steam boat passenger for each and every trip or voyage such passengers may be conveyed upon the Hudson River, on board of any steam boat, over one hundred miles, and half that sum for any distance less than one hundred miles, and over thirty miles; the proceeds of all lotteries which shall be drawn in this state, after the sums now granted upon them shall be paid; all the net proceeds of this state from the Western Inland Lock Navigation Company; all the net proceeds of the said canals, and each part thereof, when made: all grants and donations made or to be made for the purpose of making the said canals; all the duties upon sales at auction, after deducting thereof twenty-three thousand and five hundred dollars annually appropriated to the hospital, the Economical School and the Orphan Asylum Society, and ten thousand dollars hereby appropriated annually for the support of foreign poor in the City of New York.

VI.   *And be it further enacted,* That from and after the first Tuesday of August next, there shall be paid and collected, in the manner now directed by law, upon all salt to be manufactured in the County of Onondaga, a duty of twelve and a half cents per bushel, instead of the present duties; and the like tax or duty of twelve and a half cents per bushel upon all other salt to be manufactured in the western district of this state, which shall be collected by the superintendent of the salt springs, until otherwise directed by the Legislature; and for that purpose he shall have a responsible deputy, residing at each place where salt is or may be manufactured, with the like powers, and subject in the like duties, as his present deputies; and that all the provisions, forfeitures, penalties and restrictions, contained in the laws relative to the duties upon Onondaga salt, so far as the same may be applicable, shall be in force for the purposes of enforcing the payment and collection of the tax or duties upon salt, hereby levied and imposed; And further, that the said superintendent, instead of a yearly report to the legislature, shall make a quarter yearly report to the Commissioners of the Canal Fund, and pay into the Treasury of this state on the first Tuesday of February, May, August, and November, in each year, all the monies collected by him during the quarter preceding each of those days; deducting, in addition to what by law is now allowed to be deducted, five per cent of the duties collected at all other salt works not situated in the County of Onondaga, and two per cent of the duties upon Onondaga salt, as a compensation for collecting and paying over the same.

VII.   *And be it further enacted,* That it shall be the duty of the said Canal Commissioners, to raise the sum of two hundred and fifty thousand dollars, to be appropriated towards the making and completing of the said canals, from the Mohawk River to the Seneca River, and from Lake Champlain to Hudson's River, by causing to be assessed and levied, in such manner as the said Commissioners may determine and direct, and said sum of two hundred and fifty thousand dollars, upon the lands and real estate, lying along the route of the said canals, and within twenty-five miles of the same, on each side thereof; which sum, so to be assessed and levied, shall be assessed on the said lands and real estate adjacent to the said several canals, in such proportion for each as the said Commissioners shall determine; and the said Commissioners shall have power to make such rules and regulations, and adopt such measures for the assessing, levying and collecting the sum or sums of money, either by sale of the said lands or otherwise, as they shall deem meet; and the said assessment shall be made on said lands according to the benefit which they shall be considered, by the said Commissioners, as deriving from the making of the said canals respectively; Provided that such rules, regulations, and measures shall, before they are carried into effect, be sanctioned and approved by the chancellor and judges of the Supreme Court, or a majority of them; and provided further, that if any company or individual, subject to such tax, shall subscribe any money or other property towards the completion of the said canals, the amount of such donation or voluntary subscription shall, if the same is less than the amount of the tax, be deducted therefrom, and if more, he or they shall be entirely discharged from the said tax.

VIII.   *And be it further enacted,* That from and after the first day of May next, the aforesaid tax upon steam boat passengers shall be demanded, taken and received by each captain or master of every steam boat navigating the Hudson river; and that during each month thereafter, in which such boat shall be employed for the conveyance of passengers, it shall be the duty of such captain or master to cause to be delivered to the Comptroller of this state, a return or account, sworn to before some officer authorised to administer oaths, stating the name of the boat, the number of trips made by such boat during such month, and the whole numbers of passengers conveyed on board such boat, at each of the said trips, over one hundred miles, and the number conveyed less than one hundred miles and over thirty miles, and pay into the Treasury of this state, the amount of such tax collected during the time mentioned in the said return, deducting three per cent, thereof as a compensation for making such return and collecting and paying over the said tax: And further, that in case of any neglect or refusal, in making such return, or collecting and paying over the tax, as directed in and by this section, the captain or master so neglecting, shall forfeit and pay the sum of five hundred dollars, besides the amount of tax so directed to be collected and paid over, to be recovered in an action of debt, in the name of the people of this state, and for the use of the aforesaid fund.

From Laws of New York, 40th sess. (1817), chap. 262, pp. 361–65.

15
The City of Singapore

Singapore

DID YOU KNOW . . . ?

➤ Singapore was founded as the brainchild of Sir Thomas Raffles.
➤ The country of Singapore occupies only 250 square miles (650 square kilometers); most of the population (4.4 million) lives in the city of Singapore.
➤ Singapore holds the record for shipping tonnage among all the world's ports.
➤ Singapore has five official languages.
➤ The "Singapore Sling" cocktail was created in the early 1900s at the Raffles Hotel by a local bartender, Ngiam Tong Boon.

Geography is destiny. Singapore's amazing growth and prosperity are due in part to administrative policies that make it a free trade port. But equally important is its advantageous position on a strait between the Indian Ocean and the South China Sea connecting China, Thailand, Malaysia, Indonesia, and India, where ships bring all manner of goods to the port of Singapore. Because of its geographic centrality, Singapore exemplifies the inevitable role of multiculturalism. There are five official languages in the country: Chinese, English, Mandarin, Malay, and Tamil, although English is spoken by the majority of the population. Multiculturalism is also obvious in its houses of worship, which date back to the 1800s: from the Hindu Sri Mariamman Temple, Buddhist Kuan Yin Temple, and Sultan Mosque to the Taoist Wak Hai Cheng Temple. From earliest times, Singapore has been a crossroads of commerce and a vibrant center of exchange.

An aerial shot of Singapore, 2002. Courtesy of Corbis.

HISTORY

The republic of Singapore is small, consisting primarily of a single island at the southern tip of the Malay Peninsula. It occupies a minuscule area (250 square miles, or 650 square kilometers), and the overwhelming majority of its approximately 4.4 million people (est. 2004) live in a single city also called Singapore. Yet Singapore is a major force in today's global economy. At the end of the twentieth century, its gross domestic product was over $100 billion. It has one of the largest ports, and one of the highest standards of living, in the world. It is a major player in economic sectors such as banking and electronics manufacturing.

Around the thirteenth century, Singapore was an important Malaysian center, but it fell into decline and its population dropped to about 150 people on the entire island. Its existence as an economic dynamo is a relatively recent development, dating back to the early nineteenth century when Britain, seeking to protect and advance its economic interests in the region, prevailed on local authorities to allow it to establish a presence on the island. With its splendid strategic location, Singapore was well situated to support British trade interests, which faced intense competition from the Dutch, among others.

In 1819, Sir Thomas Stamford Raffles, an administrator with the British East India Company and lieutenant governor of Benkoelen on Sumatra, came looking for a convenient location for a fortified trading station and port to support Britain's Southeast Asian commerce. On January 30, the day after landing on Singapore, Raffles signed a preliminary agreement with the local hereditary chief, Datoo Tummungung Sree Maharajah Abdul Rahman, allowing the British to build a factory on the island. The chief was under the authority of the sultan of Johore, who was affiliated with the Dutch. However, the agreement posed a problem: the Dutch claimed the entire region, including Singapore. Raffles dealt with this issue by choosing to recognize instead the sultan's elder brother as sultan of Johore and Singapore. In early February a treaty affirming British rights in Singapore was signed by three parties: Raffles, the newly installed Sultan Hussein Mahummud Shah, and the Datoo Tummungung (Honourable Governor) Abdul Rahman. Five years later a Dutch-British treaty saw the Netherlands give up all claim to Singapore, as the British purchased sovereignty over the island and surrounding area from the sultan and the Datoo Tummungung.

Singapore became the capital of Britain's straits colonies (which also included Penang, Malacca, and Labuan) in 1832. Singapore's growth was facilitated by the development of rubber and tin production in neighboring Malaya and the opening of the Suez Canal in 1869, as well as by construction of a major British naval base at Singapore after World War I. Singapore's trade skyrocketed, growing eightfold in the four decades following 1873. The island became a self-governing British Crown colony in 1959, and has been an independent republic since 1965.

The creation of Singapore was due in large part to the diligent Raffles. He carefully arranged and planned the British foothold on the island. Raffles's insistence that Singapore be a duty-free port contributed immensely to its emergence as a trade and trans-shipment hub. He drafted an organizational plan for Singapore in 1822, which assigned certain activities to specific areas and laid the basis for the new town's growth into a multi-ethnic city. In 1823, he founded the Singapore Institution (now the Raffles Institution) for the education of Singapore residents.

The documents included here are the agreements made by Raffles with the local authorities and the official proclamation of the British "Residency of Singapore," as well as Raffles's detailed instructions to Major William Farquhar, the aide left in charge of the outpost.

CULTURAL CONTEXT

The Lion City—for that is the meaning of *Singapore*—was named by Javanese prince Sang Nila Utama in 1299. Legend has it that Utama landed on the island after a ferocious storm, and spotted a beast that looked to him like a lion (although lions are not found in Asia). He thought this was a sign of good fortune, and he decided to stay on the island and become king. He renamed it Singapura, and King Sang Nila Utama reigned for 48 years.

Evidence of early international trade can be found in artifacts discovered in recent archaeological digs that have yielded over 30,000 pieces of ancient history in the form of glass, iron tools, and fire-blackened stones, suggesting industrial activity. A dig at the site of Parliament House unearthed eight jars from the Yuan dynasty (1279–1368) and porcelain from the Ming dynasty, clear evidence of trade with China.

In 1349, Singapore was renowned as a haven for pirates, several hundred of whose ships regularly took advantage of the port's harbor. Its geographic location meant there were many places for pirates to hide and then overtake the ships that were drawn to the port as a nexus of exchange. The British East India Company had great difficulty protecting its prosperous operation from the onslaught of these crafty robbers of the high seas. Piracy was finally subdued, but it took a cooperative effort. In 1850, the British Royal Navy sent patrol ships to ply the waters and maintain port security. The Dutch then began to patrol Sumatra, south of Singapore. The Spanish patrolled the Philippines. Even China signed a treaty against piracy in 1860.

Was it through Singapore that the great wave of Islam flowed into Southeast Asia? By 2004, Indonesia had more Muslims than any other country in the world, even those in the Middle East. How did this happen? According to the seventeenth-century chronicle the *Malay Annals*, in 1388 King Paramesvara was on the run from the Majapahit Thai regime in Palembang. Singapore, expressing its customary openness, gave asylum to the king, who then turned traitor and murdered the regent of Singapore. This attracted the Thai military to Singapore, so Paramesvara left the city and hid in Malacca. The chronicle reports that in 1414 Paramesvara converted to Islam, founded a sultanate, and brought the entire Malaysian Peninsula, as well as parts of Indonesia and, of course, Singapore, under the sway of his imperial control and newly espoused religion. The history of Islam in Asia may well be traced to this series of events.

PLANNING

The planning of Singapore actually took place over a period of many years in the mind of Thomas Stamford Raffles. Son of a poor shipmaster, the boy joined the East India Company in London in 1795, where he earned just £70 a year. He was sent to Penang in 1805, where he received a pay increase to £1,500 for his job as assistant to the new governor's secretary. In that position, Raffles was privy to state secrets. This change of status also allowed him to move up in society, and soon he married a widow whose husband had worked for the East India Company. In just two years, he advanced to chief secretary of the operation.

Fast on the move, he was sent to Malacca, where he observed the geographic advantage of the Straits of Malacca, through which passed many Chinese ships laden with silk and other treasures. Raffles conceived an idea that would change the commercial balance of the region: why not put the Straits of Malacca under the control of the East India Company? Raffles posed the idea to the governor of Penang, but nothing happened.

However, Raffles was not one to take no for the final answer. Pushing ahead as he had done since he was 14, Raffles went over the governor's head and traveled to Calcutta to see the governor general, Minto. In Minto he found a kindred spirit. Raffles reported being inspected by Minto with a combination of "scrutiny, anticipation, and kindness" (Gardner, 169). Minto too saw the potential in Raffles's plan. In fact, he'd been considering Java, and here was a territory just north of that prize that the East India Company had tried to acquire 200 years earlier! Minto sent Raffles back to Malacca as his personal envoy, to open talks with the princes of the territory. A study was undertaken, and it was determined that 10,000 troops could take over the region, and the princes would acquiesce (Gardner, 169).

With Raffles as his chief adviser, Minto launched 90 ships on June 18, 1811. It was a mild and polite entry—no lives were lost, no looting allowed. In just 25 days, the operation was complete, and Java was no longer under foreign dominion. It was free to become part of the economic prosperity of the East India Company with young Raffles in charge.

Events brought Raffles back to London. He was now 35, and known as Sir Thomas Stamford Raffles. He took some time to write A History of Java.

Unwilling to rest on his laurels earned at such a young age, he returned to the East Indies. Still burning with his proposal as originally presented to Minto, he brought the same idea to Minto's successor, Lord Hastings: the Straits of Malacca would be a superb economic windfall for whoever controlled traffic on the straits. Hastings agreed with Raffles's strategy to leverage control and reap economic benefits from the traffic in the Straits of Malacca, and Raffles was empowered to do it.

So it was that at sunset on January 28, 1819, a boat carrying Raffles and his colleague Farquhar anchored in the port of Singapura. There is a fascinating eyewitness account from a 15-year-old boy, Wa Hakim, who watched the entire scene from a small boat where he and his family lived. Hakim was 80 when he was interviewed by H. T. Haughton, and his recall of the day was so exact that he even remembered what Raffles and Farquhar were wearing and what food they were served by the ruler they visited. Apparently the meeting went well, for 12 days later Raffles pitched a tent, then moved to a little house. Wa Hakim was hired to help build the fort, which is part of the agreement presented here in the documents authorizing the founding of Singapore. It is interesting to note that Raffles represents not the British government but the East India Company.

BUILDING

A fort was the first structure built. Raffles was wise to fortify his stake, because the Dutch might well have attacked. But while the governments were preparing to negotiate, commercial entrepreneurs were wasting no time. Within six weeks of Raffles's landing, 100 Indonesian ships anchored in the harbor, along with 2 European ships and 1 Thai trading vessel. By May, the population of the

The busy wharves of Singapore, circa 1903. Courtesy of the Library of Congress.

new settlement in Singapore had mushroomed. But governments take their time in negotiations, so few official buildings were constructed. However, commercial enterprises erected docks for loading and unloading goods, ships' stores or chandleries sprung up where sailors could buy a shackle or measure sails for the next voyage, auction houses moved goods to market, and banks provided financial support. By 1846, Singapore was bustling with commerce: six Jewish merchant enterprises worked side by side with five Chinese and five Arab businesses, not to mention two Armenian, one American, and one Indian commercial center. Most business owners lived right above their docking houses, only later moving to elaborate mansions.

Why did such an array of commercial activity spring up so quickly? Geography and the winds played a major role. From December to March, the northeast monsoon winds propelled ships from China, Thailand, and Vietnam southward, bringing dried and salted foods, tea, medicine, and silk. After a layover to await the southwest monsoon winds, the ships were propelled north again, carrying spices, gold, and opium. When the British entered the region, cotton and firearms were added to the ships' cargos. Another trading season began in September and October when rice, pepper, and the exotic bird nests and shark fins of Asian cuisine arrived, as well as rattan for furniture and mother-of-pearl from the sea. Once again, the ships left Singapore loaded with silk, opium, cotton goods, and firearms.

In 1822, Raffles began a series of administrative reforms and sketched out the development of a town. He ordered all commercial buildings to be made with brick and tiled roofs. He mandated that each building have a two-meter covered passageway where commerce could go on unhindered by monsoons or sweltering sun. Raffles also sketched out a plan to segregate new immigrants into specific areas. Although the population is no longer segregated, these ethnic neighborhoods remain today, with colorful elements of their individual cultures.

IMPORTANCE IN HISTORY

Perhaps Raffles's greatest contribution was his realization that for this diverse economic center to prosper and avoid misunderstandings, animosity, and wars there needed to be a place where Asians and Europeans could come together to appreciate each others' cultural heritage and economic insights. In October 1823, using 2,000 Spanish pounds (then the global currency), Raffles established the Singapore Institution, a school for training civil servants and teachers. It was Raffles's dream that children of great rulers around the world would come to study here, and that their mutual exchange would bring the spirit of Singapore's vision of freedom and multiculturalism to the entire world.

The evolution of government in Singapore was curious, as there were no laws per se because the city arose out of an economic agreement. Although Raffles had forbidden the sale of land, he saw the need for some form of real estate acquisition, so he required that land be sold at public auction with permanent lease and registration. Raffles also required that Singapore always remain a free port, and ordered that no taxes on trade or industry would ever impede its growth. English common law was suggested as the practical standard, but Muslim law would rule the Malay people. There were strict anticrime regulations, with an emphasis on rehabilitation, including payment to the injured as an important part of punishment for the offender. However, Raffles decided that opium and alcohol were too profitable to outlaw, and instead he raised the tariffs on these. In 1823, he forbade slavery and even tried to end debt bondage, although many immigrants served out the contracts they had agreed to when booking passage to this land of opportunity (Gardner, 174). No study of Singapore's contributions to history could be complete without mention of the "Singapore Sling," a cocktail invented in 1915 in the famous Raffles Hotel by Ngiam Tong Boon, a bartender of multicultural background. The mélange of ingredients in the exotic drink turns the potion pink to please the ladies. The recipe is available online (see "For

Further Reference"). It reminds readers that Singapore was a place where many sailors and travelers have sought rest and relaxation amid the multicultural atmosphere of this great port.

FOR FURTHER REFERENCE

Books and Articles

Furber, H. *John Company at Work*. Cambridge, MA: Harvard University Press, 1948. "John Company" is the nickname of the English East India Company.

Gardner, Brian. *The East India Company: A History*. New York: Dorset Press, 1971.

Lee, Kuan Yew. *The Singapore Story: Memoirs of Lee Kuan Yew*. New York: Prentice Hall, 1999.Sutherland, Lucy. *The East India Company in 18th Century Politics*. Oxford: Oxford University Press, 1952.

Yen, Ching-Hwang. *A Social History of the Chinese in Singapore and Malaya 1800–1911*. Singapore: Oxford University Press, 1986.

Internet

For recent statistics on Singapore, see http://www.singstat.gov.sg/.

For an overview of Singapore see the Library of Congress Country Studies at http://www.workmall.com/wfb2001/singapore and http://memory.loc.gov/frd/cs/sgtoc.html.

Interested in cuisine? To find out how the mix of Chinese and Malay traditions resulted in a new kind of fare called Peranakan cuisine, see http://www.go-singapore-hotels.com/dining/multicultural.htm.

Cheers! For the recipe for the famous Singapore Sling, see http://www.drinkboy.com/Cocktails/recipes/SingaporeSling.html.

For pictures of the ancient temples and mosques of Singapore, as well as those of the present day, see http://www.orientalarchitecture.com/singapore.htm.

For the work of Singapore artists, including Chen Chong Swee, who combined local landscapes and Chinese and European multicultural traditions, see http://www.living2000.com.sg/chenchongswee/biography.htm.

Documents of Authorization—I

I. AGREEMENT

Agreement made by the Dato Tummungung Sree Maharajah, Ruler of Singapore, who governs the country of Singapore and all the islands which are under the government of Singapore in his own name and in the name of Sree Sultan Hussein Mahummud Shah, Rejah of Johore, with Sir Thomas Stamford Raffles, Lieutenant Governor of Bencoolen and its dependencies on behalf of the Most Noble the Governor General of Bengal.

On account of the long existing friendship and commercial relations between the English Company and the countries under the authority of Singapore and Johore it is well to arrange these matters on a better footing never to be broken.

Article 1. The English Company can establish a factory (logi) situated at Singapore or other place in the Government of Singapore-Johore.

Article 2. On account of that the English Company agree to protect the Dato Tummungung Sree Maharajah.

Article 3. On account of the English Company having the ground on which to make a factory they will give each year to the Dato Tummungung Sree Maharajah three thousand dollars.

Article 4. The Dato Tummungung agrees that as long as the English Company remain and afford protection according to this Agreement he will not enter into any relations with or let any other nation into his country other than the English.

Article 5. Whenever the Sree Sultan, who is on his way, arrives here, all matters of this Agreement will be settled, but the English Company can select a place to land their forces and all materials and hoist the English Company's flag. On this account we each of use put our hands and chops on this paper at the time it is written on the 4th day of Rabil Ahkir in the year 1234.

Seal of the East Indian Company.   (Signed) T.S. RAFFLES.
Chop of the Tummungung.

Documents of Authorization—II

THE TREATY

Treaty of Friendship and Alliance concluded between the Honourable Sir Thomas Stamford Raffles Lieutenant Governor of Fort Marlborough and its Dependencies, Agent to the Most Noble Francis Marquess of Hastings Governor General of India &c., &c., &c., for the Honourable English East India Company on the one part and their Highnesses Sultan Hussein Mahummud Shah Sultan of Johore and Datoo Tummungung Sree Maharajah Abdul Rahman Chief of Singapoora and its Dependencies, on the other part.

ARTICLE 1ST

The Preliminary Articles of Agreement entered into on the 30th of January 1819 by the Honourable Sir Stamford Raffles on the part of the English East India Company; and by Datoo Tummungung Sree Maharajah Abdul Rahman Chief of Singapoora and its Dependencies, for himself and for Sultan Hussein Mahummud Shah Sultan of Johore, is hereby entirely approved, ratified and confirmed by His Highness the aforesaid Sultan Mahummud Shah.

ARTICLE 2ND

In furtherance of the objects contemplated in the said preliminary agreement; and in compensation of any and all the advantages which may be foregone now or hereafter by His Highness Sultan Hussein Mahummud Shah Sultan of Johore, in consequence of the stipulations of this Treaty; the Honourable English East India Company agree and engage to pay to His aforesaid Highness the sum of Spanish Dollars Five Thousand Annually; for and during the time that the said Company may, by virtue of this treaty, maintain a Factory or Factories on any part of His Highness's hereditary Dominions; and the said company further agree to afford their protection to His Highness aforesaid as long as he may continue to reside in the immediate vicinity of the places subject to their authority. It is however clearly explained to and understood by His Highness that the English Government in entering into this alliance and in thus engaging to afford protection to His Highness is to be considered in no way bound to interfere with the internal politics of his States, or engaged to assert or maintain the authority of His Highness by force of Arms.

ARTICLE 3RD

His Highness Datoo Tummungung Sree Maharajah Abdul Rahman Chief of Singapoora and its Dependencies having by Preliminary Articles of Agreement entered into on the 30th of January 1819 granted his full permission to the Honourable English East India Company to establish a Factory or Factories at Singapoora or on any other part of His Highness's Dominions; And, the said Company having in recompence and in return for the said Grant settled on His Highness the yearly sum of Spanish Dollars Three Thousand and having received His Highness into their Alliance and protection, all and every part of the said Preliminary Articles is hereby confirmed.

ARTICLE 4TH

His Highness the Sultan Hussein Mahummud Shah Sultan of Johore and His Highness Datoo Tummungung Sree Maharajah Abdul Rahman Chief of Singapoora engage and agree to aid and assist the Honourable English East India Company against all enemies that may assail the Factory or Factories of the said Company established or to be established in the Dominions of their said Highnesses respectively.

ARTICLE 5TH

His Highness the Sultan Hussein Mahummud Shah Sultan of Johore and His Highness Datoo Tummungung Sree Maharajah Abdul Rahman Chief of

Singapoora agree, promise and bind themselves their heirs and successors, that for as long time as the Hon'ble the English East India Company shall continue to hold a Factory or Factories on any part of the Dominions subject to the authority of their Highnesses aforesaid, and shall continue to afford to their Highnesses support and protection, they their said Highnesses will not enter into any treaty with any other Nation and will not admit or consent to the Settlement in any part of their Dominions of any other power European or American.

ARTICLE 6TH

All persons belonging to the English Factory or Factories or who shall hereafter desire to place themselves under the protection of its flag, shall be duly registered, and considered as subject to British authority.

ARTICLE 7TH

The mode of administering Justice to the native population shall be subject to future discussion and arrangement between the contracting parties, as this will necessarily in a great measure depend on the Laws and usages of the various tribes who may be expected to settle in the vicinity of the English Factory.

ARTICLE 8TH

The port of Singapoora is to be considered under the immediate protection and subject to the regulation of the British Authorities.

ARTICLE 9TH

With regard to the duties which it may hereafter be deemed necessary to levy on Goods, Merchandize, Boats or Vessels, His Highness Datoo Tummungung Sree Maharajah Abdul Rahman is to be entitled to a moiety or full half of all the amount collected from Native Vessels. The expenses of the Port and the collection of duties to be defrayed by the British Government.

Done and concluded at Singapoora this 6th day of February in the year of Our Lord 1819, answering to the 11th day of the Month Rubbelakhir and Year of the Hujira 1234.

Seal of the East India Company	T.S. RAFFLES *Agent to the Most Noble the Gov. Genl. with the States of Rhio Lingin and Johor.*
Seal of the Tummungung	Seal of the Sultan

Documents of Authorization—III

PROCLAMATION

A treaty having been this day concluded between the British Government and the native authorities, and a British establishment having been in consequence founded at Singapore, the Honourable Sir T.S. Raffles, Lieutenant-Governor of Bencoolen and its Dependencies, Agent to the Governor-General, is pleased to certify the appointment by the Supreme Government of Major William Farquhar, of the Madras Engineers, to be Resident, and to command the troops at Singapore and its Dependencies; and all persons are hereby directed to obey Major Farquhar accordingly.

It is further notified, that the Residency of Singapore has been placed under the Government of Fort Marlborough [Bencoolen], and is to be considered a dependency thereof; of which, all persons concerned are desired to take notice.

Dated at Singapore, this 6th day of February, 1819.
By order of the Agent of the Most Noble
the Governor-General.
(Signed) F. CROPLY, *Secretary.*

Documents of Authorization—IV

SINGAPORE, 6th *February* 1819.
TO MAJOR WILLIAM FARQUHAR,
Resident and Commandant,
Singapore.

SIR:

Herewith I have the Honour to transmit to you one of the copies of the treaty this day concluded between the Honourable the East India Company, and their Highnesses the Sultan of Johore, and the Tummungong of Singapore and its dependencies.

2. As the object contemplated by the Most Noble the Governor General in Council, namely the establishment of a station beyond Malacca, and commanding the southern entrance of the Straits, has thereby been substantially accomplished, I proceed to give you the following general instructions for the regulation of your conduct in the execution of the duties you will have to perform as Resident and Commandant of the station which has been established.

3. As you have been present at and assisted in the previous negotiations, and are fully apprised of the political relations existing between the states in the immediate vicinity of this island, it is only necessary for

me to direct your particular attention to the high importance of avoiding all measures which can be construed into an interference with any of the states where the authority of His Netherlands Majesty may be established. Whatever opinion may be formed with regard to the justice or nature of the proceedings of the Dutch authorities in these seas, it is not consistent with the views of his Lordship in Council to agitate the discussion of them in this country; and a station having been obtained which is properly situated for the securing the free passage of the Straits, and for protecting and extending the commercial enterprises both of the British and native merchant, all questions of this nature will necessarily await the decision of the higher authorities in Europe.

4.   It is impossible, however, that the object of our establishment at Singapore can be misunderstood or disregarded, either by the Dutch or the native authorities; and while the former may be expected to watch with jealousy the progress of a settlement which must check the further extension of their influence throughout these seas; the latter will hail with satisfaction the foundation and the site of a British establishment, in the central and commanding situation once occupied by the capital of the most powerful Malaysian empire then existing in the East, and the prospect which it affords them of the continuance, improvement and security of the commercial relations by which their interests have been so long identified with those of the British merchant. It is from the prevalence of this feeling among the natives and the consequences which might possibly arise from it, that I am desirous of impressing on your mind the necessity of extreme caution and delicacy, not only in all communications which you may be obliged to have with the subjects of any power under the immediate influence of the Dutch, but also in your intercourse with the free and independent tribes who may resort to the port of Singapore either for the purposes of commerce or for protection and alliance. The offer which is understood to have been made to the Sultan of the Bugguese, is a sufficient proof that in all communications regarding the proceedings of the Netherlands Government we should carefully guard against the expression of any sentiment of dislike or discontent, however justly those feelings might be excited, lest our motives be misconstrued, not only by the Dutch but by the natives themselves.

5.   With regard, however, to those states that have not yet fallen under their authority, it is justifiable and necessary that you exert your influence to preserve their existing state of independence. If this independence can be maintained without the presence of an English authority it would be preferable, as we are not desirous of extending our stations; but as from the usual march of the Dutch policy, the occupation of Tringano, and the extension of their views to Siam, may be reasonably apprehended, a very limited establishment in that quarter may become ultimately necessary. It is at all events of importance to

cultivate the friendship of these powers, and to establish a friendly intercourse with them; and as the recent application from the Sultan of Tringano for a small supply of arms affords us a favorable opportunity of advancing towards this object, you will avail yourself of the first opportunity to comply with his request.

6. A similar line of policy in relation to the states of Pahang and of Lingin will be conducive to the maintenance of the influence and just weight which the English nation ought properly to possess in these seas. As it is my intention to return to this island after the completion of the arrangements at Acheen, I shall then be able to avail myself of the information you may have collected in the intervening period, relative to the political state of Borneo Proper, Indragiri and Jambi. In the meantime, it is probable that a knowledge of our establishment at this station will have considerable weight in preventing these powers from falling under the influence of the Dutch.

7. With reference to the native authorities residing under our immediate protection, it is only necessary for me to direct your attention to the conditions of the treaty concluded with these chiefs; which it will be incumbent on you to fulfil, under any circumstances that may arise, in a manner consistent with the character and dignity of the British Government. In the event of any question of importance being agitated by the Dutch Government at Batavia, or the authorities subordinate to it, you will refrain from entering into any discussion that can be properly avoided, and refer them to the authority under which you act.

8. To enable you to conduct the civil duties of the station with efficiency, I have appointed Lieutenant Croply your assistant; and that officer will conduct the details of the Pay Department, Stores and Commissariat with such other duties as you may think proper to direct. The allowances for your assistant have been fixed at Spanish dollars 400 per month, subject to the confirmation of the Supreme Government.

9. As the services of Lieutenant Croply as my acting Secretary, will be for some time required under my immediate authority, Mr. Garling of the Bencoolen Establishment will officiate until his return. In the event of its being necessary for you to leave the station or of any accident depriving the Company of your services, your assistant is appointed to succeed to the temporary charge until further orders.

10. Mr. Bernard has also been appointed to take charge provisionally of the duties of the port as Acting Master Attendant and Marine Storekeeper, and in consideration of the active duties that may be required in this department, and the general services which this office may be required

to perform, he is allowed provisionally to draw a monthly salary of 300 dollars per month.

11. As the convenience and accommodation of the port is an object of considerable importance, you will direct your early attention to it, and to the formation of a good watering place for the shipping. You will also be pleased to establish a careful and steady European at St. John's with a boat and small crew, for the purpose of boarding all square sailed vessels passing through the Straits and of communicating with you either by signals or by a small canoe as you may find most advisable.

12. It is not necessary at present to subject the trade of the port to any duties; it is yet inconsiderable, and it would be impolitic to incur the risk of obstructing its advancement by any measure of this nature.

13. In determining the extent and nature of the works immediately necessary for the defense of the port and station, my judgment has been directed in a great measure by your professional skill and experience. With this advantage and from a careful survey of the coast by Captain Ross, aided by my own personal inspection of the nature of the ground in the vicinity of the Settlement, I have no hesitation in conveying to you my authority for constructing the following works with the least delay practicable:

On the hill overlooking the Settlement, and commanding it and a considerable portion of the anchorage, a small Fort, or a commodious blockhouse on the principle which I have already described to you, capable of mounting 8 or 10 pounders and of containing a magazine of brick or stone, together with a barrack for the permanent residence of 30 European artillery, and for the temporary accommodation of the rest of the garrison in case of emergency.

Along the coast in the vicinity of the Settlement one or two strong batteries for the protection of the shipping, and at Sandy Point a redoubt and to the east of it a strong battery for the same purposes.

The entrenchment of the Cantonment by lines and a palisade, as soon as the labor can be spared from works of more immediate importance.

14. These defenses, together with a Martello tower on Deep Water Point, which it is my intention to recommend to the Supreme Government, will in my judgment render the Settlement capable of maintaining a good defense. The principle on which works were charged for a Malacca, is to be considered as applicable to this station, and it is unnecessary for me to urge on you the necessity of confining the cost of these works within the narrowest limits possible. As the construction of them, however, will necessarily demand a greater portion of care and superintendence than the performance of your duties will permit you to devote to them, I have appointed Lieutenant Ralfe of the Bengal Artillery to be the assistant

Engineer. This officer will likewise have charge of the ordnance and military stores, and for the duties attendant on both these appointments conjoined I have fixed his salary at Spanish dollar 200 per mensem, to commence from the 1st instant, and subject to the confirmation of the Supreme Government.

15.   As you will require the aid of a Staff officer to conduct the duties of the garrison, I have thought proper to authorize the appointment of a cantonment adjutant on the same allowance lately authorized at Malacca. As this officer may be considered your personal staff, I shall not make any permanent arrangement regarding it, but have appointed Lieutenant Dow to the temporary performance of its duties.

16.   The indent for ordnance and stores which you have handed to me shall be transmitted to Bengal without delay, and I request you will lose no time in the erection of store-houses for their reception. An application for the number and description of troops, which you have recommended to form the garrison of the residency will accompany the indent, together with an application for provisions equal to their supply for 12 or 15 months.

17.   I should not think myself justified at the present moment in authorizing the erection of a house for the accommodation of the chief authority, but I shall take an early opportunity of recommending the adoption of that measure, or in the event of the Supreme Government declining to authorize it, the grant of a monthly allowance sufficient to compensate for the inconveniences to which, in the infancy of the Settlement, the Resident is necessarily liable. A storehouse for the Commissariat department is at present of indispensable necessity, and you will accordingly be pleased to erect a house of this description, of such materials as can be procured, and as soon as you may find practicable. A magazine built of such material, for the military store, would be subject to some risk; and I therefore confide to your professional judgment the adoption of such measures for their security as you may judge most expedient under the circumstances.

18.   For a very short period it may be necessary to retain the brig *Ganges* as a store vessel, but I rely on your discharging her the moment her services can be dispensed with.

19.   In the event of your adopting this arrangement, you will be pleased immediately to tranship to that vessel the public property now on board the H.C. hired ship *Mercury*, whose charter expires on the 24th instant, previously to which you will accordingly be pleased to discharge her from the public service. You will inform the commander, that I am entirely satisfied with his conduct while he was under my authority, and that as tonnage will probably be required to convey troops and stores from P. of W. Island, I shall be happy, in the event of his early arrival at that port, to

consider his request for the further employment of his ship to be entitled to some considerations.

20. You are already apprised that the H.C. ship *Nearchus* has been put under your orders, and the services of the schooner *Enterprise* will be also available by you, during the remainder of the period of two months for which she was engaged.

21. The accounts of the residency are those that detail the receipt and disbursement of the public money. These are principally:

    (i.) An account particulars of military disbursements in which every military abstract and disbursement is clearly and correctly entered.

    (ii.) A general account particulars, which will comprise the particulars of every disbursement of whatever nature, and containing also, under the head of "Military Establishment," a correct copy of No. 1, and,

    (iii.) A general treasury account, showing on the one side the general amount of the disbursement made on each particular account or head, with the balance remaining on hand; and on the other, the balance which remained on the 1st of the month, together with all the sums which may be received during the course of it.

22. The accounts of the commissariat cannot at present be arranged according to the established forms, they can however be kept with correctness by Mr. Garling, and I shall take care to procure and to forward from Pinang the necessary forms under which the first assistant will probably be able to arrange them on his taking charge of his appointment. You will of course exercise a strict superintendence over this department, no disbursements from which are to be made without your authority; and you will be pleased to examine the accounts rendered to you previously to transmitting them to Fort Marlborough.

23. A quarterly account of expenditure and remains of military stores will be transmitted to me. You will also be pleased to forward the usual returns to the Presidency of Fort William [Bengal] agreeably to the regulations of the service.

24. It does not occur to me that there is any other point of importance on which it is necessary at present to give you any instructions. I shall probably return to this residency after a short absence, and if in the meantime any important matter should occur, which I have not anticipated in this letter, I have the satisfaction afforded me by a perfect reliance on your acknowledged zeal, in the advancement and protection of the honour and interests of our country, moderated by the prudence and judgement which the infancy of our present establishment so particularly demands.

I have, &c.,
(Signed) T.S. RAFFLES

From Charles Burton Buckley, *An Anecdotal History of Old Times in Singapore, from the Foundation of the Settlement under the Honourable the East India Company on February 6th, 1819 to the Transfer to the Colonial Office as Part of the Colonial Possessions of the Crown on April 1st, 1867* (Kuala Lumpur: University of Malaya Press, 1965).

16
The Suez Canal

Egypt

DID YOU KNOW . . . ?

➤ The Suez Canal was under intermittent construction for 3,700 years, and completed in 1869.
➤ The canal is 101 miles (163 kilometers) long.
➤ 1.5 million people worked on the construction; 120,000 died.
➤ The canal construction cost 432,807,882 francs.
➤ The concession to build the canal was granted immediately following an amazing feat of horsemanship by the canal's progenitor, Ferdinand de Lesseps, in the presence of the Pasha of Egypt.
➤ The canal enabled shipping to pass from the Mediterranean Sea to the Indian Ocean, via the Red Sea.

The Suez Canal, from first dig to final christening, took around 3,700 years to build. Its actual completion in 1869 officially capped what was in many respects a paradigmatic endeavor. The canal's construction set new standards in use of machine power and international negotiations, and channeled its way through the desert to connect the Mediterranean and Red Seas and facilitate world trade. It stretches 101 miles (163 kilometers) on a north–south route through expansive desert, snaking through nearby lakes and continuing across the Isthmus of Suez, between Port Said on the Mediterranean and Suez on the Red Sea. It was not the first link ever made between the two seas, but it was the grandest.

A statue of De Lesseps looks over the Suez Canal at Port Said. Courtesy of the Library of Congress.

HISTORY

In 1850 B.C., the first canal in the region was dug. Its purpose was primarily for trade, and ensuing generations made their marks on it. The Romans extended it, the Byzantines neglected it, and the Arabs reopened it—until military interests filled it up in A.D. 770.

Previous efforts at canal building in the region, both for purposes of irrigation as well as transportation, led to connecting the Red Sea with the Nile delta via the Great Bitter Lake, a route that was simpler to implement but more onerous for navigation. Those waterways were undertaken under the pharaohs, reconstructed under the Roman emperor Trajan, and reconstructed again under the Arab ruler Amr ibn al-Aas in the seventh century, but all were eventually abandoned.

The notion of a canal across the isthmus was broached by the Venetians in the fifteenth century and later by the French, but the project failed to move beyond the talk stage. One obstacle was an analysis by French engineers during the reign of Napoleon Bonaparte, which concluded, erroneously, that the levels of the two seas differed by about 30 feet (9 meters), which meant locks would be needed. Eventually it was realized that the two seas had about the same level.

The canal project finally got underway thanks to the determination of Ferdinand de Lesseps, who pressed forward undeterred through a welter of political, financial, and technical roadblocks. De Lesseps was not an engineer, but he possessed remarkable drive and imagination and had long dreamed of "piercing" the Isthmus. Rather than digging the open-cut canal straight across the isthmus, which is only about 75 miles (120 kilometers) wide, the builders adopted a more roundabout route that made use of several existing lakes—Lake Manzala, Lake Timsah, Great Bitter Lake, and Little Bitter Lake. Construction began in 1859. The project was begun with hundreds of thousands of manual workers under forced labor—overall, more than 1.5 million people worked on the job, and over 120,000 are said to have lost their lives—and it was completed with the help of dozens of specially designed, then state-of-the-art (steam-powered) dredgers, excavators, and other machines.

The barren isthmus, where Suez had been the only populated point of any size, was refashioned. Besides the canal itself, the project produced new towns, notably Port Said and Ismailia. Ismailia—located at the halfway point, on Lake Timsah—was designed by de Lesseps as a base of operations for the canal builders. Today the city has a population of a few hundred thousand and serves as the capital of a governorate, as does Port Said. Supporting infrastructure was constructed, including roads and irrigation and water-supply canals. In addition, thousands of acres of land were brought under cultivation. Construction took somewhat longer than the six years planned, owing not only to the difficulties of the job but also to such unanticipated developments as a cholera epidemic in 1865.

The ceremonial opening of the canal, on November 17, 1869, was an elaborate affair, with foreign dignitaries and royalty in attendance. Unfortunately, composer Giuseppi Verdi had not yet finished the opera that is said to have been commissioned for the occasion, but the work, *Aida*, did not receive its premiere until 1871, at the Cairo Opera.

Britain gained a controlling interest in the Suez Canal Company in 1875, and the waterway was nationalized by Egypt in 1956. Over the years, the canal has been enlarged. Originally measuring 26 feet (8 meters) deep, 72 feet (22 meters) wide at the bottom, and at least 190 feet (58 meters) wide at the surface, by 1963 it had reached a width of 179 feet (55 meters) and a depth of 33 feet (10 meters).

From the beginning, Pasha al-Said's *firman* of 1854 mandated that the canal be open on equal terms to ships of all nations. The principle of open access to all—in war as well as peace—was legalized in the international Convention of Constantinople of 1888 (not signed by Britain until 1904). Even after the convention, however, on several occasions, passage through the canal was partially or fully blocked for military or political reasons.

CULTURAL CONTEXT

The Suez Canal was not a continuous project. The waterway, perhaps initiated by the legendary Sesostris, was used by Egyptians during the time of Seti I in 1380 B.C. The ancient channel—still visible in areas near Wadi Tumilat and used by the constructors of the modern Suez Canal—did not connect all the way to the Red Sea. At the time of Darius in 520 B.C., ships that entered from the Mediterranean came up the Pelusiac section of the Nile to Bubastis, where they continued to Heropolis. However, there was a break at that point, and all cargo had to be reloaded onto ships on the Red Sea. This was no doubt an excellent business for the area of transfer, but was inconvenient and expensive for shippers.

It was not until 385 B.C. that the inventive Ptolemy Philadelphus finally connected the canal to the Red Sea, building a new town called Arsinoë at the juncture. When the Arabs conquered Egypt in the seventh century A.D., Amr began to build another canal to pass through Cairo. But while the canal was a boon to shipping and commerce, it was altogether too strategically valuable. The fierce warrior Caliph al-Mansur, founder of Baghdad, ordered the canal closed in A.D. 770 to cut off the supply route being used by his Arabian enemies.

PLANNING

Planning for the canal's nineteenth-century rebirth began in 1844 with the arrival in Cairo of the Saint-Simonians, a clan of Frenchmen who aimed to rejuvenate the Mediterranean through macroengineering projects. The leader, Prosper Enfantin, received permission from the viceroy of Egypt, Mohammed Ali Pasha, to survey a canal route across the Isthmus of Suez. Although the survey was completed in two weeks, no work was begun, and the pasha died two years later.

Ali was succeeded by Mohammed Pasha al-Said, a 300-pound man who had been boyhood friends with a boy named Ferdinand de Lesseps, the son of the French consul in Cairo. The two used to have breakfast together because the pasha's son very much liked the cereal served in Ferdinand's house. After breakfast, the boys played together and became close childhood pals.

Many years later, when Ferdinand de Lesseps was a well-known retired diplomat, he revisited his boyhood home in Egypt in 1854 to find that his pal was now the khedive and viceroy of Egypt. Still good friends, the two went on a hunting expedition with a group of al-Said's associates. The party pitched their tents the first night by a high wall to protect themselves from fierce winds. At some point, Ferdinand jumped his horse over the wall, although no one else in the pasha's entourage dared attempt such a feat. Al-Said was so delighted that he told de Lesseps that he would offer him anything it was in his power to grant. It was at this moment that de Lesseps realized his dream. He asked Al-Said to give him a concession to build the Suez Canal. Al-Said kept his word. Thus the Suez Canal may be the only macroengineering project in history that resulted from a combination of superb horsemanship and breakfast cereal. More likely, it is the childhood friendship and trust that such a long amity creates that is the real basis of the contract.

Reproduced here is the khedive's original *firman* granting the concession. The terms under which the canal was built and run were amplified and modified in a subsequent 1856 act, also given here, which stipulated that the Suez Canal Company would operate the canal for 99 years, after which time ownership would revert to the Egyptian government.

The first document offers evidence of the friendship between de Lesseps and the pasha. In the authorizing agreement of November 30, 1854, the contract begins with a most unusual phrase not found in any other macroengineering text, "Our friend Mons. Ferdinand de Lesseps," and thereafter grants full authority over the canal construction to de Lesseps. There is further evidence of friendship and cooperation in many aspects of the authorizing document of the Suez Canal. In the first paragraph, the pasha reveals that his friend envisions not just a construction project but a "company for this purpose composed of capitalists of all nations." In fact, the expansive nature of the endeavor is noted in its name of a "Universal Company."

BUILDING

Land acquisition, labor management, and innovative machinery were de Lesseps's three primary concerns in building the Suez Canal—and there was the little matter of money.

Land rights were spelled out in the first contract. In article IV, it is stipulated that "all necessary land not belonging to private persons shall be granted to it free of cost." There is more discussion of land in article VII, where it is agreed that "the Egyptian Government will give up to the Company the uncultivated lands belonging to the public domain, which shall be irrigated and cultivated at the expense of the Company." In the deal, Egypt would gain tremendous value by improving the land through irrigation. While the company got the land for 10 years free of taxes, for the next 89 years it had to pay tithes, and after that, if it wanted to retain the land, it had to pay costs comparable to those levied on lands of similar quality, that is, irrigated, cultivated, and next to one of the greatest shipping channels in the world.

Labor management was a sore point from the very ancient days. The historian Herodotus notes that over 100,000 people perished in the building of the 609 B.C. effort to build a canal. But the work done in A.D. 1856 was also a challenge for workers. The agreement stated that four-fifths of the jobs had to go to Egyptians, and there were also Turks employed, and the French sent laborers as well. Forced labor was involved. But when the former pasha died and was succeeded by Pasha al-Said in 1863, he declared forced labor to be against the values of the country. Cutting off a huge source of labor supply caused disagreements among many of the participating nations, until Emperor Napoleon III finally settled the dispute in 1864 in an agreement that allocated land and money.

When manual labor is readily available—forced labor at that—the oldest and most primitive construction methods are often used. But when labor is in short supply, innovations begin to crop up. Such was the case with the Suez Canal. As soon as workers became scarce after forced labor was abolished, innovative new engineering methods and machinery were soon developed that accelerated the progress of construction.

In addition, construction took on a more modern project-management approach, with the work separated into four contracts. First came building materials: a contract for delivery of 230,000 cubic meters of concrete blocks. The next contract called for digging the channel to Port Said. The third contract was for digging the channel farther, this time through the more difficult high ground, which required advanced methods. The fourth contract, given to Paul Borel and Alexandre Levalley, brought the channel from Lake Timsa to the Red Sea.

Dredging equipment representing the latest technology of the times was used, and the shortage of labor produced highly advanced engineering and machines that no doubt saved many lives. Much of the dredging equipment had been developed in France, and the contract provided that the needed equipment would be imported free of any taxes.

Always the diplomat, de Lesseps put his communication skills to use to raise money. In 1865, when construction had progressed sufficiently to show the world, he invited 100 representatives from countries around the world to a preview tour. These visitors were from chambers of commerce—future customers of the canal. While future revenues were thus assured, it was the costs of the day that were challenging de Lesseps. Originally de Lesseps had gone to Rothschild, the renowned investment-banking establishment, to obtain funding for the project. The normal fee of perhaps five percent was the going rate. But when de Lesseps heard the figure, he stormed out of Rothschild's office, refusing to pay a finder's fee, even though it was normal investment banking practice. Instead, de Lesseps was certain he could persuade the French peasants to fund the dream. With that, de Lesseps set up his own offices and raised the money without an investment banker! When the first public offering was opened on November 5, 1858, 400,000 shares were sold at 500 francs per share.

But finances remained a challenge throughout the process. When the canal was opened for business, the cost of construction by then was 432,807,882 francs. The financial situation was still shaky, but on November 16, 1869, just 10 years after construction had begun, 65 ships from nations around the world, headed by a French vessel with the empress Eugénie on board, sailed down the route, immediately followed by commercial vessels. The next year 500 ships used the Suez Canal. Traffic increased exponentially, and soon the canal was both an engineering and a financial success.

By 1875, Rothschild had gained back what he had long ago lost when de Lesseps stormed out of his office. "The khedive of Egypt found himself compelled by his heavy debts to offer for sale his 177,000 shares in the Suez Canal. [Prime Minister Benjamin] Disraeli, on receiving early news that the shares were to be disposed of, promptly borrowed the four millions needed from the Rothschilds in the name of the cabinet and waited for parliament to sanction his audacious stroke" (*Encyclopedia Britannica*, 1948, vol. 3, 250). While not many people owned shares of the Suez Canal and derived direct financial benefit, the world also gained a great musical work as a result of the engineering triumph. When the Suez Canal opened in November 1868, the khedive of Egypt built a new opera house for Cairo to mark the event. It is believed that the khedive commissioned Guiseppe Verdi to write a new opera for the occasion, one celebrating Egyptian culture. Verdi took the khedive's commission and began to think about a work that could symbolize the Suez Canal and the culture of Egypt. Verdi's collaborator on *Don Carlos*, Camille du Locle, came to him with a scenario sketched by a French Egyptologist who worked for the khedive. The Frenchman, Auguste Mariette, had created a story that du Locle refined and gave to Verdi, who laid out the basic operatic elements of recitative and scena, working with his wife, Guiseppina. The libretto was written by Antonio Ghislanzoni. It is ironic that this great work did not premiere in Cairo as it was supposed to, because the scenery and costumes that were being crafted in Paris could not be shipped out of Paris due to the Franco-Prussian War and the siege of Paris in November 1870, just two months before the scheduled January 1871 Cairo debut. The delay allowed Verdi to make needed refinements, and the opera finally opened in Cairo on December 24, 1871, to rave reviews. Verdi was awarded the honor of Commendatore of the Ottoman Order.

The Suez Canal has been a boon to shipping for nearly one and a half centuries. The shortcut provided by this engineered waterway, which connected markets in Europe with those in India, was an obvious selling point. Even though the canal cannot yet accommodate the largest supertankers of the early twenty-first century, it remains a crucial transportation link. In May 2002, for example, it accommodated 1,135 vessels with a total tonnage of 37.6 million.

IMPORTANCE IN HISTORY

The Suez Canal is the result of a diplomatic vision. The first agreement between Ferdinand de Lesseps and Pasha al-Said calls, in article VI, for "tariffs of dues for passage . . . [which] shall be always equal for all nations, no particular advantage can ever be stipulated for the exclusive benefit of any one country." The Suez Canal has made a major contribution to peace in the world.

In 1952, Egyptian president Nasser sought international funds to improve the Suez Canal. At first, Nasser solicited funding from the United Nations,

"Opening of the Suez Canal," printed in Frank Leslie's illustrated newspaper, 1870. Courtesy of the Library of Congress.

World Bank, and Western democratic nations. But money continued to be a problem, so Nasser reached out more widely, cutting arms deals with communist countries, especially Czechoslovakia. An outraged U.S. State Department withdrew American funds, and Nasser retaliated by nationalizing the Suez Canal on July 26, 1959. A war of ownership ensued, but eventually the Suez Canal was returned to Egyptian control. Through the brilliant efforts of a negotiating team that included Jean-Paul Calon, general counsel of the Suez Canal Company, the World Bank was persuaded to compensate company shareholders. Instead of going broke, the Suez Canal Company became one of the leading financial investment houses in Europe.

The Suez Canal Company was initially established as a privately owned enterprise. Its executive offices were located in Paris, and the French and British governments shared jointly in stock ownership through government agencies, banks, and private individuals. A similar formula was later adopted for the Chunnel between France and England.

What can be learned from the Suez Canal and the importance of control of strategic assets in times of peace and in times of war? What of future construction that may not be on owned land? For example, how would a submerged transport system across the Atlantic Ocean be controlled? Or solar power satellites in space? The Suez Canal stands as a model for future macroengineering projects with its vision of openness to all nations with equal treatment, and raises hope and questions for the future on how peace can be achieved and maintained.

FOR FURTHER REFERENCE

Books and Articles

Adams, Michael. *Suez and After: Year of Crisis*. Boston: Beacon Press, 1958.

Baker, A. J. *Suez: The Seven Day War*. New York: Praeger, 1965.

Banaja, A. A. *Red Sea, Gulf of Aden and Suez Canal: A Bibliography on Oceanographic and Marine Environmental Research*. Edited by Selim A. Morcos and Allen Varley. Compiled by A. A. Banaja, A. L. Beltagy, and M. A. Zahran, with scientific contributions by M. Kh. Ed-Sayed. Jeddah, Saudi Arabia, and Paris: Alexco-Persga and UNESCO, 1990.

Beaufre, Andre. *The Suez Expedition*. 1956. Reprint, New York: Praeger, 1969.

Bowie, Robert Richardson. *Suez*. 1956. Reprint, London and New York: Oxford University Press, 1974.

de Lesseps, Ferdinand. *Recollections of Forty Years*. Translated by C. B. Pitman. 2 vols. London: Chapman and Hall, 1887.

Encyclopedia Britannica. Chicago, University of Chicago Press. 1948 (no edition for this year). See article on Beaconsfield, Benjamin Disraeli, Earl of (1804–1881), 246.

Karabell, Zachary. *Parting the Desert: The Creation of the Suez Canal*. New York: Vintage, 2004.

Kunz, Diane B. *The Economic Diplomacy of the Suez Crisis*. Chapel Hill: University of North Carolina Press, 1991.

Lloyd, Selwyn, Lord. *Suez 1956: A Personal Account*. New York: Mayflower Books, 1978.

Mutting, Anthony. *No End of a Lesson: The Story of the Suez*. New York: C. N. Potter, 1967.

Robertson, Terence. *Crisis: The Inside Story of the Suez Conspiracy*. New York: Atheneum, 1965.

Schonfield, Hugh Joseph. *The Suez Canal in Peace and War, 1869–1969*. Coral Cables, FL: University of Miami Press, 1969.

Tesson, Thierry. *Ferdinand de Lesseps*. Paris: J.-C. Lattès, 1992.

U.S. Department of State. *The Suez Canal Problem, July 26–September 22, 1956: A Documentary Publication*. Washington, DC: U.S. Department of State, 1956.

Internet

For a select bibliography on the Suez Crisis from the Dwight D. Eisenhower Presidential Library in Abilene, Texas, see http://www.eisenhower.utexas.edu/suez.htm.

For the story of *Aida*'s commission and creation, see http://www.r-ds.com/opera/verdiana.aida.htm.

For Benjamin Disraeli's speech on the acquisition of the Suez Canal shares, Feb. 21, 1876, see http://www.historyhome.co.uk.

Music

John, Elton, and Tim Rice. *Aida*. Island Records 524628, 1999.

Verdi, Giuseppe. *Aida*. Claudio Abbado, conductor; with Placido Domingo, Martina Arroyo, and others. 2 discs. CD Opera D'Oro 1167, 1998.

Documents of Authorization

This original Firman of Concession was replaced by the Charter of Concession of 1856, and a subsequent Text of Convention followed in 1888. Related documents followed thereafter: the Anglo-Egyptian Treaty of Alliance of 1936, the Suez Canal Company and Egyptian Government Agreement of 1949, the Anglo-Egyptian Agreement on Suez Canal of 1954, and the Suez Canal Company Nationalisation Law of 1956.

FIRMAN OF CONCESSION

(November 30, 1854)

Granted by the Khedive Mohammed Pasha al-Said to Ferdinand de Lesseps.

Our friend Mons. Ferdinand de Lesseps, having called our attention to the advantages which would result to Egypt from the junction of the Mediterranean and the Red Seas, by a navigable passage for large vessels, and having given us to understand the possibility of forming a company for this purpose composed of capitalists of all nations; we have accepted the arrangements which he has submitted to us, and by these presents grant him exclusive power for the establishment and direction of a Universal Company, for cutting through the Isthmus of Suez, with authority to undertake or cause to be undertaken all the necessary works and erections, on condition that the Company shall previously indemnify all private persons in case of dispossession for the public benefit. And all within limits, upon the conditions and under the responsibilities, settled in the following Articles.

ARTICLE I

Mons. Ferdinand de Lesseps shall form a company, the direction of which we confide to him, under the name of the UNIVERSAL SUEZ MARITIME CANAL COMPANY, for cutting through the Isthmus of Suez, the construction of a passage suitable for extensive navigation, the foundation of appropriation of two sufficient entrances, one from the Mediterranean and the other from the Red Sea, and the establishment of one or two ports.

ARTICLE II

The Director of the Company shall be always appointed by the Egyptian Government, and selected, as far as practicable, from the shareholders most interested in the undertaking.

ARTICLE III

The term of the grant is ninety-nine years, commencing from the day of the opening of the Canal of the two Seas.

ARTICLE IV

The works shall be executed at the sole cost of the Company, and all the necessary land not belonging to private persons shall be granted to it free of cost. The fortifications which the Government shall think proper to establish shall not be at the cost of the Company.

ARTICLE V

The Egyptian Government shall receive from the Company annually fifteen per cent of the net profits shown by the balance sheet, without prejudice to the interest and dividends accruing from the shares which the Government reserves the right of taking upon its own account at their issue, and without any guarantee on its part either for the execution of the works or for the operations of the Company; the remainder of the net profits shall be divided as follows: Seventy-five per cent to the benefit of the Company; ten per cent to the benefit of the members instrumental in its foundation.

ARTICLE VI

The tariffs of dues for the passage of the Canal of Suez, to be agreed upon between the Company and the Viceroy of Egypt, and collected by the Company's agents, shall be always equal for all nations; no particular advantage can ever be stipulated for the exclusive benefit of any one country.

ARTICLE VII

In case the Company should consider it necessary to connect the Nile by a navigable cut with the direct passage of the Isthmus, and in case the Maritime Canal should follow an indirect course, the Egyptian Government will give up to the Company the uncultivated lands belonging to the public domain, which shall be irrigated and cultivated at the expense of the Company, or by its instrumentality.

The Company shall enjoy the said lands for ten years free of taxes, commencing from the day of the opening of the canal; during the remaining eighty-nine years of the grant, the Company shall pay tithes to the Egyptian Government, after which period it cannot continue in possession of the lands above mentioned without paying to the said Government an impost equal to that appointed for lands of the same description.

ARTICLE VIII

To avoid all difficulty on the subject of the lands which are to be given up by the Company, a plan drawn by M. Linant Bey, our Engineer Commissioner attached to the Company, shall indicate the lands granted both for the line and the establishments of the Maritime Canal and for the alimentary Canal from the Nile, as well as for the purpose of cultivation, conformably to the stipulations of Article VII.

It is moreover understood that all speculation is forbidden from the present time, upon the lands to be granted from the public domain, and that the lands previously belonging to private persons and which the proprietors may hereafter wish to have irrigated by the waters of the alimentary Canal, made at the cost of the Company, shall pay a rent of._____ per feddan cultivated (or a rent amicably settled between the Government and the Company).

ARTICLE IX

The Company is further allowed to extract from the mines and quarries belonging to the public domain, any materials necessary for the works of the canal and the erections connected therewith, without paying dues; it shall also enjoy the right of free entry for all machines and materials which it shall import from abroad for the purposes of carrying out this grant.

ARTICLE X

At the expiration of the Concession the Egyptian Government will take the place of the Company, and enjoy all its rights without reservation, the said Government will enter into full possession of the Canal of the two Seas, and of all the establishments connected therewith. The indemnity to be allowed the Company for the relinquishment of its plant and moveables, shall be arranged by amicable agreement or by arbitration.

ARTICLE XI

The statutes of the Society shall be moreover submitted to us by the Director of the Company, and must have the sanction of our approbation. Any modifications that may be hereafter introduced must previously receive our sanction. The said statutes shall set forth the names of the founders, the list of whom we reserve to ourselves the right of approving. This list shall include those persons whose labours, studies, exertions, or capital have previously contributed to the execution of the grand undertaking of the Canal of Suez.

ARTICLE XII

Finally, we promise our true and hearty co-operation, and that of all the functionaries of Egypt in facilitating the execution and carrying out of the present powers.

To my attached friend

FERDINAND DE LESSEPS

Of high birth and elevated rank.
Cairo, 30th November, 1854.

The grant made to the Company having to be ratified by his Imperial Majesty the Sultan, I send you this copy that you may keep it in your possession. With regard to the works connected with the excavation of the Canal of Suez, they are not to be commenced until after they are authorised by the Sublime Porte.

3 Ramadan, 1271

(The Viceroy's Seal.)

A true translation of the Turkish text.

KOENIG BEY
Secretary of Mandates to
His Highness the Viceroy.
Alexandria, May 19, 1855.

From Hugh J. Schonfield, *The Suez Canal in Peace and War: 1869–1969* (Coral Gables, FL: University of Miami Press, 1969), 174–77.

CHARTER OF CONCESSION AND BOOK OF CHARGES

For the Construction and Working of

THE SUEZ CANAL MARITIME CANAL AND DEPENDENCIES

We, Mohammed-Said Pasha, Viceroy of Egypt, considering our charter bearing date the 30th November, 1854, by which we have granted to our friend M. Ferdinand de Lesseps exclusive power to constitute and direct a *Universal Company* for cutting the Isthmus of Suez, opening a passage suitable for large vessels, forming or adapting two sufficient entrances, one on the Mediterranean, the other on the Red Sea, and establishing one or two ports, as the case may be:

M. Ferdinand de Lesseps, having represented to us that in order to constitute a company as above described under the forms and conditions generally adopted for companies of that nature, it is expedient to stipulate beforehand by a fuller and more specific document, the burdens, obligations, and services to which that company will be subjected on the one part, and the concessions, immunities, and advantages to which it will be entitled, as also the facilities which will be accorded to it for its administration, on the other part:

Have decreed as follows the conditions of the concession which is the subject matter of these presents.

I. CHARGES

ARTICLE I

The Company founded by our friend M. Ferdinand de Lesseps in virtue of our charter of the 30th November, 1854, shall execute at its own cost, risk, and damage all the necessary works and constructions for the establishment of:

1st   A canal navigable by large vessels between Suez on the Red Sea, and the Gulf of Pelusium on the Mediterranean;

2nd   A canal of irrigation adapted to the river traffic of the Nile, joining that river to the above-mentioned Maritime Canal;

3rd   Two branches for irrigation and supply, striking out of the preceding canal, and in the direction respectively of Suez and Pelusium.

The works will be completed within the period of six years, unavoidable hindrances and delays excepted.

ARTICLE II

The Company shall have the right to execute the works they have undertaken, themselves and under their own management, or to cause them to be executed by contractors by means of public tender or private contract under penalties. In all cases, four-fifths of the workmen employed upon these works shall be Egyptians.

ARTICLE III

The Canal navigable by large vessels shall be constructed of the depth and width fixed by the scheme of the International Scientific Commission.
Conformably with this scheme, it will commence at the port of Suez; it will pass through the basin of the Bitter Lakes and Lake Timsah, and will debouche into the Mediterranean at whatever point in the Gulf of Pelusium may be determined in the final plans to be prepared by the engineers of the Company.

ARTICLE IV

The canal of Irrigation adapted to the river traffic, according to the terms of the said scheme, shall commence in the vicinity of the city of Cairo, follow the Wadi Tumilat (ancient land of Goshen), and will fall into the Grand Maritime Canal at Lake Timsah.

ARTICLE V

The branches from the above Canal shall strike out from it above the debouchure into Lake Timsah, from which point they shall proceed, on one side of to Suez, and on the other to Pelusium, parallel to the Grand Maritime Canal.

ARTICLE VI

Lake Timsah shall be concerted into an inland harbour capable of receiving vessels of the highest tonnage.
The Company shall moreover be bound, if necessary:

1st   To construct a harbour of refuge at the entrance of the Maritime Canal into the Gulf of Pelusium;

2nd   To improve the port and roadstead of Suez so that it shall equally afford a shelter to vessels.

ARTICLE VII

The Maritime Canal, the ports connected therewith, as also the Junction Canal of the Nile and the branch Canals, shall be permanently maintained in good condition by the Company and at their expense.

ARTICLE VIII

The owners of contiguous lands desirous of irrigating their property by means of water-courses from the Company's canals shall obtain permission so to do in consideration of the payment of an indemnity or rent, the amount whereof shall be fixed according to Article 17 hereinafter recited.

ARTICLE IX

We reserve the right of appointing at the official headquarters of the Company a special commissioner, whose salary they shall pay and who shall represent at the Board of Direction the rights and interests of the Egyptian Government in the execution of these presents.

If the principal office of the Company be established elsewhere than in Egypt, the Company shall be represented at Alexandria by a superior agent furnished with all necessary powers for securing the proper management of the concern and the relations of the Company with our Government.

II. CONCESSIONS

ARTICLE X

For the construction of the Canals and their dependencies mentioned in the foregoing articles, the Egyptian Government grants to the Company, free of impost or rent, the use and enjoyment of all lands not the property of individuals which may be found necessary.

It likewise grants to the Company the use and enjoyment of all uncultivated lands not the property of individuals which shall have been irrigated and cultivated by their care and at their expense, with these provisos:

1st   That lands comprised under the latter head shall be free of impost during those years only, to the date from their being put in a productive condition;

2nd   That after that period, they shall be subject for the remainder of the term of concession to the same obligations and imposts to which are subjected under like circumstances, the lands in other provinces of Egypt;

3rd   That the Company shall afterwards, themselves or through their agents, continue in the use and enjoyment of these lands and the water-courses necessary to their fertilisation, subject to payment to the Egyptian Government of the imposts assessed upon lands under like conditions.

ARTICLE XI

For determining the area and boundaries of the lands conceded to the Company under Article X, reference is made to the plans hereunto annexed, in which plans the lands conceded for the construction of the Canals and their dependencies free of impost or rent, conformably to Clause 1 is coloured black, and the land conceded for the purpose of cultivation, on paying certain duties conformably with Clause 2 is coloured blue.

All acts and deeds done subsequently to our charter of the 30th November, 1854, the effect of which would be to give to individuals as against the Company either claims to compensation which were not then vested in the ownership of the lands, or claims to compensation more considerable than those which the owners could then justly advance, shall be considered void.

ARTICLE XII

The Egyptian Government will deliver to the Company, should the case arise, all lands the property of private individuals, whereof possession should be necessary for the execution of the works and the carrying into effect of the concession, subject to the payment of just compensation to the parties concerned.

Compensation for temporary occupation or definitive appropriation shall as far as possible be determined amicably; in case of disagreement the terms shall be fixed by a court of arbitration deciding summarily and composed of:

1st   An arbitrator chosen by the Company;
2nd   An arbitrator chosen by the interested parties;
3rd   A third arbitrator appointed by us.

The decisions of the court of arbitration shall be executed without further process, and subject to no appeal.

ARTICLE XIII

The Egyptian Government grants to the leasing Company, for the whole period of the concession, the privilege of drawing from the mines and quarries belonging to the public domain, without paying duty, impost or compensation, all necessary materials for the construction and maintenance of the works and buildings of the undertaking. It moreover exempts the Company from all duties of customs, entrance dues and others, on the importation into Egypt of all machinery and materials whatsoever which they shall bring from foreign countries for employment in the construction of the works or working the undertaking.

ARTICLE XIV

We solemnly declare for our part and that of our successors, subject to the ratification of His Imperial Majesty the Sultan, that the Grand Maritime Canal from Suez to Pelusium and the ports appertaining thereto, shall always remain open as a neutral passage to every merchant ship crossing from one sea to another, without any distinction, exclusion, or preference of persons or nationalities, on payment of the dues and observance of the regulations established by the *Universal Company* lessee for the use of the said Canal and its dependencies.

ARTICLE XV

In pursuance of the principle laid down in the foregoing Article, the *Universal Company* can in no case grant to any vessel, company, or individual, any advantage or favour not accorded to all other vessels, companies, or individuals on the same conditions.

ARTICLE XVI

The term of the Company's existence is fixed at 99 years reckoning from the completion of the works and the opening of the Maritime Canal to large vessels.

At the expiration of the said term, the Egyptian Government shall enter into possession of the Maritime Canal constructed by the Company, upon condition, in that event, of taking all the working stock and appliances and stores employed and provided for the naval department of the enterprise, and paying to the Company such amount for the same as shall be determined either amicably or by the decision of sworn appraisers.

Nevertheless, if the Company should retain the concession for a succession of terms of 99 years, the amount stipulated to be paid to the Egyptian Government by Article XVIII, hereinafter recited, shall be raised for the second term to 20 per cent, for the third term to 25 per cent, and so on augmenting at the rate of 5 per cent for each term, but so as never to exceed on the whole 35 per cent of the net proceeds of the undertaking.

ARTICLE XVII

To indemnify the Company for the expenses of construction, maintenance and working, charged upon them by these presents, we authorise the Company henceforth, and during the whole term of their lease, as determined by Clauses 1 and 3 of the preceding Article, to levy and receive for passage through and entrance into the canals and ports thereunto appertaining, tolls and charges for navigation, pilotage, towage or harbour dues, according to tariffs which they shall be at liberty to modify at all times, upon the following express conditions.

1st That these dues be collected, without exception or favour, from all ships under like conditions;

2nd That the tariffs be published three months before they come into force, in the capitals and principal commercial ports of all nations whom it may concern;

3rd That for the simple right of passage through the Canal, the maximum toll shall be ten francs per measurement ton on ships and per head on passengers, and that the same shall never be exceeded.

The company may also, for granting the privilege of establishing water-courses, upon the request of individuals by virtue of Article VIII, receive dues, according to tariffs to be hereafter settled, proportionable to the quantity of water diverted and the extent of the lands irrigated.

ARTICLE XVIII

Nevertheless in consideration of the concessions of land and other advantages accorded to the Company by the preceding Articles, we reserve on behalf of the Egyptian Government a claim of 15 per cent on the net profits of each year, accord to the dividend settled and declared by the General Meeting of Shareholders.

ARTICLE XIX

The list of Foundation Members who have contributed by their exertions, professional labours, and capital to the realisation of the undertaking before the establishment of the Company, shall be settled by us.

After the said payment of the Egyptian Government, according to Article XVIII above recited, there shall be divided out of the net annual profits of the undertaking, one share of 10 per cent among the Foundation Members or their heirs or assigns.

ARTICLE XX

Independently of the time necessary for the execution of the works, our friend and authorised agent, M. Ferdinand de Lesseps, shall preside over and direct the Company, as original founder, during ten years from the first day on which the term of concession for 99 years shall begin to run, by the terms of Article XVI above contained.

From Hugh J. Schonfield, *The Suez Canal in Peace and War, 1869–1969* (Coral Gables, FL: University of Miami Press, 1969), 178–83.

17

The Transcontinental Railroad

United States

DID YOU KNOW . . . ?

➤ The railroad was completed in 1869 with the driving of the Golden Spike in Utah.

➤ The railroad was the largest government construction project in U.S. history to that date.

➤ More than 1,800 miles (2,900 kilometers) of track were laid.

➤ On July 1, 1862, Abraham Lincoln signed key legislation authorizing the project.

➤ It was difficult to find laborers; eventually 6,000 Chinese immigrants were employed, equaling 80 percent of the workforce.

➤ By 1880, $50 million worth of cargo was traveling the completed line, and 200 million acres were being cultivated.

➤ Before the railroad, it took five to six months to cross the country; now the journey took 10 days.

In October 1860, in Dutch Flat, California, an engineer named Theodore Judah met with Daniel Strong, a shopkeeper and expert on the terrain of the Sierra Nevada mountains, and a fellow who could function on land like a harbor pilot might on the water. The mountains loomed as the greatest obstacle to building a transcontinental railroad. Strong knew that most of the crossing points over the mountains were double—that is, it would be necessary to go up a mountain, down across a valley, up another mountain, and down before the road was once again flat. But Strong also knew one place, called Donner, where it might

The final spike has been driven, completing the U.S. Transcontinental Railroad.
Courtesy of the Library of Congress.

be possible to traverse the range over just one summit by taking a route that follows up the American River and down the Truckee River. This would give the builders two advantages: a single summit to overcome, and a gradual slope down—like a natural ramp. It took Judah and Strong two days to reach the top of the Donner crossing point, but as soon as Judah saw the vista, he knew immediately what a coup this choice could be. What Daniel Strong already knew, Theodore Judah discovered on that mountaintop. Right then and there, Judah and Strong drew up the articles of incorporation for the Central Pacific Railroad Company and began to look for investors.

HISTORY

The medieval idea of moving heavily loaded carts by using rails to reduce friction was introduced to England in the seventeenth century, and without the application of this simple engineering concept, railroads would never have flourished. In America, this innovation was originally called a "gravity road." In 1764 in Lewiston, New York, a military road using rails was constructed by a British engineer and mapmaker named Captain John Montressor.

The U.S. desire to reach the West Coast fueled the quest for a quick and more convenient way to traverse the land and reap the alluring harvests of the West: gold and silver. It was not long until the idea of connecting the two American coasts occurred to the country's leaders. The desire to unite the eastern and western part of the country had been expressed earlier by President George Washington, who wanted to extend the Potomac River westward as a means of greater communication. President Lincoln tried to accomplish the union of a country whose western borders were far out of reach administratively. The urge to push west was apparent from the earliest days of the country, beginning with the Pilgrims' landing on the eastern shore, the Lewis and Clark expedition (1804–6), the Potomac Canal's failure, and the Erie Canal's success. All were steps to open the West and facilitate full transcontinental communication, commerce, and exchange.

The first U.S. railroad to carry passengers and general freight was the Baltimore and Ohio, chartered in Maryland in 1827. At midcentury, the federal government began granting land to certain railroads in exchange for reduced carriage charges for government use. By 1853 one could travel by rail from New York City to Chicago. The following year, Chicago was connected with the Mississippi River at Rock Island, Illinois. By 1856, the first rail line was opened on the West Coast, the Sacramento Valley Railroad, which ran between Sacramento and Folsom, California. But the immense intervening territory remained traversable only by some combination of stagecoach, prairie schooner, boats, horseback, and walking.

In January 1845, and then again in January 1848, Asa Whitney, a successful New York mercantile trader, began to promote the idea of a transcontinental railway, having grown accustomed to traveling long distances while pursuing his successful China trade (see Documents I and II). Although Whitney's proposals were not directly adopted, he did raise the key issue.

In July 1862, on behalf of Whitney, Zadock Pratt introduced a proposal to the U.S. Congress to charter a 65-mile section for a railroad, paying the wages in land, which was in abundant supply, and enticing immigrants who were looking for a new start. In 1853, Congress granted funds to the Army Topographic Corps to explore a route from the Mississippi River to the Pacific Ocean. Under the provisions of the Army Appropriation Act of 1853, Secretary of War Jefferson Davis considered and charted some routes to the Pacific. Ironically, the routes followed those earlier envisioned by Whitney.

The first transcontinental railway was made possible by the Pacific Railway Act, signed into law by President Lincoln on July 2, 1862. That measure (presented here as Document III) provided land grants and subsidies in the form of loans to the new Union Pacific Railroad and the Central Pacific Railroad (founded the preceding year). To help the railroads meet the immense costs of the project, another Pacific Railway Act was passed in 1864, which expanded the land grants and authorized the railroads to sell bonds.

The Central Pacific Railroad, which would cross the Sierra Nevada Mountains, began building its portion of the link from Sacramento in 1863. The Union Pacific, which would cross the Great Plains and the Rocky Mountains, started out from the 100th meridian of longitude—a site near Omaha, Nebraska.

However, construction by the Union Pacific was delayed by the Civil War, and work did not get under way until 1865.

The magnitude of this macro project is illustrated by the materials and labor required. More than 1,800 miles (2,900 kilometers) of new track were laid for the project. Thousands of workers were recruited. The Central Pacific relied primarily on Chinese migrants recruited from Canton, and the Union Pacific on European immigrants (especially Irish) as well as Civil War veterans. The workers typically put in 12- to 16-hour days, braving harsh conditions that included Indian attacks, blazing summer heat, frostbite, mountain avalanches, and accidental detonation of explosives.

Since both companies wanted to win the profitable Salt Lake City business, and since the miles of track laid affected the size of the government subsidies they received, the Central Pacific and Union Pacific found themselves unable to agree on a meeting point; each company surveyed and graded land considerably beyond the eventual meeting point. Congress intervened with a compromise, and the two lines were linked, making it possible to ride the 2,000 miles from the Missouri River to California in just 6 days; the rail journey between New York and San Francisco took 10 days. On the morning of May 10, 1869, the United States was united geographically by the driving of the last spike, the legendary "golden spike," on the new transcontinental railroad. The event was celebrated in an official ceremony at Promontory Point, Utah, with live telegraph coverage provided by Western Union. (A little-noted fact is that at the same time the railroad was being built, the contractors were also to place telegraph lines. Sections 18 and 19 of the contract call for the installation of both rail and telegraph lines, and put them in the government's control in times of war, as well as outlining military and postal use.)

After the transcontinental railroad was completed in 1869, the trip between the eastern states and the West Coast no longer required months of perilous overland journey across deserts and over mountains, or a sea voyage around the lower tip of South America.

This critical link between East and West helped to heal the North–South rupture caused by the Civil War. While the union between North and South required overcoming philosophical obstacles and unification by force of arms, the East–West linkup required the demolition of different kinds of barriers.

CULTURAL CONTEXT

The expansion from east to west had a terrible impact on Native Americans. Red Cloud, chief of the largest tribe of the Teton Sioux Nation, made war on the United States to protest the railroad and its destruction of the Sioux habitat and its sacred buffalo and the imbalance being created between humans and the environment. Red Cloud's actions resulted in a settlement by treaty, which awarded the Sioux Nation a huge land reservation in perpetuity. This may be the only war the Native Americans ever won against the U.S. government.

Native Americans numbering 350,000 lived in the western United States in the mid-nineteenth century. When gold was discovered, swarms of new settlers raced westward. At that time, however, it took five months to cross the continent by overland wagon. The most difficult stretch—one littered with discarded household objects that became too burdensome to carry—was the 40 miles from the Humboldt River across the desert to the Truckee River. The last opportunity to obtain drinking water was the Humboldt River. From that point west, thirst was the challenge that made the drive almost unbearable for humans and for the animals that dragged the carts and covered wagons. Besides chairs and furniture, people discarded heavy cookstoves and resorted to eating raw food. Many succumbed, but there was nowhere to bury the bodies. Corpses were simply interred in shallow graves in the roadway, which was rendered secure from foraging wolves only by the weight of wagons that rolled continually over the path and hardened the surface (Bowers and Bain).

The West offered a new promise for the New England and Jamestown pilgrims up and down the East Coast who were farmers. Some had significant land holdings and prospered; some used slave labor, which made farming even more prosperous, but most grew wealthy gradually. The lure of the gold rush was the tantalizing possibility of instant wealth. The appetite to get rich quick sparked something in the American psyche that has never left it; it may have fueled the ceaseless American fascination with the new, with instant gratification, with inventions and innovations. When the railroads moved into the territories, people immediately recognized a much faster route to getting rich, if only they could get a piece of the action. The prospect, however remote, of finding gold was consistent with the mushrooming optimism of a new country and remains today a hallmark of the American psyche.

The year 1869 marked a point in time when technology was dramatically overcoming geography. In that year, both the U.S. transcontinental railroad and the Suez Canal opened, and it was also in 1869 that the *Great Eastern,* the largest ship ever built to that date, laid the Atlantic cable from Newfoundland to Ireland, thus making it possible to send a telegram from Los Angeles to Limerick.

PLANNING

President Lincoln signed the bill authorizing the transcontinental railroad partly because of the dream of a 13-year-old boy. Theodore Judah was a civil engineer who had begun working on eastern railroads as a teenager. Other boys in history were inspired by a childhood glimpse of a grand vision: for example, Pierre-Paul Riquet, who sat wide-eyed at a meeting in Paris with his father as the board presented a map of what Riquet later built—the Canal des Deux Mers, linking the Atlantic coast of France with its Mediterranean shores; and Mohammed Pasha al-Said of Egypt and his childhood friend Ferdinand de Lesseps, who as adults built the Suez Canal. As an adult Judah took the first job he could find on the western rails in 1854. It was Judah who saw the possibilities of a coast-to-coast railroad. Judah's wife said he took the western job not because it was especially distinguished but rather because of its strategic positioning.

Judah's planning and persuasion skills were sufficiently strong to convince the U.S. Congress to fund a land survey to determine a potential route for a transcontinental railroad. After Judah completed the land survey in October 1861, he went back east to Washington, DC. There he spread out a 60-foot-long map in a little office on a long corridor of the Capital Office Building. As congressmen passed, Judah would approach them, bring them into his tiny room, and show them the map and survey close-ups of important passages on the route, including the advantageous Donner crossing point. By dint of his personal force and vision, he successfully lobbied for the project.

The plan came to the attention of President Lincoln, who strongly favored such a railroad. At the time, he was seeking greater communication and union with the new states of California (1850), Oregon (1859), and Nevada (soon to come, in 1864). California was of urgent concern because it was threatening to secede. On July 1, 1862, Lincoln signed the authorization document presented here. The agreement allowed the Central Pacific Railroad to build east from Sacramento, California, and permitted the Union Pacific to build west from the Missouri River. The railroad builders were granted 6,400 acres of land for every mile of track laid and $48,000 in government bonds for each mile completed. But the contract had strictures. The U.S. government would withhold 20 percent of the bonds until the entire railroad line was in working order, and would not release any money to either company until 40 miles of operative railroad had been completed. If the railroad between Missouri and Sacramento was not completed within 12 years, all the assets would be forfeited.

Another aspect of planning that is not widely known was the creation and implementing of what may be the first limited liability contract in America. Thomas Clark Durant, vice president of the Union Pacific Railroad, had been an ophthalmologist, and he proved to be far-sighted in financial as well as medical affairs. Durant obtained control of 2.2 million shares of Union Pacific stock and then set up a financial structure on the side called Crédit Mobilier, the shares of which he made available to a select circle of friends and congressmen who were "influential in approving federal subsidies for the cost of railroad construction . . . enabling railroad builders to make huge profits" (*Reader's Companion to American History,* 1991). In a mere 18 months, shares increased in value by 341 percent, according to historian Stanley Hirshson. On January 6, 1873, a select committee of the U.S. House of Representatives made an inquiry into the affairs of the Union Pacific

Railroad Company and Crédit Mobilier, which led to the Crédit Mobilier scandal of 1872–73.

Crédit Mobilier had two remarkable features: it banked on the railroad being able to eventually obtain land it did not yet own but would if it finished the job, and it allowed investors, even if the operation failed, to lose only the monies they had put in and nothing more. Each investor's personal holdings, such as his home and bank accounts, were out of reach of any liability.

BUILDING

In 1863, just two days after the Gettysburg Address, Lincoln made a far-reaching decision. He looked at the 450 miles across Nebraska's Platte River Valley, which was promising and geographically advantageous. But geography was not the only factor he weighed; political considerations eventually persuaded him to bypass the easy route through Nebraska and instead agree to an eastern terminus at Council Bluffs, Iowa. Why? Three years earlier, when he was seeking the presidential nomination (which had gone into a third ballot against his opponent, Seward), behind-the-scenes deals were arranged with the Iowa delegates who agreed to switch their votes to Lincoln. Now, as president, Lincoln felt honor-bound to give something back to Iowa. It proved to be substantial repayment for favors given—the biggest government construction project in U.S. history at the time.

However, the Union Pacific was so involved in raking in enormous profits, inflating prices, and working deals with leaders of both construction and the body politic via Crédit Mobilier that the company soon fell behind schedule. When the Civil War ended in 1865, the Union Pacific had not completed a single length of track! Congress almost cut off its funding. Thomas Durant suddenly realized what was about to happen and sent telegrams to the construction foremen insisting that they get to work at once. Durant wired the crew heads in Nebraska daily until the work began to speed up.

Manpower was a problem. It was not easy to find people willing to undertake the arduous work of building a railroad. A few immigrant crews participated in construction of the Central Pacific line, but for the most part workers were locals, people who wanted to be near their homes or near the gold mines. No one wanted to be out in the middle of the prairie. So recruiters for Central Pacific tried distributing handbills to attract more workers. They advertised for 5,000 workers, but only 200 showed up. The recruiters were desperate. They had all the money they needed; the problem was, no one wanted to do the work.

The solution? The Chinese, who were arriving in California to escape a recent famine in the Kwangtung region of China, presented an attractive, potential labor force. Labor boss James Harvey Strobridge was reluctant, to which his boss, Charlie Crocker replied, "No, no, they can do it. They built the Great Wall. They can do it." (http://www.pbs.org/wgbh/ amex/tcrr/filmmore/pt.html)

Crocker convinced Strobridge to try 50 Chinese workers; they were quite successful. Another 100 were hired, with equal success. Finally the company recruited only Chinese laborers. By 1866, during the height of the railroad building, the Central Pacific had more than 6,000 Chinese workers, about 80 percent of their total workforce.

Diet had a major impact on the health and survival of the workers. In the Irish camps, the bill of fare was boiled beef, boiled beans, and coffee; in the Chinese camps, the meals were stir-fried fare served on rice with tea. The Chinese were careful about the health of their food and water, refusing to use water from ditches. As a result, few Chinese fell ill with dysentery. Nevertheless, many died from the hazards of the work itself.

The style and organizational discipline of work changed after the Civil War, when war veterans began working on the railroad. Accustomed to taking orders and moving military supplies with logistical efficiency, these new rail workers transformed the management and systems of operations. For veterans of the Union and Confederate armies, the railroads offered a constructive alternative to war and the opportunity to put war-acquired skills to economic use.

Innovative technology was required to get through the rock of the mountains. Using traditional black-powder explosive, it took 15 months to push a tunnel through a mountain. Owing to the Union Pacific's faster progress compared to the Central Pacific's delayed construction, the Central Pacific made a desperate (or was it heroic?) decision to use liquid nitroglycerine, which was powerful but unstable. Indeed, in 1867 transport of the volatile substance was outlawed in California after a Panamanian steamer was blown up by the fuel. In San Francisco, an explosion of nitroglycerine blasted body parts to the tops of buildings blocks away from the site of the explosion. James Howden, a British chemist, found a way to brew the explosive on-site, thereby getting around the transport ban. Charlie Crocker, the labor boss, brought Howden to the summit to work his magic. And magic it was in the nimble hands of the Chinese, who had few mishaps and much success with this new tool for tunneling.

But it was not all work, for men toiling all day in the middle of the prairie required some rest and relaxation. Amusements began to appear, especially notable in the movable feast called "Hell on Wheels." Supplied with everything that a hard-working fellow could want, the portable welcome wagon could be packed onto railroad flatbed cars and transported to the next work site, providing gambling, alcohol, and the pleasure of ladies who came to the West to ply their trade. But Hell on Wheels was small compared to Bull's Big Tent, a precursor of Las Vegas, offering many games of chance. Carnal pleasure was available in cubicles at the back of the tent. Sexually transmitted diseases became a major problem, so the owner of Bull's Big Tent introduced another innovation. Bull's tent had a doctor whose office was near the cubicles. So after a man had enjoyed gambling, at least he did not have to gamble

The construction crew of a Union Pacific Railroad line pose in front of a supply train, 1867. © MPI / Getty Images.

on his health, for he could visit the good Doctor Allen and get protection for his amatory pursuits.

There was administrative work to be done as well. In 1867 Grenville Dodge, head of engineering for Union Pacific, established a new town 90 miles north of Denver, at the confluence of two creeks that supplied potable water. The town was situated between Omaha and Salt Lake City so it could serve as a change point for westbound locomotives that needed extra cars; at the same point, eastbound trains could drop unneeded equipment. This center of administration, with a military headquarters, machine shops, water, and dining and sleeping facilities, was called Cheyenne. The price for a standard lot of land in the town ranged from $600 to $4,000. Sadly, the town was named after a Native American tribe that did not benefit from the town's prosperity.

By 1880 $50 million worth of cargo was traveling the completed line, and 200 million acres of land were being cultivated by people who had traveled west via the railroad. Before the railroad, it took five to six months to cross the country; now the journey took just over a week. Commerce increased exponentially, just as it did when the Erie Canal had opened earlier. This coast-to-coast link could be thought of as the American Silk Road, a crucial new route for exchanging ideas as well as goods.

The railroad also wrought new advances in mapmaking. Before this time, little was known about details of the U.S. interior. According to the Smithsonian

Museum, "Surveying and mapping activities flourished in the United States as people began moving inland over [what had once been an] inadequately mapped continent." (http://lcweb2.loc.gov/ammem/gmdhtml/rrhtml/rrintro.html) The technological advances of mapmaking, especially lithography, which had been invented in Bavaria, began arriving about the time America was building into the west. Lithography made it possible to make maps, and soon mapmaking and railroad building began to coincide. Suggestions arose for all kinds of possibilities, including the founding of Yellowstone National Park with a proposed rail line right to its entrance, just in time for the 1872 opening of the park. By the turn of the century, there was great demand for maps of all kinds, and this new industry went hand-in-hand with the railroad.

IMPORTANCE IN HISTORY

Reducing the journey across the United States from months by covered wagon to days via the transcontinental railroad made a monumental difference in the commercial and social interactions of the United States. This success holds added promise that similar leaps may occur in the use of trains in evacuated tubes (meaning tubes with the air pumped out, thus eliminating resistance), making it possible to traverse the distance between Boston and San Francisco in less than an hour, or perhaps a train in such a tube laid on the floor of the Atlantic Ocean from Boston to Buenos Aires.

FOR FURTHER REFERENCE

Books and Articles

Achenbach, Joel. *The Grand Idea: George Washington's Potomac and the Race to the West.* New York: Simon and Schuster, 2004.

Ambrose, Stephen E. *Nothing Like It in the World: The Men Who Built the Transcontinental Railroad, 1863–1869.* New York: Simon and Schuster, Touchstone Books, 2000.

Boorstin, Daniel J. *The Americans: The Democratic Experience.* New York: Random House, 1965.

Haney, Louis H. *A Congressional History of Railways.* 2 vols. 1908–10. Reprint, New York: Augustus M. Kelley, 1968.

McDonnell, Greg. *Canadian Pacific: Stand Fast, Craigellachie!* Erin, Canada: Boston Mills Press, 1954.

Poor, Henry Varnum. *Manual of the Railroads of the United States for 1870–71.* New York: H. V. and J. W. Poor, 1870.

Reader's Companion to American History. Houghton Mifflin Online Study Center. http://college.hmco.com/history/readerscomp/rcah/html/rc_021900_crditmobilie.htm

Thompson, Slason. *A Short History of the American Railways.* Chicago: Bureau of Railway News and Statistics, 1925.

Twain, Mark. *Roughing It.* 1892. Reprint, New York: Penguin Classics, 1985.

Whitney, Asa. *A Project for a Railroad to the Pacific.* New York: George Ward, 1849.

Internet

For history and a select bibliography of mapmaking and the railroads, see http://lcweb2. loc.gov/. Select "Railroad Maps."

For a complete transcript of the PBS *American Experience: Transcontinental* Railroad, see http://www.pbs.org/wgbh/amex/tcrr/filmmore/pt.html.

For the complete text of the *Report of the Select Committee of the House of Representatives Appointed under the Resolution of January 6, 1873, to Make Inquiry in Relation to the Affairs of the Union Pacific Railroad Company, the Credit Mobilier of America, and Other Matters Specified in Said Resolution and in Other Resolutions Referred to Said Committee*, see http://cprr.org/Museum/Credit_Mobilier.html.

For a summary time line from 1769 to 1889, see http://www.pbs.org/wgbh/amex/tcrr/ timeline/index.html.

For western railroad songs with historical narration, including "Hell on Wheels," see http://mcneilmusic.com/railroad.html.

For Greg Schindel, official singer on California Western Railroad's "Skunk Train," including a list of songs and a new CD, see http://www.trainsinger.com/.

Film and Television

Transcontinental Railroad. PBS. Boston: WGBH, 2003. Part of PBS's *American Experience* series, this film and its Web site (http://www.pbs.org/wgbh/amex/tcrr/) offer many features helpful to understanding the railroad, including an interactive route and a teacher's guide. The editors acknowledge a great debt and appreciation of research contributed by authorities including Phil Roberts, Wendell Huffman, Stanley Hirshson, Sue Fawn Chung, Frank Chin, and Donald Fixico (historians); Fred Gamst (anthropologist); David Bain (writer); and many other scholars. A complete transcript is available at http://www.pbs.org/wgbh/amex/ tcrr/filmmore/pt.html.

Documents of Authorization—I

28th Congress, Doc. No. 72 *Ho. of Reps.*
2nd Session.
RAILROAD FROM LAKE MICHIGAN TO THE PACIFIC.
MEMORIAL
Of
ASA WHITNEY, OF NEW YORK CITY,
Relative To
The construction of a railroad from lake Michigan to the Pacific ocean
January 28, 1845.
Read, and referred to the Committee on Roads and Canals.
To the honorable the Senate and House of Representatives of the United States in Congress assembled:

Your memorialist begs respectfully to represent to your honorable body, that by rivers, railroads, and canals, all the States east and north of the Potomac connect directly with the waters of the great lakes. That there is a chain of railroads in projection and being built from New York to the southern shores of lake Michigan, crossing all the veins of communication to the ocean, through all the states south and east of the Ohio river, producing commercial, political, and national results and benefits which must be seen and felt through all our vast confederacy.

Your memorialist would further represent to your honorable body, that he has devoted much time and attention to the subject of a railroad from lake Michigan, through the Rocky mountains, to the Pacific ocean; and that he finds such a route practicable—the results from which would be incalculable, far beyond the imagination of man to estimate. To the interior of our vast and widely-spread country, it would be as the heart to the human body; it would, when all completed, cross all the mighty rivers and streams which send their way to the ocean throughout vast and rich valleys from Oregon to Maine, a distance of more than three thousand miles.

The incalculable importance of such a chain of roads will readily be seen and appreciated by your honorable body. It would enable us, in the short space of eight days, and perhaps less, to concentrate all the forces of our vast country at any point from Maine to Oregon, in the interior, or on the coast. Such easy and rapid communication, with such facilities for exchanging the different products of the different parts, would bring all our immensely wide-spread population together as one vast city; the moral and social effects of which must harmonize all together as one family, with but one interest—the general good of all.

Your memorialist respectfully represents further to your honorable body, that the roads from New York to lake Michigan (a distance of 840 miles) will no doubt be completed by the States through which they pass, or by individuals; that from lake Michigan to the mouth of the Columbia river is 2,160 miles, making from New York to the Sandwich islands 5,100 miles; from the Columbia river to Japan is 5,600 miles, making from New York to Japan 8,600 miles; from the Columbia river to Amoy, in China, (the port nearest to the tea and silk provinces,) is 6,200 miles, making from New York to Amoy only 9,200 miles; which, with a railroad to the Pacific, thence to China by steam, can be performed in 30 days; now being a sailing distance of nearly 17,000 miles, requiring from 100 to 150 days for its performance. Then the drills and sheetings of Connecticut, Rhode Island, and Massachusetts, can be transported to China in 30 days; and the teas and rich silks of China, in exchange, come back to New Orleans, to Charleston, to Washington, to Baltimore, to Philadelphia, to New York, and to Boston, in 30 days more. Comment is unnecessary. Your honorable body will readily see the revolution by this to be wrought in the entire commerce of the world; and that this must inevitably be its channel, when the rich freights from the waters of the Mississippi and the Hudson will fill to overflowing with the products of all the earth the storehouses of New York and New Orleans, the great marts dividing the commerce of the world; while each State, and every town in our vast confederacy, would receive its just proportion of influence and benefits, compared with its vicinity to, or facility to communicate with, any of the rivers, canals, or railroads crossed by this great road.

Your memorialist would respectfully represent to your honorable body its political importance; that, affording a communication from Washington to the Columbia river in less than eight days, a naval depot, with a comparatively small navy, would command the Pacific, the South Atlantic, and the Indian oceans, and the China Sea.

Your memorialist begs respectfully to represent further to your honorable body, that he can see no way or means by which this great and important work can be accomplished for ages to come, except by a grant of a sufficient quantity of the public domain; and your memorialist believes that, from the proceeds of such a grant, he will be enabled to complete said road in a reasonable time, and at the same time settle the country through which it passes, so far as the lands may be found suited to cultivation, with an industrious and frugal people; and this, in a comparatively short space of time, accomplish what will otherwise require ages, and thus at once giving us the power of dictation to those who will not long remain satisfied without an attempt to dictate to us. Our system of free government works so well, diffusing so much intelligence, dispensing equal justice, and assuring safety to all, and producing so much general comfort and prosperity, that its influence must, like a mighty flood, sweep away all other systems. Then let us not flatter ourselves that this overwhelming current is not to meet resistance; for to us directly will that resistance be applied, and your memorialist believes that we must yet meet that desperate and final struggle which shall perpetuate our system, and religious and civil liberty.

Your honorable body are aware of the over-population of Europe; and your memorialist would respectfully represent, that by the application of machinery, and its substitution for manual labor, the latter no longer receives its just or sufficient reward, and thousands in the fear of starvation at home are driven to our shores, hoping from our wide-spread and fertile soil to find a rich reward for their labor. Most of them ignorant, and all inexperienced, having been herded together in large numbers at home, they dread separation—they fear the wilderness or prairie—refuse to separate from their associates, or to leave the city; their small means soon exhausted, they see abundance around them, almost without price, but that small price they can no longer pay; necessity plunges them into vice, and often crime, and they become burdensome to our citizens, and which evil is increasing to an alarming extent; and your memorialist believes it must increase, unless there can be some great and important point in our interior to which they can be attracted immediately on their landing; where their little means, with their labor, can purchase lands; where they will escape the tempting vices of our cities, and where they will have a home with their associates, and where their labor from their own soil will not only produce their daily bread, but, in time, an affluence of which they could never have dreamed in their native land.

Your memorialist believes that this road will be the great and desirable point of attraction; that it will relieve our cities from a vast amount of misery, crime, and taxation; that it will take the poor unfortunates to a land where they will be compelled to labor for a subsistence; and as they will soon find that their labor and efforts receive a just and sufficient reward—finding themselves and their little ones surrounded with comfort and plenty, the recompense for their own toil—their energies will kindle into a flame of ambition and desire, and we shall be enabled to educate them to our system, to industry, prosperity, and virtue. Your memorialist confidently expects all this, and more.

Your memorialist would further respectfully represent to your honorable body, that from an estimate (as near accurate as can be made, short of an actual survey) the cost of said road, to be built in a safe, good, and substantial manner, will be about $50,000,000; and as the road cannot (from the situation of the uninhabited country through which it will pass) earn anything, or but little before its completion, therefore a further sum will be required to keep it in operation, repair, &c., of $15,000,000; making the total estimated cost of said road, when competed, $65,000,000. It may require some years before the earnings of said road (at the low rate of tolls necessary for its complete success) can be much, if anything, beyond its current expenses for repairs, &c.; but after a period of _____ years, and at the very lowest possible rate of tolls, it must earn more than ample for its repairs and expenses. It would be the only channel for the commerce of all the western coast of Mexico and South America, of the Sandwich islands, of Japan, of all China, Manila, Australia, Java, Singapore, Calcutta, and Bombay. Not only all ours, but the commerce of all Europe, to the most of those places, must pass this road. Your memorialist says *must;* because the saving of time (so all important to the merchant) from the long and hazardous voyage around either of the capes, would force it; and in a few years would be built up cities, towns, and villages, from the lake to the ocean, which would alone support the road. Being built from the public lands, the road should be free, except so far as sufficient for the necessary expenses of operation, repairs, &c.; and your memorialist believes that, at a very low rate of tolls, a sum would be gained sufficient, after all current expenses, to make a handsome distribution for public education; and as a part of the earning of said road will be from foreign commerce, your memorialist begs to respectfully submit the subject to your wise consideration.

Your memorialist respectfully further represents to your honorable body, that, from the knowledge he can procure of them, he finds that the lands for a long distance east of the mountain are bad—of little or no value for culture; that through, and for some distance beyond the mountains, would also be of but very little, if any, value; therefore your memorialist is satisfied that it will require an entire tract of 60 miles in width, from as near to Lake Michigan as the unappropriated lands commence, to the Pacific Ocean. Therefore, in view of all the important considerations here set forth, your memorialist is induced to pray that your honorable body will grant to himself, his heirs, and assigns, such tract of land, the proceeds of which to be strictly and faithfully applied to the building and completing the said road—always with such checks and guarantees to your honorable body as shall secure a faithful performance of all the obligations and duties of your memorialist; and that after the faithful completion of this great work, should any lands remain unsold, or any moneys due for lands, or any balance of moneys received for land sold, and which have not been required for the building of the said road, then all and every one of them shall belong to your memorialist, his heirs, and assigns, forever.

Your memorialist further prays that your honorable body will order a survey of said route, to commence at some point to be fixed upon as most desirable on the shore of Lake Michigan, between the 42nd and 45th degree of north latitude; thence west to the gap or pass in the mountains; and thence by the most practicable route to the Pacific Ocean.

Your memorialist would respectfully represent one further consideration to your honorable body: that, in his opinion, Oregon must fast fill up with an industrious, enterprising people from our States; that they will soon attract and draw to them large numbers from the states of Europe, all expecting to share in the benefits of our free government, claiming its care and protection. But the difficulty of access to them, either by land or water, will forbid such a hope; and your memorialist believes that the time is not far distant when Oregon will become a State of such magnitude and importance, as to compel the establishment of a separate government—a separate nation—which will have cities, ports and harbors, all free—inviting all the nations of the earth to a free trade with them; when they will control and monopolize the valuable fisheries of the Pacific; control the coast trade of Mexico and South America, of the Sandwich Islands, Japan, and China; and be our most dangerous and successful river in the commerce of the world. But your memorialist believes that this road will unite them to us; enabling them to receive the protecting care of our government; sharing in its blessings, benefits, and prosperity, and imparting to us our share of the great benefits from their local position, enterprise, and industry. But your honorable body will see all this and more.

And, as in duty bound, will every pray

ASA WHITNEY
January 28, 1845.
Blair & Rives, printers

Documents of Authorization—II

30th Congress, [SENATE.] Miscellaneous
1st Session No. 28

Memorial

Of

ASA WHITNEY,

Praying

For a grant of land to enable him to construct a railroad from lake Michigan to the Pacific ocean.

January 17, 1848.

Referred to the Committee on Public Lands, and ordered to be printed.

To the Senate and House of Representatives of the United States in Congress assembled:

Your memorialist begs respectfully to represent to your honorable body, that he presented a memorial to the last session of the 28th Congress, praying that a tract of the public lands, 60 miles in width, from Lake Michigan to the Pacific Ocean, might be set apart and granted expressly to furnish means, by sale and settlement, to enable him and his associates to construct a railroad to connect with the above named points.

Said memorial was referred to the Committee on Roads and Canals, and a unanimous report adopted "recommending the subject to the deliberate attention of Congress and the people, and the public lands as the only means for such a work, which should not long be delayed, as the lands were rapidly being taken up." Said memorial is now respectfully submitted.

During the summer of 1845 your memorialist, with a company of young men from different States, explored and examined a part of the proposed route. The object and result of said exploration were declared and expressed by your memorialist to the 29th Congress. His memorial was referred to the Committee on Public Lands in the Senate, Hon. Mr. Breese chairman. His able report was unanimously adopted by the committee, and, with a bill introduced and passed to a second reading, ordered to be printed; your memorialist begs to submit said report with his memorial to your honorable body.

Your memorialist, viewing the great importance of this great work to our whole country, has devoted his whole time and attention to it. He believes he has examined the subject in all its bearings, and made himself master of it, and fixed upon the only plan by which this work can ever be accomplished.

Your memorialist would now represent and explain the plan by which he proposes to carry out this great work. He prays that your honorable body will be pleased to set apart 60 miles wide of the public lands (and an equivalent for any which may have been taken up) from Lake Michigan to the Pacific Ocean, for this especial purpose. He has explored and examined a part of the route, and from the lake onward for 800 miles the land is of the very best quality, but nearly 500 miles of this 800 without timber, and then no timber on to the Rocky Mountains.

That after this 800 miles, onward nearly to the ocean, the land is represented as very poor—too poor to sustain settlement; therefore the whole work is based on the 800 miles of the first part, with the belief that the facilities which the road would create and give to settlement, intercourse and communications with markets, would rend a part of the poor lands useful and available.

Your memorialist does not ask your honorable body for the appropriation of one dollar in money, or even for a survey of the route He proposed to make the surveys, commence the work, with machinery, preparations, and arrangements for its continuance, and complete ten miles of road, at his own expense; and when the ten miles is completed to the satisfaction of the commissioner, (appointed as your honorable body shall direct,) and with the satisfaction that the work will be continued, then your memorialist would receive five miles, or one-half of the lands to the lie of the ten miles of road completed), with which to reimburse himself. The other five miles, or half of the lands, to be held by the government, and so on for each and every ten miles for the 800 miles of good land, or so far as the one-half of the land set apart will furnish means to complete ten miles of road. Thus the road would be completed for the 800 or more miles, and in operation with one-half (the alternates five miles) settled with towns, villages, and cities, while the other half (or alternates) held by the government would be enhanced in value more than fourfold what all is now worth, and held or sold as the demand for actual settlement many require; but when sold, to be sold as Congress shall direct, and the proceeds held as a fund to continue and complete the road through the poor lands all to the ocean; and the road and machinery also held by the government as further security that the work will be continued and completed. Beyond the 800 miles of good lands, and through the poor lands, when each and every ten miles of road shall have been completed, and the entire ten miles by sixty of lands do not furnish means to reimburse for the actual outlay, then the fund which may have been accumulated from the reserved half of good lands, or the lands, shall be applied to this purpose; but in all cases, the ten miles of road must be completed to the satisfaction of the Commissioner, before any lands, or money from lands sold, can be touched by your memorialist and associates.

When the road is so far advanced that security can be given to the government that it will be completed, then your memorialist and associates shall pay to the government _____ cents per acre for all the lands set apart for this work; but the balance, with the fund from the half of the good lands if any, after the road is completed, shall be held subject to keep the road in repair and operation while it may be considered as an experiment, and until by its earnings it can provide for itself; then the surplus land, and funds if any, with the road and machinery, shall belong to and be the property of your memorialist and his associates and their heirs and assigns, but leaving with Congress, if necessary, the power of prescribing the mode of sale for any surplus lands at public auction to the highest bidder, and leaving with the Congress the power of fixing and regulating the tolls of said road forever, after sufficient only for repairs, operation, and necessary expenses; with power also to fix and regulate the transportation of United States mails, troops, munitions of war, &c., belonging to the government; thus, making it a national road, still built and carried on purely as an individual enterprise, without any government, political, or party machinery or influence.

Your memorialist would further represent, that one mile by sixty wide would give 38,400 acres; and when the lands are good, one mile of land, at $1.25 per acre would furnish means sufficient only to build two miles of good road, (as this must be,) with heavy rail, with bridges and the necessary machinery; and having the double quantity of land on the first part is the sure and only guaranty to the people that the road will be completed, and without which it would be impossible and idle to attempt it.

Your memorialist would further represent, that the distance from the lake to the ocean, on a straight line, is but 1,780 miles; that from the lake to the pass in the mountains is 1,098 miles; and a road may be constructed on a straight line; but allow for detour 50 miles, is 1,148 miles; thence to the mouth of the Columbia river, or to Puget's sounds, is 682; but allow for detour 200 miles, is 882 miles— making the estimated distance from the lake to the ocean 2,030 miles.

It is estimated that it will cost for a good road and turnouts,

$20,000 per mile; for 2,030 miles- - - - - - - - - - - - - $40,600,00

And as the road, except for this side of the Missouri, cannot

earn any income until all is completed, a further

sum for repairs, operation, and machinery, will be

required of: - *$20,000,000*

Probable cost of road, when completed and ready

for use - $60,600,000;

but it has been estimated much higher.

The 2,030 miles by 60 wide, which your memorialist has prayed for, for this work, would amount to 77,952,000 acres.

The 800 miles of good lands would give 30,720,000 acres; from which deduct waste land and usual expenses of sale, and allow the facilities of the road to enhance the value so as to average $1.25 per acre, would yield $32,832,000. Thence to the ocean is 1,230—47,232,000 acres—the greater part of which is represented as being too poor to sustain settlement; but allow the facilities, which the road would undoubtedly create, to cause it to average one-half of the present government price, ($1.25 per acre,) and deduct expenses of sale, and we have $27,044,000 more—making together $59,879,000, or less than the estimated cost of the road. But your memorialist believes that, by connecting the sale and settlement of the lands with the building of the road, and the great advantages which the road would render to settlement, he will be enabled to realize the means for the full and completed accomplishment of the work; *but if the commencement is delayed even for a few months, the lands on the first part of the route (on which all depends) will be so far taken up as to defeat it forever.*

Your memorialist believes the lands which he has prayed for are of no value, (except for a small part of the first part of the route,) and believes it impossible for settlement to take place without the road first, as there are nearly 1,200 miles without timber, and no navigable streams to communicate with civilization, and no possible means to transport materials for buildings and fences; therefore settlement would be impossible, and the land of no use to man, or value to the nation; but by taking settlement and materials on with the road, connecting the two together, the hopes and expectations of your memorialist can be realized, but not otherwise. The estimates and calculations which he has presented to your honorable body is not to show the present or future value of the land, but to show that he has full confidence in the effect which the road will produce on them.

Your memorialist believes that the nation at large will receive benefits far beyond any present or future value of the lands; and as it is a work so directly and decidedly national, that a price should be fixed for the lands—not at what the government are now selling the best at—not at what even it is proposed to reduce the price by graduation, but at the actual cost to the government—that though 16 cents per acre has been named as the price your memorialist to pay for *all* the lands, he considers that price as too high, believing the government can never, in any other way or time, realize so large a sum, and believing the government should not speculate upon a work promising such vast and beneficial national results. Your memorialist taking upon himself the entire risk and responsibility, should the enterprise fail, the government lose nothing, while he must lose all; he therefore feels that the price for the lands should be fixed at not above their actual cost to the government, and it cannot be expected that your memorialist and associates will pay out some 12–1/2 millions of dollars for lands to build this road without expecting a return for it; therefore if the lands set apart do not furnish means to complete the road, and reimburse the 12–1/2 millions, than so much more must be added to the tolls as will pay for the use of this investment; so the government and the people are interested in fixing the price to be paid for the lands at their actual cost.

Your memorialist believes that he has fixed upon the only route across our continent where such a road can be built, where the streams can be bridged, so as to make an uninterrupted intercourse from ocean to ocean; the only route where the wilderness lands can be made to produce the means for the work; the only route where so vast an extent of wilderness country can be opened to settlement, production, and communication with all the markets of the world, creating and producing the only means to increase and sustain commerce, as well as all other branches of industry; the only route where the climate would not destroy our animal and vegetable products, thereby closing to us forever the vast markets of Japan, China, and all Asia; the only route which would give all our Atlantic and gulf cities a fair opportunity to participate in all its vast benefits. It will be found, from actual calculation, that the starting point, on the lake or at the crossing of the Mississippi, is nearer to Charleston and Savannah by 50 miles, than to New York, and 250 miles nearer than Boston; and the Charleston and Georgia roads are *now* completed nearly to Tennessee, and will be the first from the Atlantic to reach this; that Mobile and New Orleans, by proposed railroad route, are nearer than New York by 311 miles, and nearer than Boston by 511 miles; Richmond and Baltimore 200 miles, and Philadelphia 100 miles nearer than New York; and there is no other route across our continent which would change the present route for the commerce of Europe with Asia.

A canal at Panama, Nicaragua, or Tehuantepec, has been mooted for nearly two hundred years, surveys and explorations made; but all rests where it commenced, and will undoubtedly remain so. No one has examined and calculated to see if anything in distance could be gained, and your memorialist begs to present to your honorable body the actual distances from London to Asia, via the present sea voyage, and via a proposed canal, as well as via railroad to Oregon. His calculations are for a canal at Panama, though Nicaragua and Tehuantepec are a few degrees north and west, would not increase or diminish the distances, but the navigation and access to which from Europe would be far more dangerous and difficult than Panama.

From London to Panama 81° of longitude and 42° of

latitude must be overcome, and which, in a straight line,

would vary little from - - - - - - - - - - - - - - - - - - - 5,868 mi.

From Panama to Canton is 170° of longitude, measuring

60 miles to the degree, and is on a line - - - - - - - - - - - *10,200* "

Making from London to Canton, on a line via any canal— - - - - 16,068"

From Canton to England, via the cape of Good Hope, in the season of the northeast monsoon, as follows:

From Canton, through the China Sea, to the equator, is - - - - - 1,320 mi.

From the equator to Lunda straits, to 12° south latitude - - - - - - - 750 "

Through the region of the southeast trades to 27° south

latitude and 50° east longitude— - - - - - - - - - - - - - - - - 3,200 "

Thence to the cape - 1,560 "

From the cape to London- *6,900* "

13,730 "

Again: From Canton to England, via the Cape of Good Hope, in the season of the southwest monsoon—

From Canton to the straits of Formosa- - - - - - - - - - - - 480 miles.

Thence to Pill's straits, passing near the Pelew islands - - - - - - 1,300 "

Thence to Alla's straits - 1,200 "

Thence to 27° south latitude and 50° east Longitude - - - - - - - 3,900 "

Thence to the cape - 1,560 "

And thence to London - *6,900* "

15,340 "

In the first instance, the route by canal would increase the distance between London and Canton 2,338 miles; and in the latter, 728 miles.

The distances both for a canal and for via the cape, are calculated for a straight line from point to point; but owing to trades and currents, a sail vessel could not make either voyage on a straight line, and the voyage from London to China is estimated at not less than 17,000 miles, and it would be increased in the same manner and proportion by the canal route.

From London to New York is 74° of longitude; at 45

miles each - 3,330 mi

Thence to Puget's sound or Columbia river, via proposed

railroad, is - 2,963 "

Thence to Shang-Hae, in China, is 115° of longitude;

at 47 miles each, is - *5,405* "

Making from London to China, via New York and via

railroad -11,698 "

For the railroad part of this route, the actual railroad distance is taken to the Mississippi; thence to the ocean 250 miles is allowed for detour. The seaport may be made by steam on a line, and a saving from London to China, over the canal route, of 4,370 miles on a straight line, and equal to more than 6,000 miles under influence of trades and currents, and for sail vessels, the distance being so great from point to point that steam could not be used, except at an enormous expense; and there would also be a saving of more than half in time.

Again: From England to Singapore, via the proposed canal, during the northeast monsoon –

From London through the canal at Panama - - - - - - - - - - 586 miles

Thence to Singapore on a line, 180° longitude, at 60 miles

each - *10,800* "

16,668 "

From England to Singapore, via the Cape of Good Hope, northeast monsoon –

From England to the cape - - - - - - - - - - - - - - - - - - 7,730 mi

Thence past the island of Amsterdam and St. Paul's, to

105° east longitude, and between 39 and 30° south latitude - - - - 4,320 "

Thence to Auger point - 1,740 "

And thence to Singapore - - - - - - - - - - - - - - - - - - - *560* "

14,350 mi

or 2,318 miles against a canal.

From London to New York, as before - - - - - - - - - - - - - 3,330mi

Thence to Puget's sound - - - - - - - - - - - - - - - - - - - 2,963 "

Thence to Singapore 132°, at 55 miles each - - - - - - - - - - *7,260* "

13,553 mi

or 3,115 miles less than straight lines by a proposed canal, and may be accomplished by steam, saving more than half the time.

Again: from England to Valparaiso, via Cape Horn, is - - - - - 9,400 mi

From England to Valparaiso, via proposed canal - - - - - - - - *8,978* "

422 miles

difference in favor of canal, but would not change route.

By reference to a globe, it will be seen that a vessel anywhere on the coast from Panama, bound to China, would gain more than 2,000 miles in distance by first proceeding to Oregon, and thence to China. For steam, this route is the only means of supply for fuel; and it will be seen that in crossing the globe within the tropics, the degree of longitude measures full 60 miles, while on a course from a point at 30° on a line to 46° latitude measures but 47 miles to the degree. Comment is unnecessary; but your memorialist begs to submit an article from DeBows Commercial Review for October, fully explaining this subject of route. It was prepared with great care and labor, and your memorialist believes its statements and calculations of distances to be correct, and doubts not it will satisfy all who read it.

Your honorable body will see, from the map [*Editors' note:* Map is unavailable] herewith submitted, that our continent is placed in the centre of the world; Europe, with 250 millions of population, on one side, and all Asia on the other side of us, with 700 millions of souls. The Atlantic, 3,000 miles across, separating us from Europe, while the calm Pacific rolls 5,000 miles between us and Asia, and no part over 25 days from us; and it will be seen that this proposed road will change the present route for all the vast commerce of all Europe with Asia, bring it across our continent, make it and the world tributary to us, and, at the lowest tolls, give us 25 millions of dollars per annum for transit alone. It would bind Oregon and the Pacific coast to us, and forever prevent the otherwise inevitable catastrophe of a separate nation growing up west, to rise at our decline, and control us and the world. It would open the vast markets of Japan, China, Polynesia, and all Asia to our agricultural, manufacturing, and all other products. It would open the wilderness, to the husbandman, and take the products of the soil to all the markets of the world. It would make available and bring into market lands now too remote from civilization, and add millions of wealth to the nation. The labor of the now destitute emigrant would grade the road, and purchase him a home, where comfort and plenty would surround all. Man's labor would receive its proper reward, and elevate him from inducement to vice and crime. It would unite and bind us together as one family, and the whole world as one nation, giving us the control over all, and making *all* tributary to us.

Your memorialist would further represent to your honorable body that his memorial and plan, presented to the 28th Congress, was the first matured plan ever presented to Congress, or to the world, for a railroad to connect the Atlantic with the Pacific, across our continent; that in his memorial to the 29th Congress, the origin of that plan is dated back to 1830; that your memorialist in urging this plan embraced all others, by declaring "the work could not be done by the government; could not be done by States not yet formed; and could not be done by individual enterprise; because no man would invest money in a work which could not produce any income during his life time;" therefore your memorialist believes there can be no plan for this work of which his is not the origin and foundation. His plan has now been before the public more than three years, and the expression throughout the country is universally in its favor; and the press has, almost without exception urged its adoption, and the legislatures of twelve States, by almost unanimous votes, have passed resolutions approving and declaring it "the only feasible plan by which this great work can and be accomplished," recommending its adoption by your honorable body, and instructing and request their delegates "to give it their prompt attention and support."

Your memorialist would further represent that he has devoted exclusively more than three years in this country, and nearly two years in Asia, to this great subject; that he can commence the work without any delay, and is fully satisfied that he can carry it out to its full and perfect completion on the plan he has proposed; but any material alteration would defeat the whole. He therefore prays that your honorable body will take this great subject into early and deliberate consideration and action; and, as in duty bound; will every pray.

ASA WHITNEY
Of New York
Washington, D.C., January 17, 1848.
Tippin & Streeper, printers.

Documents of Authorization—III

An Act to aid in the Construction of a Railroad and Telegraph Line from the Missouri River to the Pacific Ocean, and to secure the Government the Use of the same for Postal, Military, and Other Purposes.

Be it enacted by the Senate and House of Representatives of the United States of America in Congress assembled, That Walter S. Burgess, William P. Blodget, Benjamin H. Cheever, Charles Fosdick Fletcher, of Rhode Island; Augustus Brewster, Henry P. Haven, Cornelius S. Bushnell, Henry Hammond, of Connecticut; Isaac Sherman, Dean Richmond, Royal Phelps, William H. Ferry, Henry A. Paddock, Lewis J. Stancliff, Charles A. Secor, Samuel R. Campbell, Alfred E. Tilton, John Anderson, Azariah Boody, John S. Kennedy,

H. Carver, Joseph Field, Benjamin F. Camp, Orville W. Childs, Alexander J. Bergen, Ben. Holliday, D. N. Barney, S. De Witt Bloodgood, William H. Grant, Thomas W. Olcott, Samuel B. Ruggles, James B. Wilson, of New York; Ephraim Marsh, Charles M. Harker, of New Jersey; John Edgar Thompson, Benjamin Haywood, Joseph H. Scranton, Joseph Harrison, George W. Cass, John H. Bryant, Daniel J. Morell, Thomas M. Howe, William F. Johnson, Robert Finney, John A. Green, E. R. Myre, Charles F. Wells, junior, of Pennsylvania; Noah L. Wilson, Amasa Stone, William H. Clement, S. S. L'Hommedieu, John Brough, William Dennison, Jacob Blickinsderfer, of Ohio; William M. McPherson, R. W. Wells, Willard P. Hall, Armstrong Beatty, John Corby, of Missouri; S. J. Hensley, Peter Donahue, C. P. Huntington, T. D. Judah, James Bailey, James T. Ryan, Charles Hosmer, Charles Marsh, D. O. Mills, Samuel Bell, Louis McLane, George W. Mowe, Charles McLaughlin, Timothy Dame, John R. Robinson, of California; John Atchison and John D. Winters, of the Territory of Nevada; John D. Campbell, R. N. Rice, Charles A. Trowbridge, and Ransom Gardner, Charles W. Penny, Charles T. Gorham, William McConnel, of Michigan; William F. Coolbaugh, Lucius H. Langworthy, Hugh T. Reid, Hoyt Sherman, Lyman Cook, Samuel R. Curtis, Lewis A. Thomas, Platt Smith, of Iowa; William B. Ogden, Charles G. Hammond, Henry Farnum, Amos C. Babcock, W. Seldon Gale, Nehemiah Bushness and Lorenzo Bull, of Illinois; William H. Swift, Samuel T. Dana, John Bertram, Franklin S. Stevens, Edward R. Tinker, of Massachusetts; Franklin Gorin, Laban J. Bradford, and John T. Levis, of Kentucky; James Dunning, John M. Wood, Edwin Noyes, Joseph Eaton, of Maine; Henry H. Baxter, George W. Collamer, Henry Keyes, Thomas H. Canfield, of Vermont; William S. Ladd, A. M. Berry, Benjamin F. Harding, of Oregon; William Bunn, junior, John Catlin, Levi Sterling, John Thompson, Elibu L. Phillips, Walter D. McIndoe, T. B. Stoddard, E. H. Brodhead, A. H. Virgin, of Wisconsin; Charles Paine, Thomas A. Morris, David C. Branham, Samuel Hanna, Jonas Votaw, Jesse L. Williams, Isaac C. Elston, of Indiana; Thomas Swan, Chauncey Brooks, Edward Wilkins, of Maryland; Francis R. E. Cornell, David Blakely, A. D. Seward, Henry A. Swift, Dwight Woodbury, John McKusick, John R. Jones, of Minnesota; Joseph A. Gilmore, Charles W. Woodman, of New Hampshire; W. H. Grimes, J. C. Stone, Chester Thomas, John Kerr, Werter R. Davis, Luther C. Challis, Josiah Miller, of Kansas; Gilbert C. Monell and Augustus Kountz, T. M. Marquette, William H. Taylor, Alvin Saunders, of Nebraska; John Evans, of Colorado; together with five commissioners to be appointed by the Secretary of the Interior, and all persons who shall or may be associated with them, and their successors, are hereby created and erected into a body corporate and politic in deed and in law, by the name, style, and title of "The Union Pacific Railroad Company"; and by that name shall have perpetual succession, and shall be able to sue and to be sued, plead and be impleaded, defend and be defended, in all courts of law and equity within the United States, and may make and have a common seal; and the said corporation is hereby authorized and empowered to lay out, locate, construct, furnish, maintain, and enjoy a continuous railroad and telegraph, with the appurtenances, from a point on the one hundredth meridian of longitude west from Greenwich, between the south margin of the valley of the Republican River and the north margin of the valley of the Platte River, in the Territory of Nebraska, to the western boundary of Nevada Territory, upon the route and terms hereinafter provided, and is hereby vested with all the powers, privileges, and immunities necessary to carry into effect the purposes of this act as herein set forth. The capital stock of said company shall consist of one hundred thousand shares of one thousand dollars each, which shall be subscribed for and held in such manner as the by-laws of said corporation shall provide. The persons hereinbefore named, together with those to be appointed by the Secretary of the Interior, are hereby constituted and appointed commissioners, and such body shall be called the Board of Commissioners of the Union Pacific Railroad and Telegraph Company, and twenty-five shall constitute a quorum for the transaction of business. The first meeting of said board shall be held at Chicago at such time as the commissioners from Illinois herein named shall appoint, not more than three nor less by them to the other commissioners by depositing a call thereof in the post office at Chicago, post paid, to their address at least forty days before said meeting, and also by publishing said notice in one daily newspaper in each of the cities of Chicago and Saint Louis. Said board shall organize by the choice from its number of a president, secretary, and treasurer, and they shall require from said treasurer such bonds as may be deemed proper, and may from time to time increase the amount thereof as they may deem proper. It shall be the duty of said board of commissioners to open books, or cause books to be opened, at such times and in such principal cities in the United States as they or a quorum of them shall determine, to receive subscriptions to the capital stock of said corporation, and a cash payment of ten per centum on all subscriptions, and to receipt therefor. So soon as two thousand shares shall be in good faith subscribed for, and ten dollars per share actually paid into the treasury of the company, the said president and secretary of said board of commissioners shall appoint a time and place for the first meeting of the subscribers to the stock of said company, and shall give notice thereof in at least one newspaper in each State in which subscription books have been opened at least thirty days previous to the day of meeting, and such subscribers as shall attend the meeting so called, either in person or by proxy, shall then and there elect by ballot not less than thirteen directors for said corporation; and in such election each share of said capital shall entitle the owner thereof to one vote. The president and secretary of the board of commissioners shall act as inspectors of said election, and shall certify under their hands the names of the directors elected at said meeting; and the said commissioners, treasurer, and secretary shall then deliver over to said directors all the properties, subscription books and other books in their possession, and thereupon the duties of said commissioners and the officers previously appointed by them shall cease and determine forever, and thereafter the stockholders shall constitute said body politic and corporate. At the time of the first and each triennial election of directors by the stockholders two additional directors shall be appointed by the President of the United States, who shall act with the body of directors, and *to* be denominated directors on the part of the government; any vacancy happening in the government directors at any time may be filled by the President of the United States. The directors to be appointed by the President shall not be stockholders in the Union Pacific Railroad Company. The directors so chosen shall, as soon as may be after their election, elect from their own number a president and vice-president, and shall also elect a treasurer and secretary. No person shall be a director in said company unless he shall be a bona fide owner of at least five shares of stock in the said company, except the two directors to be appointed by the President as aforesaid. Said company, at any regular meeting of the stockholders called for that purpose, shall have power to make by-laws, rules, and regulations as they shall deem needful and proper, touching the disposition of the stock, property, estate, and effects of the company, not inconsistent herewith, the transfer of shares, the term of office, duties, and conduct of their officers and servants, and all matters whatsoever which may appertain to the concerns of said company; and the said board of directors shall have power to appoint such engineers, agents, and subordinates as may from time to time be necessary to carry into effect the object of this act, and to do all acts and things touching the location and construction of said road and telegraph. Said directors may require payment of subscriptions to the capital stock, after due notice, at such times and in such proportions as they shall deem necessary to complete the railroad and telegraph within the time in this act prescribed. Said president, vice-president, and directors shall hold their office for three years, and until their successors are duly elected and qualified, or for such less time as the by-laws of the corporation may prescribe; and a majority of said directors shall constitute a quorum for the transaction of business. The secretary and treasurer shall give such bonds, with such security, as the said board shall from time to time require, and shall hold their offices at the will and pleasure of the directors. Annual meetings of the stockholders of the said corporation, for the choice of officers (when they are to be chosen) and for the transaction of annual business, shall be holden at such time and place and upon such notice as may be prescribed in the by-laws.

SEC. 2. And be it further enacted, That the right of way through the public lands be, and the same is hereby granted to said company for the construction of said railroad and telegraph line; and the right, power, and authority is hereby given to said company to take from the public lands adjacent to the line of said road, earth, stone, timber, and other materials for the construction thereof; said right of way is granted to said railroad to the extent of two hundred feet in width on each side of said railroad where it may pass over the public lands, including all necessary grounds for stations, buildings, workshops, and depots, machine shops, switches, side tracks, turntables, and water stations. The United States shall extinguish as rapidly as may be the Indian titles to all lands falling under the operation of this act and required for the said right of way and grants hereinafter made.

SEC. 3. And be it further enacted, That there be, and is hereby, granted to the said company, for the purpose of aiding in the construction of said railroad and telegraph line, and to secure the safe and speedy transportation of the mails, troops, munitions of war, and public stores thereon, every alternate section of public land, designated by odd numbers, to the amount of five alternate sections per mile on each side of said railroad, on the line thereof, and within the limits of ten miles on each side of said road, not sold, reserved, or otherwise disposed of by the United States, and to which a preemption or homestead claim may not have attached, at the time the line of said road is definitely fixed: Provided, That all mineral lands shall be excepted from the operation of this act; but where the same shall contain timber, the timber thereon is hereby granted to said company. And all such lands, so granted by this section, which shall not be sold or disposed of by said company within three years after the entire road shall have been completed, shall be subject to settlement and preemption, like other lands, at a price not exceeding one dollar and twenty-five cents per acre, to be paid to said company.

SEC. 4. And be it further enacted, that whenever said company shall have completed forty consecutive miles of any portion of said railroad and telegraph line, ready for the service contemplated by this act, and supplied with all necessary drains, culverts, viaducts, crossings, sidings, bridges, turnouts, watering places, depots, equipment, furniture, and all other appurtenances of a first class railroad, the rails and all the other iron used in the construction and equipment of said road to be American manufacture of the best quality, the President of the United States shall appoint three commissioners to examine the same and report to him in relation thereto; and if it shall appear to him that forty consecutive miles of said railroad and telegraph line have been completed and equipped in all respects as required by this act, then, upon certificate of said commissioners to that effect, patents shall issue conveying the right and title to said lands to said company, on each side of the road as far as the same is completed, to the amount aforesaid; and patents shall in like manner issue as each forty miles of said railroad and telegraph line are completed, upon certificate of said commissioners. Any vacancies occurring in said board of commissioners by death, resignation, or otherwise, shall be filled by the President of the United States: Provided, however That no such commissioners shall be appointed by the President of the United States unless there shall be presented to him a statement, verified on oath by the president of said company, that such forty miles have been completed, in the manner required by this act, and setting forth with certainty the points where such forty miles begin and where the same end; which oath shall be taken before a judge of a court of record.

SEC. 5. And be it further enacted, That for the purposes herein mentioned the Secretary of the Treasury shall, upon the certificate in writing of said commissioners of the completion and equipment of forty consecutive miles of said railroad and telegraph, in accordance with the provisions of this act, issue to said company bonds of the United States of one thousand dollars each, payable in thirty years after date, bearing six per centum per annum interest, (said interest payable semi-annually,) which interest may be paid in United States treasury notes or any other money or currency which the United States have or shall declare lawful money and a legal tender, to the amount of sixteen of said bonds per mile for such section of forty miles; and to secure the repayment to the United States, as hereinafter provided, of the amount of said bonds so issued and delivered to said company, together with all interest thereon which shall have been paid by the United States, the issue of said bonds and delivery of the company shall ipso facto constitute a first mortgage on the whole line of the railroad and telegraph, together with the rolling stock, fixtures and property of every kind and description, and in consideration of which said bonds may be issued; and on the refusal or failure of said company to redeem said bonds, or any part of them, when required so to do by the Secretary of the Treasury, in accordance with the provisions of this act, the said road, with all the rights, functions, immunities, and appurtenances thereunto belonging, and also all lands granted to the said company by the United States, which, at the time of said default, shall remain in the ownership of the said company, may be taken possession of by the Secretary of the Treasury, for the use and benefit of the United States: Provided, This section shall not apply to that part of any road now constructed.

SEC. 6. And be it further enacted, That the grants aforesaid are made upon condition that said company shall pay said bonds at maturity, and shall keep said railroad and telegraph line in repair and use, and shall at all times transmit dispatches over said telegraph line, and transport mails, troops, and munitions of war, supplies, and public stores upon said railroad for the government, whenever required to do so by any department thereof, and that the government shall at all times have the preference in the use of the same for all the purposes aforesaid, (at fair and reasonable rates of compensation, not to exceed the amounts paid by private parties for the same kind of service;) and all compensation for services rendered for the government shall be applied to the payment of said bonds and interest until the whole amount is fully paid. Said company may also pay the United States, wholly or in part, in the same or other bonds, treasury notes, or other evidences of debt against the United States, to be allowed at par; and after said road is completed, until said bonds and interest are paid, at least five per centum of the net earnings of said road shall also be annually applied to the payment thereof.

SEC. 7. And be it further enacted, That said company shall file their assent to this act, under the seal of said company, in the department of the Interior, within one year after the passage of this act, and shall complete said railroad and telegraph from the point of beginning as herein provided, to the western boundary of Nevada Territory before the first day of July, one thousand eight hundred and seventy-four: Provided, That within two years after the passage of this act said company shall designate the general route of said road, as near as may be, and shall file a map of the same in the Department of the Interior, whereupon the Secretary of the Interior shall cause the lands within fifteen miles of said designated route or routes to be withdrawn from preemption, private entry, and sale; and when any portion of said route shall be finally located, the Secretary of the Interior shall cause the said lands hereinbefore granted to be surveyed and set off as fast as may be necessary for the purposes herein named: Provided, That in fixing the point of connection of the main trunk with the eastern connections, it shall be fixed at the most practicable point for the construction of the Iowa and Missouri branches, as hereinafter provided.

SEC. 8. And be it further enacted, That the line of said railroad and telegraph shall commence at a point on the one hundredth meridian of longitude west from Greenwich, between the south margin of the valley of the Republican River and the north margin of the valley of the Platte River, in the Territory of Nebraska, at a point to be fixed by the President of the United States, after actual surveys; thence running westerly upon the most direct, central, and practicable route, through the territories of the United States, to the western boundary of the Territory of Nevada, there to meet and connect with the line of the Central Pacific Railroad Company of California.

SEC. 9. And be it further enacted, That the Leavenworth, Pawnee, and Western Railroad Company of Kansas are hereby authorized to construct a railroad and telegraph line, from the Missouri River, at the mouth of the Kansas River, on the south side thereof, so as to connect with the Pacific railroad of Missouri, to the aforesaid point, on the one hundredth meridian of longitude west from Greenwich, as herein provided, upon the same terms and conditions in all respects as are provided in this act for the construction of the railroad and telegraph line first mentioned, and to meet and connect with the same at the meridian of longitude aforesaid; and in case the general route or line of road from the Missouri River to the Rocky Mountains should be so located as to require a departure northwardly from the purpose line of said Kansas railroad before it reaches the meridian of longitude aforesaid, the location of said Kansas road shall be made so as to conform thereto; and said railroad through Kansas shall be so located between the mouth of the Kansas River, as aforesaid, and the aforesaid point, on the one hundredth meridian of longitude, that the several railroads from Missouri and Iowa, herein authorized to connect with the same, can make connection within the limits prescribed in this act, provided the same can be done without deviating from the general direction of the whole line to the Pacific coast. The route in Kansas, west of the meridian of Fort Riley, to the aforesaid point, on the one hundredth meridian of longitude, to be subject to the approval of the President of the United States, and to be determined by him on actual survey. And said Kansas company may proceed to build said railroad to the aforesaid point, on the one hundredth meridian of longitude west from Greenwich, in the territory of Nebraska. The Central Pacific Railroad Company of California, a corporation existing under the laws of the State of California, are hereby authorized to construct a railroad and telegraph line from the Pacific coast, at or near San Francisco, or the navigable waters of the Sacramento River, to the eastern boundary of California, upon the same terms and conditions, in all respects, as are contained in this act for the construction of said railroad and telegraph line first mentioned, and to meet and connect with the first mentioned railroad and telegraph line on the eastern boundary of California. Each of said companies shall file their acceptance of the conditions of this act in the department of the Interior within six months after the passage of this act.

SEC. 10. And be it further enacted, That the said company chartered by the State of Kansas shall complete one hundred miles of their said road, commencing at the mouth of the Kansas River as aforesaid, within two years after filing their assent to the conditions of this act, as herein provided, and one hundred miles per year thereafter until the whole is completed; and the said Central Pacific Railroad Company of California shall complete fifty miles of their said road within two years after filing their assent to the provisions of this act, as herein provided, and fifty miles per year thereafter until the whole is completed; and after completing their roads, respectively, said companies, or either of them, may unite upon equal terms with the first-named company in constructing so much of said railroad and telegraph line and branch railroads and telegraph lines in this act hereinafter mentioned through the Territories from the State of California to the Missouri River, as shall then remain to be constructed, on the same terms and conditions as provided in this act in relation to the said Union Pacific Railroad Company. And the Hannibal and St. Joseph Railroad, the Pacific Railroad Company of Missouri, and the first-named company, or either of them, on filing their assent to this act, as aforesaid, may unite upon equal terms, under this act, with the said Kansas company, in constructing said railroad and telegraph, to said meridian of longitude, with the consent of the said State of Kansas; and in case said first-named company shall complete their line to the eastern boundary of California before it is completed across said State by the Central Pacific Railroad Company of California, said first-named company is hereby authorized to continue in constructing the same through California, with the consent of said State, upon the terms mentioned in this act, until said roads shall meet and connect, and the whole line of said railroad and telegraph is completed; and the Central Pacific Railroad Company of California, after completing its road across said State, is authorized to continue the construction of said railroad and telegraph through the Territories of the United States to the Missouri River, including the branch roads specified in this act, upon the routes hereinbefore and hereinafter indicated, on the terms and conditions provided in this act in relation to the said Union Pacific Railroad Company, until said roads shall meet and connect, and the whole line of said railroad and branches and telegraph is completed.

SEC. 11. And be it further enacted, That for three hundred miles of said road most mountainous and difficult of construction, to wit: one hundred and fifty miles westwardly from the eastern base of the Rocky Mountains, and one hundred and fifty miles eastwardly from the western base of the Sierra Nevada mountains, said points to be fixed by the President of the United States, the bonds to be issued to aid in the construction thereof shall be treble the number per mile hereinbefore provided, and the same shall be issued, and the lands herein granted be set apart, upon the construction of every twenty miles thereof, upon the certificate of the commissioners as aforesaid that twenty consecutive miles of the same are completed; and between the sections last named of one hundred and fifty miles each, the bonds to be issued to aid in the construction thereof shall be double the number per mile first mentioned, and the same shall be issued, and the lands herein granted be set apart, upon the construction of every twenty miles thereof, upon the certificate of the commissioners as aforesaid that twenty consecutive miles of the same are completed: Provided, That no more than fifty thousand of said bonds shall be issued under this act to aid in constructing the main line of said railroad and telegraph.

SEC. 12. And be it further enacted, That whenever the route of said railroad shall cross the boundary of any State or Territory, or said meridian of longitude, the two companies meeting or united there shall agree upon its location at that point, with reference to the most direct and practicable through route, and in case of difference between them as to said location the President of the United States shall determine the said location; the companies named in each State and Territory to locate the road across the same between the points so agreed upon, except as herein provided. The track upon the entire line of railroad and branches shall be of uniform width, to be determined by the President of the United States, so that, when completed, cars can be run from the Missouri River to the Pacific coast; the grades and curves shall not exceed the maximum grades and curves of the Baltimore and Ohio railroad; the whole line of said railroad and branches and telegraph shall be operated and used for all purposes of communication, travel, and transportation, so far as the public and government are concerned, as one connected, continuous line; and the companies herein named in Missouri, Kansas, and California, filing their assent to the provisions of this act, shall receive and transport all iron rails, chairs, spikes, ties, timber, and all materials required for constructing and furnishing said first-mentioned line between the aforesaid point, on the one hundredth meridian of longitude and western boundary of Nevada Territory, whenever the same is required by said first-named company, as cost, over that portion of the roads of said companies constructed under the provisions of this act.

SEC. 13. And be it further enacted, That the Hannibal and Saint Joseph railroad Company of Missouri may extend its roads from Saint Joseph, via Atchison, to connect and unite with the road through Kansas, upon filing its assent to the provisions of this act, upon the same terms and conditions, in all respects, for one hundred miles in length next to the Missouri River, as are provided in this act for the construction of the railroad and telegraph line first mentioned, and may for this purpose, use any railroad charter which has been or may be granted by the legislature of Kansas; Provided, That if actual survey shall render it desirable, the said company may construct their road, with the consent of the Kansas Legislature, on the most direct and practicable route west from St. Joseph, Missouri, so as to connect and unite with the road leading from the western boundary of Iowa at any point east of the one hundredth meridian of west longitude, or with the main trunk road at said point; but in no event shall lands or bonds be given to said company, as herein directed, to aid in the construction of their said road for a greater distance than one hundred miles. And the Leavenworth, Pawnee, and Western Railroad Company of Kansas may construct their road from Leavenworth to unite with the road through Kansas.

SEC. 14. And be it further enacted, That the said Union Pacific Railroad Company is hereby authorized and required to construct a single line of railroad and telegraph from a point on the western boundary of the State of Iowa, to be fixed by the President of the United States, upon the most direct and practicable route, to be subject to his approval, so as to form a connection with the lines of said company at some point on the one hundredth meridian of longitude afore-said, from the point of commencement on the western boundary of the State of Iowa, upon the same terms and conditions, in all respects, as are contained in this act for the construction of the said railroad and telegraph first mentioned; and the said Union Pacific Railroad Company shall complete one hundred miles of the road and telegraph in this section provided for, in two years after filing their assent to the conditions of this act, as by the terms of this act required, and at the rate of one hundred miles per year thereafter, until the whole is completed: Provided, That a failure upon the part of said company to make said connection in the time aforesaid, and to perform the obligations imposed on said company by this section and to operate said road in the same manner as the main line shall be operated shall forfeit to the government of the United States all the rights, privileges, and franchises granted to and conferred upon said company by this act. And whenever there shall be a line of railroad completed through Minnesota or Iowa to Sioux City, then the said Pacific Railroad Company is hereby authorized and required to construct a railroad and telegraph from said Sioux City upon the most direct and practicable route to a point on, and so as to connect with, the branch railroad and telegraph in this section hereinbefore mentioned, or with the said Union Pacific Railroad, said point of junction to be fixed by the President of the United States, not further west than the one hun-dredth meridian of longitude aforesaid, and on the same terms and conditions as provided in this act for the construction of the Union Pacific Railroad as afore-said, and to complete the same at the rate of one hundred miles per year; and should said company fail to comply with the requirements of this act in relation to the said Sioux City railroad and telegraph, the said company shall suffer the same forfeitures prescribed in relation to the Iowa branch railroad and telegraph hereinbefore mentioned.

SEC. 15. And be it further enacted, That any other railroad company now incorporated, or hereafter to be incorporated, shall have the right to connect their road with the road and branches provided for by this act, at such places and upon such just and equitable terms as the President of the United States may prescribe. Wherever the word company is used in this act it shall be construed to embrace the words their associates, successors, and assigns, the same as if the words had been properly added thereto.

SEC. 16. And be it further enacted, That at any time after the passage of this act all of the railroad companies named herein, and assenting hereto, or any two or more of them, are authorized to form themselves into one consolidated company; notice of such consolidation, in writing, shall be filed in the Department of the Interior, and such consolidated company shall thereafter proceed to construct said railroad and branches and telegraph line upon the terms and conditions provided in this act.

SEC. 17. And be it further enacted, That in case said company or companies shall fail to comply with the terms and conditions of this act, by not completing said road and telegraph and branches within a reasonable time, or by not keeping the same in repair and use, but shall permit the same, for an unreasonable time, to remain unfinished, or out of repair, and unfit for use, Congress may pass any act to insure the speedy completion of said road and branches, or put the same in repair and use, and may direct the income of said railroad and telegraph line to be thereafter devoted to the use of the United States, to repay all such expenditures caused by the default and neglect of such company or companies; Provided, That if said roads are not completed, so as to form a continuous line of railroad, ready for use, from the Missouri River to the navigable waters of the Sacramento River, in California, by the first day of July eighteen hundred and seventy-six, the whole of all of said railroads before mentioned and to be constructed under the provisions of this act, together with all their furniture, fixtures, rolling stock, machine shops, lands, tenements, and hereditament, and property of every kinds and character, shall be forfeited to and be taken possession of by the United States: Provided, that of the bonds of the United States in this act provided to be delivered for any and all parts of the roads to be constructed east of the one hundredth meridian of west longitude from Greenwich, and for any part of the road west of the west foot of the Sierra Nevada mountain, there shall be reserved of each part and installment twenty-five per centum, to be and remain in the United States treasury, undelivered, until said road and all parts thereof provided for in this act are entirely completed; and of all the bonds provided to be delivered for the said road, between the two points aforesaid, there shall be reserved out of each installment fifteen per centum, to be and remain in the treasury until the whole of the road provided for in this act is fully completed; and if the said road or any part thereof shall fail of completion at the time limited therefor in this act, then and in that case the said part of said bonds so reserved shall be forfeited to the United States.

SEC. 18. And be it further enacted, that whenever it appears that the net earnings of the entire road and telegraph, including the amount allowed for services rendered for the United States, after deducting all expenditures, including repairs, and the furnishing, running, and managing of said road, shall exceed ten per centum upon its cost, exclusive of the five per centum to be paid to the United States, Congress may reduce the rates of fare thereon, if unreasonable in amount, and may fix and establish the same by law. And the better to accomplish the object of this act, namely, to promote the public interest and welfare by the construction of said railroad and telegraph line, and keeping the same in working order, and to secure to the government at all times (but particularly in time of war) the use and benefits of the same for postal, military and other purposes, Congress may, at any time, having due regard for the rights of said companies named herein, add to, alter, amend, or repeal this act.

SEC. 19. And be it further enacted, That the several railroad companies herein named are authorized to enter into an arrangement with the Pacific Telegraph Company, the Overland Telegraph Company, and the California State Telegraph Company, so that the present line of telegraph between the Missouri River and San Francisco may be moved upon or along the line of said railroad and branches as fast as said roads and branches are built; and if said arrangement be entered into, and the transfer of said telegraph line be made in accordance therewith to the line of said railroad and branches, such transfer shall, for all purposes of this act, be held and considered a fulfillment on the part of said railroad companies of the provisions of this act in regard to the construction of said line of telegraph. And, in case of disagreement, said telegraph companies are authorized to remove their line of telegraph along and upon the line of railroad herein contemplated without prejudice to the rights of said railroad companies named herein.

SEC. 20. And be it further enacted, That the corporation hereby created and the roads connected therewith, under the provisions of this section, shall make the Secretary of the Treasury an annual report wherein shall be set forth—

First. The names of the stockholders and their places of residence, so far as the same can be ascertained;.

Second. The names and residences of the directors, and all other officers of the company;

Third. The amount of stock subscribed, and the amount hereof actually paid in;

Fourth. A description of the lines of road surveyed, of the lines thereof fixed upon for the construction of the road, and the cost of such surveys;

Fifth. The amount received from passengers on the road;

Sixth. The amount received for freight thereon;

Seventh. A statement of the expense of said road and its fixtures;

Eighth. A statement of the indebtedness of said company, setting forth the various kinds thereof. Which report shall be sworn to by the president of the said company, and shall be presented to the Secretary of the Treasury on or before the first day of July in each year.

APPROVED, July 1, 1862.

From U.S. Statutes at Large 12 (1862).v

18
The Brooklyn Bridge

United States

DID YOU KNOW . . . ?

➤ The bridge's construction began in 1870 and was completed in 1883.
➤ The bridge's length is 0.3 miles (0.5 kilometers); its main span measures 1,595 feet (486 meters).
➤ Its total length, including approach roads, is 6,016 feet (1,834 meters), it is 85 feet (26 meters) wide, and in the middle it clears high water by 135 feet (41 meters).
➤ The bridge's wooden caissons were the largest ever built: 119 by 129 feet (36.3 by 39.3 meters) at the base, and 104 by 117 feet (31.7 by 35.7 meters) on top.
➤ Two dozen workers lost their lives in its construction, including John Roebling, the designer and builder.
➤ It has inspired more poetry than any other bridge in history.
➤ The Brooklyn Bridge quite possibly employed the first female field engineer.

Could it have been the philosopher Hegel who first turned the mind of John Roebling, builder of the Brooklyn Bridge, to thoughts of spanning two disparate sides of a point, whether virtual or real? Hegel once said Roebling was his favorite pupil, one who truly recognized the importance of following one's own intuition and thinking. While Roebling studied engineering, hydraulics, and architecture at the Royal Polytechnic Institute of Berlin, perhaps it was his philosophical background that gave his engineering the innovative spark needed to build a bridge that many called "the eighth wonder of the world."

A shot of the Brooklyn Bridge. Courtesy of Getty Images / PhotoDisc.

Why a bridge to Brooklyn from Manhattan? At the time, Brooklyn was an independent city, actually larger than New York City, with a port even larger than the Port of Boston. New York was still the center of finance, however, so it made sense that linking the two cities would create a powerful, reciprocating engine of commerce.

HISTORY

In the early 1800s, Brooklyn was a populous, independent, but slightly isolated town. Without a bridge, the only way to travel to and from Manhattan was by slow ferry; storms caused delays, and in winter the crossing was hampered by ice. The New York state legislature discussed building a bridge over the East River as early as 1802, and there was even talk of building a tunnel.

In the mid-1850s, John Augustus Roebling, a German-born engineer and wire-cable manufacturer, appeared among the bridge advocates with his own proposal. He was already known for designing major bridges built in Pittsburgh, Niagara Falls, and Cincinnati, but he will be forever remembered as a founding father of modern suspension-bridge design, and the Brooklyn Bridge was his masterpiece. In a report to the New York Bridge Company on September 1, 1867, about his grand design (reproduced here as a Related Cultural Document),

Roebling foresaw the role of engineering as public art: "Its most conspicuous feature—the great towers—will serve as landmarks to the adjoining cities, and they will be entitled to be ranked as national monuments."

Despite an economic downturn and the interruption of the Civil War, eventually Roebling's plan for a bridge with an unprecedented (at the time) span of about 0.3 miles (0.5 kilometers) was adopted. In April 1867 the legislature passed a bill to incorporate the New York Bridge Company for the purpose of building the bridge. Roebling was named the company's chief engineer. Approval came from Congress in 1869 with legislation establishing the bridge as a postal road.

Both the state and federal measures stressed that the bridge must not interfere with navigation. The state bill set the minimum elevation above the middle of the river at high tide at 130 feet (40 meters). Construction began in early 1870. It was a monumental project that stretched the boundaries of the technology of the day. The wooden caissons, for example, were the largest ever built: "The dimensions of the New York anchorage and that of the one in Brooklyn were 119 by 129 feet at the base, and 104 by 117 feet on top" (McCullough, 329), with a weight of 60,000 tons.

The project was dogged by tragedy. More than two dozen workers lost their lives. Roebling himself died of tetanus in 1869, which developed after he suffered a crushed foot in an accident while making observations for siting the Brooklyn tower. He was succeeded as chief engineer by his son, Washington. In 1872, Washington Roebling was crippled by decompression sickness (more commonly known as the bends) while working in the Manhattan caisson. Thereafter he supervised the work from his nearby apartment, looking at the work site through a spyglass. His liaison at the construction site was none other than his wife, Emily, who may have been the first female field engineer. Other accidents slowed progress, including an explosion in the caisson on the New York side and a prolonged fire in the Brooklyn caisson.

But the eighth wonder of the world, as the behemoth was advertised, was eventually completed in 1883, after 14 years of construction. The Brooklyn Bridge, connecting New York City's Manhattan Island with Brooklyn on Long Island, was the longest suspension bridge in the world. But that was not its sole distinction. It was also the first suspension bridge to use cable wire made of steel instead of iron, and the bridge's construction marked the first use of a pneumatic caisson (a watertight compressed-air-filled shell). Its main span measured 1,595 feet (486 meters), the longest in the world until the Firth of Forth Bridge in Scotland (a cantilever structure) was built in 1890. Today, Japan's Akashi Kaikyo suspension bridge, completed in 1998, has a span of 6,532 feet (1,991 meters). The Brooklyn Bridge's total length, including approaches, is 6,016 feet (1,834 meters); it is 85 feet (26 meters) wide, and in the middle it clears high water by 135 feet (41 meters). It has four wire cables, each 15 3/4 inches (40 centimeters) in diameter, each composed of 5,434 wires. The granite towers stand 276 feet (84 meters) above mean high water. When the bridge was opened, the only building taller in the city was Trinity Church, including its steeple.

At the ceremonial dedication of the bridge on May 24, 1883, Abram Hewitt, an iron maker and political reformer (and future New York mayor) likened it to the pyramids of Egypt. In fact, the bridge developed into an icon of technological progress famous around the world, and an inspiration for poets such as the American, Hart Crane, and the Russian Vladimir Mayakovsky. By providing a physical link between Brooklyn and New York City, the bridge lent impetus to the movement supporting a merger of the two cities, which took place just before the end of the nineteenth century.

CULTURAL CONTEXT

A vacation in Bavaria may have inspired John Roebling and opened the door to his destiny. Roebling went for a hike to Bamberg, where he first saw what was referred to in those days as "the miracle bridge"—a suspension bridge over the Regnitz River that used iron chains. Roebling sat on a rock and began sketching various aspects of the bridge. All his education came together in one electric realization, and he knew he had found his vocation as a builder of innovative bridges.

Roebling came to the United States in 1831, intending to be a farmer in Saxonburg, Pennsylvania. By 1840, he had applied for a patent for *in-situ* spinning of wire rope, an innovation that could be traced back to his early insights inspired by the bridge in Bamberg. Once the patent had been granted, Roebling moved to Trenton, New Jersey, to build a factory to manufacture the wire. In an unlucky twist of fate, just after Roebling sold the family farm in Saxonburg, oil was discovered on the land; had he been the owner he would have become wealthy, but instead he went on to become famous.

Another woeful incident was the crushing of his left hand while working with the wire-rope machinery at his Trenton factory. Roebling had been an accomplished pianist, but he could play no more and instead turned his attention to using wire-rope technology in innovative bridge and aqueduct designs. He built a cable-suspended wooden aqueduct over the Allegheny River in 1845. Then he won the contract for a bridge in Niagara, New York; the design was a tremendously important innovation because trains could go over the bridge. An earlier bridge constructed for this purpose had fallen down, so Roebling's success was well regarded.

PLANNING

Ferries plied the straits of the so-called East River (actually a passage from the ocean, and therefore subject to tidal turbulence amplified by the narrowing of the straits) to service the burgeoning trade and commerce between the bustling port of the city of Brooklyn (originally named Breuckelen and chartered in 1646 by the Dutch West India Company), and the financial center of New York City, also founded by the Dutch West India Company. New York was becoming especially important as the terminus of traffic for the new Erie Canal, carrying goods from the western part of the state, including valuable salt from Onondaga.

Although Roebling was paid only $8,000 per year for his services as chief engineer, the Brooklyn Bridge was a financial bonanza. "According to the law, the entire corporation, though representing $4.5 million of the people's money (from the two cities), was actually controlled by the private stockholders. So . . . the man with ten shares of stock ($1,000 worth) had as much say as the City of New York or the City of Brooklyn with all their millions tied up in the venture" (McCullough, 132).

Much went on behind the scenes, and many politicians became rich in the process. "And if you believe that, I've got a bridge I can sell you," is a standard American joke; the bridge it refers to is the Brooklyn Bridge. The bridge's financing and shareholder process became a famous scandal because of the remorseless padding of expenses by Boss Tweed and his gang. It was Tweed who had managed to turn the construction contract for the new county courthouse on New York's Chamber Street from a modest project of $250,000 to a monumental $13 million (twice the price paid for Alaska).

The Brooklyn Bridge went public in 1874 and began to pay interest to its shareholders at that time. The bridge did not actually open until 1883 (three years after the time stipulated in the authorizing incorporation contract, reproduced here as Document I), but the project was profitable before it was even fully built. Bridge profits were restricted by law to 15 percent net (see the incorporation contract, section 6), and it was stipulated that if tolls produced more than the agreed level of profit, tolls must be reduced. There was also a specified limit on how much profit could be taken in a sale (see the incorporation contract, section 7): if the bridge was sold to either the city of New York or the city of Brooklyn, payment was to be comprised of the cost of building and property acquired plus a profit equal to an additional 33.3 percent, and a provision that all tolls would cease and the bridge would be free to all users.

In the documents of authorization presented here, the reader will find two separate contracts. The first is the act of incorporation for the New York Bridge Company, an agreement between New York State and the associate partners of the enterprise. The incorporation document discusses the rights and responsibilities of the company. One notable portion is the clause at the conclusion of section 1, which stipulates that the bridge will be open for public use by June 1870.

The second document, an act of Congress, is very different. The parties to this agreement are the cities of Brooklyn and New York. The national agreement includes details about navigation of the waterway the bridge spanned and stipulates the height of the span (i.e., tall enough for ships to pass under the bridge at all tide levels). The waterway was also considered a postal thoroughfare and a strategic transport route. Hence the secretary of war is mentioned throughout the document, and is named as the approval authority.

BUILDING

Prior to the construction of the Brooklyn Bridge, iron was the primary material used. In fact, steel was regarded as a suspect material not yet proven over time as was iron. Nevertheless, Roebling had complete confidence in his own judgment that the crucible-cast steel he had designed would be six times stronger than necessary. When a scandal was uncovered—a wire contractor had been substituting a cheaper Bessemer steel—because of Roebling's prudent specifications, the lesser steel proved to be more than adequate because it was still five times stronger than necessary.

One innovative decision made early in the process was to construct an extremely heavy roadway. Previously, bridges were built with lightweight roadways in the belief that it would be easier for the cables to hold up a lighter structure. But such lightweight bridges were easily damaged by wind. Soon engineers learned that bridges need a very heavy roadway, and such a heavy load required cables that would not snap—precisely the kind Roebling adopted.

Another innovation was the use of pneumatic caissons, the chambers that were submerged in the river so work could be carried on underwater. The seams of the caissons were made watertight by caulking them with oakum (loose hemp or jute fiber from unraveled rope, impregnated with tar). The caisson was then painted with airtight varnish and totally covered in an armor of boilerplate, and hot tar was poured over the entire structure like syrup. To prevent fire, the inside of the caisson was lined with boilerplate. McCullough gives a description of the caissons:

> The easiest way to explain how the caisson would work . . . was to describe it as a huge diving bell that would be built of wood and iron, shaped like a gigantic box, with a heavy roof, strong sides, and no bottom. Filled with compressed air, it would be sent to the bottom of the river by building up layers of stone on its roof. The compressed air would keep the river out, help support the box against the pressure of water and mud, and make it possible for men to go down inside to dig out the riverbed. As they progressed and as more stone was added, the box would sink slowly, steadily, deeper and deeper, until it hit a firm footing. Then the excavation could stop, the interior of the box would be filled in solid with concrete, and that would be the foundation for the bridge tower. (173)

Such a structure was dark and dangerous. The dark was mitigated somewhat by painting the interior of the caisson white, but the danger remained. Like deep-sea divers, those working in the caissons suffered decompression sickness, or the bends. Three workers died and 15 were disabled. What is especially sad is that the bends could perhaps have been prevented—certainly eased—if James Eads, the engineer who built the St. Louis Bridge, had shared his experience with Roebling. Unfortunately, the two had had a disagreement and their colleagueship as builders of great American bridges dissolved. Eads had had similar problems with the bends, but he had found a physician who practiced the remedy first pioneered by English and French engineers—gradual decompression rather than immediate return to the surface.

The hazards of building the Brooklyn Bridge were many, and they eventually took the lives of 30 workers, including the chief engineer. John Roebling crushed his foot while working on the bridge and died from an infection related to the injury. His son, Washington, carried on, but he too was stricken while working on the bridge; although he did not die, he was disabled from a bout with the bends, which affected him while working in the underwater caissons. Washington's wife, Emily, rented an apartment overlooking the bridge construction so that her invalid husband could continue to oversee the work while she herself acted in his stead on-site.

Upon completion of the bridge, Emily planned an enormous reception to celebrate its opening. Not willing to see her husband left out of the ceremony, she insisted on bringing 21 dignitaries right into the convalescent bedroom to

Close-up view of construction of the Brooklyn Bridge, circa 1880.
Courtesy of the Library of Congress.

congratulate the Roebling dynasty. Then she closed the place and packed the family off to their summer cottage in Newport.

IMPORTANCE IN HISTORY

The Brooklyn Bridge was a privately held and financed contribution to the public infrastructure of New York. There are pluses and minuses in such an arrangement. On the plus side, the government does not have to deploy workers or lay out funds. On the minus side, private building still requires permissions from many government agencies; therefore a huge fund is required just to pay the lawyers to obtain the needed permissions. Perhaps this is why Boss Tweed worked his way into the project, in order to smooth the governmental permissions process; without his considerable political weight, the permissions could have been significantly slowed. In more general terms, the permissions process is of critical importance to large-scale modern projects. On this point, George Shultz, former secretary of state and later an officer of Bechtel Corporation, commented that he needed more than 300 permissions to build a pipeline in the West, and eventually the cost of the lawyers was so great that Bechtel had to abandon the project even though it was financially viable in every other respect.

As a topic for further research, the process of permission-granting merits continuous study and improvement across international, national, state, and local lines. In Japan it is notably less time-consuming to receive authorizations for the construction of large-scale projects.

It is interesting to note that at one point, a tunnel was considered as a method for crossing the river between New York and Brooklyn; it might have been less expensive. As in the case of the Channel Tunnel between England and France, there was disagreement over whether to build a tunnel or bridge. French minister of the interior Jules Moch put together a study group that ultimately recommended the more expensive alternative, a bridge. In the Northumberland Straits, which links mainland Canada to Prince Edward Island, both options were considered; a bridge was chosen despite the fact that the bridge cannot remain open for much of the year due to hazardous ice that prohibits vehicular traffic.

Divers today use slow decompression, and the recent innovation of the hyperbaric chamber is now available as a therapy. If not already under consideration, perhaps future water-based engineering projects, such as underwater tunnels and bridges spanning waterways, should have hyperbaric chambers on-site.

An icon of modern times, the bridge attracted daredevils who performed feats on the behemoth; the earliest record is that of a barkeep named Steve Brodie, who on July 23, 1886, leaped from the bridge into the East River and survived. Those wishing to pursue more moderate kinds of kinesthetic engagement can walk across the span. The first day it opened, May 25, 1883, 150,300 people crossed over on foot.

FOR FURTHER REFERENCE

Books and Articles

Mann, Elizabeth. *The Brooklyn Bridge: A Wonders of the World Book*. New York: Mikaya Press, 1996. For children ages 9–12.

McCullough, David. *The Great Bridge*. New York: Simon and Schuster, 1972.

Internet

For an overview of the bridge with illustrations of all aspects, including the caissons, see http://www.endex.com/gf/buildings/bbridge/bbridgefacts.htm.

On the hyperbaric chamber and its history dating to 1662, see http://www.tbims.org/combi/ubb/Forum4/HTML/000030.html.

For information about Hart Crane's work on the Brooklyn Bridge, see http://www.poets.org/poems/Poempmt.cfm?45442B7C000C07070F77.

For Jack Kerouac's poem "The Brooklyn Bridge Blues," see http://www.geocities.com/yesterdayswine/BrooklynBridgeBlues.html?200415.

Film and Television

The Brooklyn Bridge. Videocassette. Directed by Ken Burns. Boston: PBS Home Video, 1997.

Documents of Authorization—I

AN ACT TO INCORPORATE THE NEW YORK BRIDGE COMPANY, FOR THE PURPOSE OF CONSTRUCTING AND MAINTAINING A BRIDGE OVER THE EAST RIVER, BETWEEN THE CITIES OF NEW YORK AND BROOKLYN.

Passed April 16, 1867; three fifths being present.

The People of the State of New York, represented in Senate and Assembly, do enact as follows:

Section 1. John T. Hoffman, Simeon B. Chittenden, Edward Ruggles, Smith Ely, Jr., Samuel Booth, Granville T. Jenks, Alexander McCue, Henry E. Pierrepont, Martin Kalbfleisch, John Roach, Charles A. Townsend, Henry G. Stebbins, Charles E. Bill, Chauncey L. Mitchell, T. Bailey Myers, Seymour L. Husted, William A. Fowler, William W.W. Wood, Andrew H. Green, Edmund W. Corlies, William C. Rushmore, Ethelbert S. Mills, Alfred W. Craven, Arthur W. Benson, T.B. Cornell, John N. Hayward, Isaac Van Annden, Pomeroy P. Dickinson, Alfred M. Wood, J. Carson Brevoort, William Marshall, Samuel McLean, John W. Combs, William Hunter, Jr., John H. Prentice, Edmond Driggs, John P. Atkinson, John Morton, and their associates, are hereby created a body corporate and politic, by the name of "The New York Bridge Company," for the purpose of constructing and maintaining a permanent bridge over the East river, between the cities of New

York and Brooklyn; and as such corporation, are invested with all the powers and privileges, and are subject to all the liabilities conferred and imposed by title three, chapter eighteen of part one, of the Revised Statutes; provided, however, that the said bridge shall be completed and opened to the public use on, or before the first day of June, in the year one thousand eight hundred and seventy.

Section 2. The said corporation shall have power to purchase, acquire and hold as much real estate as may be necessary for the site of said bridge, and of all piers, abutments, walls, toll-houses, and other structures proper to said bridge, and for the opening of suitable avenues of approach to said bridge, not any land under water, in the river, beyond the pier lines established by law; to borrow money from time to time, not exceeding in the aggregate, at any one time, the amount of the capital; to make and establish, from time to time, ordinances and laws, under reasonable penalties, to be recovered in the name of and on behalf of the said corporation, in any court in the city and county of New York, or county of Kings, having the jurisdiction of justices of the peace, regulating the travel over said bridge by vehicles and animals; and in case of destruction of said bridge, to reconstruct and maintain the same, and to borrow additional moneys sufficient for that purpose.

Section 3. The capital stock of said corporation shall be five millions of dollars, divided into shares of one hundred dollars each. The directors of said corporation may, at any time, with the consent in writing of stockholders holding a majority of the stock, increase such capital. The shares shall be deemed personal property, and may be transferred in such manner as shall be prescribed by the by-laws of the corporation.

Section 4. The persons named in the first section of this act, shall constitute the first board of directors of said corporation, and shall hold their places as such until the first Monday of June, one thousand eight hundred and sixty-eight, and until others shall be elected in their stead. The number of directors, after the time last mentioned, shall not be less than thirteen nor more than twenty-one, to be fixed in the by-laws of the corporation. An election of directors shall take place on the first Monday of June, in the year last aforesaid, and annually on that thereafter, at an hour and place to be designated in the by-laws; and the persons then elected by a majority of shares voted upon by stockholders in person or by proxy, shall constitute the directors for the then ensuing year, and until others shall be duly elected in their places. All vacancies which shall occur in the board of directors, by death, resignation, mental incompetency, removal from the State, or otherwise, shall be filled by appointment of a majority of the remaining members, for the balance of the term thus vacated. A majority of said board shall constitute a quorum.

Section 5. The officers of the corporation shall consist of a president, secretary and treasurer, who shall be annually elected by the incoming board of directors. Such subordinates may be appointed from time to time, as the board may direct. A record of the proceedings of the board of directors shall be kept by the secretary, and a statement of the financial condition of the corporation, the amount of money expended on account of said bridge and its appurtenances, and of all its receipts and expenditures, shall be annually prepared by the president and treasurer, verified by them under oath, and filed in the office of the Secretary of State, on or before the first day of June in each year.

Section 6. The board of directors shall have power to fix the rates of toll for persons, animals, carriages and vehicles of every kind or description passing over the same. Toll-gates shall be kept at each end of the bridge, and the toll demanded and paid upon entering on the bridge. The rates of toll shall be posted up conspicuously at the toll-gates. The said directors shall reduce the rates of toll from time to time, so that the net profits of said bridge shall not exceed the sum of fifteen per cent per annum, after deducting the expenses of repairs and improvements to said bridge, its appurtenances and approaches, and all just and proper damages against the said corporation.

Section 7. The cities of New York and Brooklyn or either of them, may at the time take the said bridge and appurtenances and acquire all property therein by the payment to the said corporation of the cost thereof, together with thirty-three and one-third per cent in addition thereto, provided the said bridge be made free, to be passed by travelers and vehicles without tolls or other charges.

Section 8. Any person willfully doing any injury to the said bridge or any of its appurtenances, shall forfeit and pay to the said corporation three times the amount of such injury, and shall be deemed guilty of a misdemeanor, and be subject to a penalty not exceeding five hundred dollars, and to imprisonment not exceeding six months, in the discretion of the courts.

Section 9. Concurrent jurisdiction shall be possessed and exercised by the courts of the city and county of New York and the county of Kings respectively, over all crimes and offences committed upon said bridge over the East river. It shall be the duty of the commissioners of the metropolitan police to keep an adequate force at all times during the day and night upon the said bridge, for the protection of the public, which force shall be in addition to the force authorized for the two cities, and shall be paid at the joint expense of the said two cities, to be raised in the same manner as the other police expenses of the said two cities and as a part thereof.

Section 10. Nothing in this act contained shall be construed to authorize nor shall it authorize the construction of any bridge which shall obstruct the free and common navigation of the East river, or the construction of any pier in the said river beyond the pier lines established by law. Such bridge shall not be at a less elevation than one hundred and thirty feet above high tide at the middle of the river. It shall not obstruct any street which it shall cross, but such street shall be spanned by a suitable arch or suspended platform as shall give a suitable height for the passage under the same for all purposes of public travel and transportation. No street running in the line of said bridge shall be closed without full compensation to the owners of land fronting on the same, for all damages they may sustain by reason thereof. The said bridge shall commence at or near the junction of Main and Fulton streets in the city of Brooklyn, and shall be so constructed as to cross the river as directly as possible to some point at or below Chatham square, not south of the junction of Nassau and Chatham streets, in the City of New York. The said bridge shall be built with a substantial railing or siding, and shall be kept fully lighted through all hours of the night.

Section 11. If the said corporation shall be unable to agree, for any reason, with the owner or owners of any real estate required for its purposes as aforesaid, for the purchase thereof, it shall have the right to acquire the same, in the manner and by the like special proceedings as are authorized and provided for obtaining title to real estate required for the purposes of a railroad corporation, under the fourteenth section of the act entitled "An act to authorize the formation of railroad corporations and to regulate the same," passed April second, one thousand eight hundred and fifty, and the other sections of the said act relative thereto, and any acts amendatory thereof or in addition thereto, and for that purpose all such acts shall be considered applicable to the corporation hereby created, as far as may be, in like manner as if the same were named therein; and such modifications may be made in the formal part of the proceedings, in order to apply the same to the corporation hereby created instead of a railroad corporation, as shall be approved by the supreme court; and the said court may make such orders and regulations as to the mode and manner of conducting the proceedings, and all things relative thereto, so as to effectuate and make the same valid for acquiring title to such real estate as the said court may deem proper; and the title thus acquired by the said corporation shall vest in it the fee simple of the said lands. The said court, on sufficient cause being shown, and on proof of payment or tender of the amount to be paid for such real estate in any manner, as may have been required by said court, may issue summary process, in such form as may be deemed proper, to the sheriff of the proper county, commanding him, without delay, to put the said corporation, by it proper agents, in the possession of such real estate, and may enforce such process in such manner as may be conformable to law.

Section 12. The cities of New York and Brooklyn, or either of them, are authorized to subscribe to the capital stock of said company such amount as two-thirds of their common councils respectively shall determine, and to issue bonds in payment of such subscriptions, payable in not less than thirty years, or may guarantee the payment of the principal and interest of the bonds of the company, in such amounts as the said common council shall respectively determine.

Section 13. This act shall take effect immediately.

FORTIETH CONGRESS. Sess. III. CH. 139, March 3, 1869.

Documents of Authorization—II

CHAPTER CXXXIX—AN ACT TO ESTABLISH A BRIDGE ACROSS THE EAST RIVER, BETWEEN THE CITIES OF BROOKLYN AND NEW YORK, IN THE STATE OF NEW YORK, A POST-ROAD.

Be it enacted by the Senate and House of Representatives of the United States of America in Congress assembled, That the bridge across the East River, between the cities of New York and Brooklyn, in the State of New York, to be constructed under and by virtue of an act of the legislature of the State of New York, entitled "An act to incorporate the New York Bridge Company, for the purpose of constructing and maintaining a bridge over the East River between the cities of New York and Brooklyn," passed April sixteenth, eighteen hundred and sixty-seven, is hereby declared to be, when completed in accordance with the aforesaid law of the State of New York, a lawful structure and post-road for the conveyance of the mails of the United States: *Provided,* That the said bridge shall be so constructed and built as not to obstruct, impair, or injuriously modify the navigation of the river; and in order to secure a compliance with these conditions, the company, previous to commencing the construction of the bridge, shall submit to the Secretary of War a plan of the bridge, with a detailed map of the river at the proposed site of the bridge, and for the distance of a mile above and below the site, exhibiting the depths and currents at all points of the same, together with all other information touching said bridge and river as may be deemed requisite by the Secretary of War to determine whether the said bridge, when built, will conform to the prescribed conditions of the act, not to obstruct, impair, or injuriously modify the navigation of the river.

Section 2. *And be it further enacted,* That the Secretary of War is hereby authorized and directed, upon receiving said plan and map and other information, and upon being satisfied that a bridge built on such plan and at said locality will conform to the prescribed conditions of this act, not to obstruct, impair, or injuriously modify the navigation of said river, to notify the said company that he approves the same; and upon receiving such notification the said company may proceed to the erection of said bridge, conforming strictly to the approved plan and location. But until the Secretary of War approve the plan and location of said bridge, and notify said company of the same in writing, the bridge shall not be built during the progress of the work thereon, such change shall be subject likewise to the approval of the Secretary of War.

Section 3. *And be it further enacted,* That Congress shall have power at any time to alter, amend, or repeal this act.

Approved, March 3, 1869.

From *Laws of New York,* 90th sess. (1867), chap. 399.

Related Cultural Documents

REPORT TO THE NEW YORK BRIDGE COMPANY, SEPTEMBER 1, 1867

PLAN AND DETAILS OF ANCHORAGE, APPROACHES, TOWER, AND STEEL CABLES:

The contemplated work, when constructed in accordance with my design, will not only be the greatest bridge in existence, but it will be the great engineering work of the Continent and of the age. Its most conspicuous feature—the great towers—will serve as landmarks to the adjoining cities, and they will be entitled to be ranked as national monuments. As a great work of art, and a successful specimen of advanced bridge engineering, the structure will forever testify to the energy, enterprise, and wealth of that community which shall secure its erection.

—John Augustus Roebling
Designer of the Brooklyn Bridge

From Report to the New York Bridge Company, Sept. 1, 1867, http://www.endex.com/gf/buildings/bbridge/bbridgefacts.htm.

19
The Canadian Pacific Railway

Canada

DID YOU KNOW . . . ?

➤ Railway construction began in 1875 and was complete in 1885, five years ahead of the mandatory completion date of 1891.

➤ The railway begins in Montreal and ends in Vancouver.

➤ 3.5 million laborers worked on the railway, including "navvies" (workers who first built Canadian canals) and Chinese immigrants.

➤ Wages ranged from Can$0.75 to Can$2.50 per day.

➤ The railway transported silk cocoons in specially designed cars, as the final stage of the Silk Road.

➤ Building the railway fulfilled a promise to British Columbia that united Canada.

➤ Sandford Fleming first proposed international time zones while the railway was being built.

➤ Railway cars were used as mobile classrooms to reach isolated rural children.

The Canadian Pacific Railway could be regarded as a technological extension of the Silk Road of China. When this Canadian railway was built, special train cars were created to bring Chinese silk cocoons from the docks in Vancouver, British Columbia, and transport them east to the mills in New Jersey and New York. Silk was extremely rare and valuable, so the train cars were carefully designed to preserve the precious cocoons. Armed guards were aboard; so great was the possibility of robbery that the so-called Silk Trains stopped infrequently, speeding quickly to reach their destination with their cargo intact.

An advertisement for travel on the Canadian Pacific Railway, circa 1940. Courtesy of the Library of Congress.

HISTORY

The early Canadian Confederation, which extended from Nova Scotia in the east to Ontario in the west, was formed in 1867; two years later it grew to include the Northwest Territories. Manitoba became a province in 1870 and British Columbia in 1871, but only after obtaining a commitment that a transcontinental railway would begin construction within 2 years and be completed within 10. (Alberta and Saskatchewan became provinces in 1905.)

Many Canadian leaders were convinced that a transcontinental rail line was critical for Canada's growth. Prime Minister John A. Macdonald and his allies believed a railway across the western mountains and central prairies would not only provide a crucial transportation link between the provinces but would also help counter the aspirations of U.S. expansionists imbued with dreams of manifest destiny, who were casting greedy eyes on their northern neighbor.

Macdonald's Conservative government fell in 1873, sidelined by the so-called Pacific Scandal in which Montreal multimillionaire Hugh Allan, head of a group that had received a contract to build the railway, was revealed to have given hundreds of thousands of dollars to the Conservatives' 1872 election campaign. In the aftermath, Allan lost the contract, and the newly installed government began working on a few portions of the line. At the same time, the Canadian economy was in the throes of depression, and construction progress slowed.

In the elections of 1878, Macdonald was returned to power on a platform with three planks: in addition to a protectionist tariff policy, he called for settlement and development of the western region and for finalization of the railway to the Pacific. One depended on the other, for without the railway, settlement would be less likely. It was Macdonald's view that the railway could unite the country because it would facilitate trade and travel among the provinces and bring all Canadians together in a common interest.

New construction contracts were signed. The most important was awarded in 1880 to a Montreal-based group that included Scottish émigré George Stephen, a manufacturer and financier who would play an instrumental role in raising private capital for the project. A construction contract with Stephen's group was ratified in 1881 in an act of Parliament reproduced here as a document of authorization. The measure incorporated the Canadian Pacific Railway Company and ordered it to complete the railway by 1891. The company was granted numerous concessions, among them a subsidy of Can$25 million and a grant of 25 million acres (10.1 million hectares) of prairie land. (All subsequent monetary figures are in Canadian dollars.)

The project moved ahead quickly, especially in the prairies, where 418 miles (673 kilometers) of track were laid between April 1882 and January 1883; the entire section from Winnipeg to Calgary was completed by August 1883. Progress over the terrain north of Lake Superior was much slower, impeded not only by difficulties of access but also by unforgiving rock and spongy peat bogs (muskeg). In British Columbia, construction was hampered by a combination of mountains, gorges, and fast-flowing streams.

Even before its completion, the new railway proved its strategic value. In March 1885, the government quickly quelled the so-called Northwest Rebellion in Saskatchewan by quickly sending troops via the Canadian Pacific. The matter helped Macdonald win approval of a loan to the financially hard-pressed (at the time) railway company.

Five years ahead of schedule, in 1885, and 16 years after completion of the U.S. transcontinental railroad, the last spike was driven in a ceremony on November 7, 1885, at Craigellachie, in western British Columbia. The first passenger train to cross the continent departed from Montreal on June 28, 1886, and arrived at Port Moody, British Columbia, on July 4, for a total distance of 2,893 miles (4,656 kilometers). The British Columbia terminus was relocated the following year to Vancouver.

As expected, the ability to move people and cargo long distances with unprecedented speed contributed to the rapid development of western Canada. New cities sprouted up around railway stations. The Canadian Pacific established its western office in Winnipeg, which quickly grew into a major center for grain commerce. Vancouver thrived as a result of railway facilities there. To boost revenues, the company began to expand by including marine trade in the Pacific through chartering and purchasing ships. Among its other ancillary activities, the company also built hotels to house weary travelers and constructed irrigation systems.

CULTURAL CONTEXT

Because the Canadian Pacific Railway was completed 20 years after the U.S. transcontinental railroad, many engineers who had worked on the American railroad went to work for the Canadian company. A short time later, the Russians began work on the Trans-Siberian Railway. Thus, at this time in history the culture was one of building connections through rail, just as the previous era had been one of constructing connections via waterways and canals.

The trains could also serve as movable schoolrooms. As settlers began to move into western Canada, families quickly formed, but there was no system in place for educating the children. The solution was to hire teachers who traveled on specially designed rail cars—rolling one-room schoolhouses, complete with desks, blackboards, and a modest library. The rail cars would pull into a district and children would come for several days of teaching, then be sent home with plenty of homework. Meanwhile, the teacher and train moved on to the next district.

Special-use cars became a signature of the Canadian Pacific Railway. There were troop cars during wartime. At the conclusion of the war, they became bridal cars, transporting newly married wives to join their husbands. When wheat prices dropped to an all-time low and many Canadians lost their jobs, food-wagon cars carried supplies out to the heartland to feed people until the

Laying the railway tracks through the Connaught Tunnel, in British Columbia on the Canadian Pacific Railway Line, circa 1916. © The Art Archive / Culver Pictures.

economy recovered and they could earn a living again. When the prime minister died, his coffin was placed in a special car that rolled across the country, stopping at towns for mourners to pay their respects. On a happier note, the British monarchy enthralled loyal subjects with the pomp and circumstance of processions by visiting Canada and making use of Royal Railcars.

PLANNING

The planning process was powered by political pressure. The government called the shots, set the terms, and provided the land and loans; it was a private company that carried out these orders. The Canadian Pacific is a private company that carried out a public program, where the public party supplied the land and the land supplied the money. The arrangement is clearly stated in section 3 of the contract reproduced as a Document of Authorization: "The government may grant a subsidy of $25,000,000 and 25,000,000 acres of land."

The Canadian Pacific Railway was initiated, then collapsed in the midst of the planning process due to the political scandal mentioned earlier, and stalled completely during the succeeding administration. Finally, in 1878 Macdonald regained the position of prime minister, and construction of the railway moved into full operation.

The most difficult task remained: finding a path over the daunting Selkirk Mountains. The job of finding a pass was assigned to surveyor Major Albert Bowman Rogers. He was given $5,000 and a promise that the pass would be named in his honor. After several expeditions and close brushes with death, in May 1883 he finally found a 150-mile shortcut through woods and meadows. The railroad named the pass Rogers Pass and gave him the check for $5,000, which he at first refused to cash, wanting to frame it instead. He was finally coaxed to cash it with the promise of an engraved watch for doing so.

Originally it was hoped that the rough terrain of the Canadian midsection could be avoided by looping down through Chicago, a great rail center in America. But ultimately the desire to keep the system totally Canadian won out; the decision was made to push ahead across the hard ground of the Canadian plains. By 1884, the project was again floundering, this time for lack of funds, and the government was forced to provide additional financial support with another $22.5 million.

BUILDING

Construction of the rail line began in 1875 with the ceremonial first sod turned at what is now Thunder Bay, Ontario, marking the start of construction by the federal government on the Lake Superior section of the railway. In 1880, the far western section of the railway was started by the federal government. In the same year, the government signed a contract with George Stephen and others to construct the eastern and central sections, and gave the group full responsibility to complete the railroad and make it operational by 1891.

However, construction was progressing too slowly, and in 1882 railway officials hired the renowned railway executive William Cornelius Van Horne to oversee construction. Van Horne stated that he would have 500 miles (800 kilometers) of main line built in 1882. Floods delayed the start of the construction season, but over 417 miles (672 kilometers) of main line, as well as 28 miles (45 kilometers) of siding and branch lines, were built that year. Work continued apace until the summer of 1884, when the company ran short of funds. Political upheaval saved the day. In March 1885, white settlers and Indians in Saskatchewan began fighting in open rebellion. To quell the trouble quickly, the government sought to move militiamen west to the area via the unfinished railway: Van Horne promised to get them there in 10 days. Perhaps because the government was grateful for this service, they subsequently reorganized the railway's debt to the government and provided a further $5 million loan, money desperately needed by the Canadian Pacific.

Despite the stipulation in the contract of a maximum gradient of 2.2 percent (equivalent to a 116-foot [35.4 meters] change in elevation per mile), the engineers were forced to build a steeper incline when they could not afford to construct a tunnel to pass through Big Hill, an eight-mile (12.9 kilometers) section of the railway. Instead they opted for a 4.5 percent gradient with safety switches installed at several points. The speed limit for descending trains was set at 6 miles (10 kilometers) per hour, and special locomotives were ordered. Despite these precautions, several serious runaways occurred.

Dynamite was needed to blast through rock in many places. However, demand exceeded the imported supply so much that three factories were built near the construction sites, enough to produce three tons of dynamite a day.

No discussion of the Canadian Pacific Railway is complete without considering the "navvies"—men who had worked on Canada's canals digging the waterbeds, shoring up the silted sides of rivers to maintain the width and depth needed by vessels bearing crops and products from across Canada. When the rails replaced canals, the navvies promptly switched from waterways to railways.

Unfortunately for the navvies, history appreciates their availability and experience far more than did their employers. Navvies were paid shockingly low wages for experienced workers, perhaps as a result of their sudden unemployment. They earned between $1.00 and $2.50 per day, and from that they had to find their own transport to the work site, buy work clothes, pay medical expenses, buy food, and pay for any mail they received. And those were the luckier ones who worked in eastern Canada. Out west, where the laborers were largely Chinese immigrants (like their counterparts on the U.S. transcontinental railroad), the going rate for daily work ranged from $0.75 to $1.25. Even worse than the pay differential was the difference in work. On eastern work sites, the land was flat and moist, but the western work areas were filled with craggy rocks that required explosives and extensive drilling.

But not all the laborers on the job sites were navvies. Because work sites were so remote, once a location was established, the entire team had to

Donald Smith, the Canadian statesman, drives in the last spike at Craigellachie, British Columbia, marking the completion of the Canadian Pacific Railway. © Hulton Archive / Getty Images.

live on-site: blacksmiths, dynamite manufacturers, animal caretakers, executives, navvies, and every kind of work specialist all camped together. Train cars loaded with supplies for each day's work were sent down the track to the work site. Overall, the transcontinental project employed 3.5 million laborers. Many died from malnutrition, disease, exposure, and construction accidents.

IMPORTANCE IN HISTORY

Once the railway was built, a dependable signaling system had to be developed. In the case of the U.S. transcontinental railroad, conductors waited for telegraphed information, noting the times of other trains on the line on small pieces of paper called *flimsies*. But this was at best a tenuous system of confirmation: given the so-called dark spots where no telegraph signals could be received, accidents posed a recurrent danger. The same dangers were present in the Canadian rail system.

Sandford Fleming, a Canadian Pacific Railway route surveyor, was concerned that the trains could not run dependably unless something was done to standardize time schedules. He proposed a system of international time zones. The Canadian Pacific and the U.S. transcontinental railroads came together in 1883 (even before the Canadian Pacific was completed) and agreed jointly on a system of time zones. Eventually the idea gathered such force that the entire world became galvanized by this innovation. In 1884, the International Prime Meridian Conference held in Washington, DC, endorsed and inaugurated a worldwide system of time zones. Key portions of the conference proceedings are reproduced in this chapter as a Related Cultural Document.

The Canadian Pacific is not the only railway in Canada: the Canadian National Railway is an impressive 23,790 miles (38,286 kilometers) long and is run by the government. While it may be the largest rail system from the standpoint of miles operated, it was pieced together from five separate railways: three privately owned and two government-built. When the private systems faced bankruptcy during World War II, the government stepped in and salvaged them as government operations.

The Canadian Pacific may have been one of the first companies to adopt a logo. Perhaps because it was such a vast country with so many workers employed on the project, some common theme was needed. What is more perfect than a representation of nature's most engineering-driven animal, the beaver—also the logo of the Massachusetts Institute of Technology in Cambridge, Massachusetts?

FOR FURTHER REFERENCE

Books and Articles

Beebe, Lucius. *Trains in Transition*. New York: Bonanza Book, 1941. Reprint, New York: Hawthorn Books, 1976.

Beebe, Lucius, and Charles Clegg. *When Beauty Rode the Rails: An Album of Railroad Yesterdays*. Garden City, NY: Doubleday, 1962.

Berton, Pierre. *The Last Spike: The Great Railway, 1881–1885*. 1971. Reprint, Toronto: Anchor Canada, 2001.

———. *The National Dream: The Great Railway, 1871–1881*. Toronto: Anchor Canada, 1970.

Hawkes, Nigel. *Structures: Man-Made Wonders of the World*. London: Marshall Editions, 1990.

Internet

For detailed information on the Canadian Pacific Railway, see the company's Web site: http://www8.cpr.ca/.

For a brief overview of the Canadian Pacific Railway, including a depiction of their logo, see http://en.wikipedia.org/wiki/Canadian_Pacific_Railway.

For the original text of the document authorizing international time zones at the International Meridian Conference, October 1884, see http://wwp.greenwich2000.com/millennium/info/conference-finalact.htm.

Film and Television

The National Dream. Eight part series. Toronto: Canadian Broadcasting Company TV series 1975.

Music

Lightfoot, Gordon. "Canadian Railroad Trilogy." *The Way I Feel*. 1967. United Artists, BH22967. The Canadian Broadcasting Company commissioned Lightfoot to write this song for Canada's centennial and aired it as part of a television special titled *100 Years Young* on January 1, 1967.

Documents of Authorization

44 Victoria
CHAPTER I

AN ACT RESPECTING THE CANADIAN PACIFIC RAILWAY

(Assented to February 15, 1881)
SCHEDULE

Preamble.	Whereas by the terms and conditions of the admission of British Columbia into Union with the Dominion of Canada, the Government of the Dominion has assumed the obligation of causing a railway to be constructed, connecting the seaboard of British Columbia with the railway system of Canada;
Preference of Parliament for Construction by a Company.	And whereas the Parliament of Canada has repeatedly declared a preference for the construction and operation of such Railway by means of an incorporated Company aided by grants of money and land, rather than by the Government, and certain Statutes have been passed to enable that course to be followed, but the enactments therein contained have not been effectual for that purpose;
Greater part still unconstructed.	And whereas certain sections of the said railway have been constructed by the Government, and others are in course of construction, but the greater portion of the main line thereof has not yet been commenced or placed under contract, and it is necessary for the development of the North-West Territory and for the preservation of the good faith of the Government in the performance of its obligations, that immediate steps should be taken to complete and operate the whole of said railway;
Contract entered into.	And whereas, in conformity with the expressed desire of Parliament, a contract has been entered into for the construction of the said portion of the main line of the said railway, and for the permanent working of the whole line thereof, which contract with the schedule annexed has been laid before Parliament for its approval and a copy thereof is appended hereto, and it is expedient to approve and ratify the said contract, and to make provision for the carrying out of the same:
	Therefore Her Majesty, by and with the advice and consent of the Senate and House of Commons of Canada, enacts as follows:—
Contract approved.	1. The said contract, a copy of which, with schedule annexed, is appended hereto, is hereby approved and ratified, and the Government is hereby authorized to perform and carry out the conditions thereof, according to their purport
Charter may be granted.	2. For the purpose of incorporating the persons mentioned in the said contract, and those who shall be associated with them in the undertaking,

and of granting to them the powers necessary to enable them to carry out the said contract according to the terms thereof, the Governor may grant to them in conformity with the said contract, under corporate name of the Canadian Pacific Railway Company, a charter conferring upon them the franchises, privileges, and powers embodied in the schedule to the said contract and to this Act appended, and such charter, being published in the Canada Gazette, with any Order or Orders in Council of the Parliament of Canada, and shall be held to be an Act of incorporation within the meaning of the said contract.

Publication and effect of charter.

Certain grants of money and land may be made to the company chartered.

3. Upon the organization of the said Company, and the deposit by them, with the Government, of one million dollars in cash, or securities approved by the Government, for the purpose in the said contract provided, and in consideration of the completion and perpetual and efficient operation of the railway by the said Company, as stipulated in the said contract, the Government may grant to the Company a subsidy of twenty-five million dollars in money, and twenty-five million acres of land, to be paid and conveyed to the Company in the manner and proportions, and upon the terms and conditions agreed upon in the said contract, and may also grant to the Company the land for right of way, stations, and other purposes, and such other privileges as are provided for in the said contract. And in lieu of the payment of the said money subsidy direct to the Company, the Government may convert the same, and any interest accruing thereon, into a fund for the payment, to the extent of such fund, of interest on the bonds of the Company, and may pay such interest accordingly; the whole in manner and form as provided for in the said contract.

Conversion of money grant authorized.

Certain materials may be admitted free of duty.

4. The Government may also permit the admission free of duty, of all steel rails, fish plates, and other fastenings, spikes, bolts and nuts, wire, timber and all material for bridges to be used in the original construction of the said Canadian Pacific Railway, as defined by the act thirty-seventh Victoria, chapter fourteen, and of a telegraph line in connection therewith, and all telegraphic apparatus required for the first equipment of such telegraph line, the whole as provided by the tenth section of the said contract.

Company to have possession of completed portions of the railway.

5. Pending the completion of the eastern and central sections of the said railway as described in the said contract, the Government may also transfer to the said Company the possession and right to work and run the several portions of the Canadian Pacific Railway as described in the said Act thirty-seventh Victoria, chapter fourteen, which are already constructed, and as the same shall be hereafter completed; and upon the completion of the said eastern and central sections the Government may convey to the Company, with a suitable number of station buildings, and with water service (but without equipment), those portions of the Canadian Pacific Railway constructed, or agreed by the said contract to be constructed by the Government, which shall then be completed; and upon completion of the remainder of the portion of the said railway to be constructed by the Government, that portion also may be conveyed by the Government to the Company, and the Canadian Pacific Railway defined as aforesaid shall become and be thereafter the absolute property of the Company; the whole, however, upon the terms and conditions, and subject to the restrictions and limitations contained in the said contract.

Conveyance to company when the contract is performed.

Security may be taken for operation of the railway.

6. The Government shall also take security for the continuous operation of the said Railway during the ten years next subsequent to the completion thereof in the manner provided by the said contract.

THIS CONTRACT AND AGREEMENT MADE BETWEEN HER MAJ-
ESTY THE QUEEN, acting in respect of the Dominion of Canada and herein
represented and acting by the Honourable SIR CHARLES TUPPER, K.C.M.G.,
Minister of Railways and Canals and George Stephen and Duncan McIntyre, of
Montreal in Canada, John S. Kennedy of New York, in the State of New York,
Richard B. Angus, and James J. Hill, of St. Paul, in the State of Minnesota,
Morton, Rose & Co., of London, England, and Kohn, Reinach & Co., of Paris,
France, Witnesses:

That the parties hereto have contracted and agreed with each other as follows, namely:—

Interpretation clause.	1. For the better interpretation of this contract, it is hereby declared that the portion of railway hereinafter called the Eastern section, shall comprise that part of the Canadian Pacific Railway to be constructed, extending from
Eastern section.	the Western terminus of the Canada Central Railway, near the East end of Lake Nipissing, known as Callander Station, to a point of junction with that
Lake Superior section.	portion of the said Canadian Pacific Railway now in course of construction extending from Lake Superior to Selkirk on the East side of Red River; which latter portion is hereafter called the Lake Superior section. That the portion of said railway, now partially in course of construction, extending
Central Section.	from Selkirk to Kamloops, is hereinafter called the Central section; and the portion of said railway now in course of construction, extending from Kamloops to Port Moody, is hereinafter called the Western section. And that
C.P. Railway, Company.	the words, "the Canadian Pacific Railway", are intended to mean the entire railway, as described in the Act thirty-seventh Victoria, chapter fourteen. The individual parties hereto are hereinafter described as the Company; and the Government of Canada is hereinafter called the Government.
Security to be given by the company.	2. The contractors, immediately after the organization of the said Company, shall deposit with the Government $1,000,000 in cash or approved securities, as a security for the construction of the railway hereby contracted for. The Government shall pay to the Company interest on the cash deposited at the rate of four per cent, per annum, half-yearly, and shall pay over to the Company
Conditions thereof.	the interest received upon securities deposited—the whole until default in the performance of the conditions hereof, or until the repayment of the deposit; and shall return the deposit to the Company on the completion of the railway, according to the terms hereof, with any interest accrued thereon.
Eastern and central sections to be constructed by company described.	3. The Company shall lay out, construct and equip the said Eastern section, and the said Central section, of a uniform gauge of 4 feet 8 ½ inches; and in order to establish an approximate standard whereby the quality and the character of the railway and of the materials used in the construction thereof, and of the equipment thereof may be regulated, the Union Pacific Railway of the United States as the same was when first constructed, is hereby selected
Standard of railway and pro-vision in case of disagreement as to conformity to it.	and fixed as such standard. And if the Government and the company should be unable to agree as to whether or not any work done or materials furnished under this contract are in fair conformity with such standard, or as to any other question of fact, excluding questions of law, the subject of disagreement shall be, from time to time, referred to the determination of three referees,

263

one of whom shall be chosen by the Government, one by the Company, and one by the two referees so chosen, and such referees shall decide as to the party by whom the expense of such reference shall be defrayed. And if such two referees should be unable to agree upon a third referee, he shall be appointed at the instance of either party hereto, after notice to the other, by the Chief Justice of the Supreme Court of Canada. And the decision of such referees, or of the majority of them, shall be final.

Commencement and regular progress of the work.

4. The work of construction shall be commenced at the eastern extremity of the Eastern section not later than the first day of July next, and the work upon the Central section shall be commenced by the Company at such point towards the eastern end thereof on the portion of the line now under construction as shall be found convenient and as shall be approved by the Government, at a date not later than the 1st May next. And the work upon the Eastern and Central sections shall be vigorously and continuously carried on at such rate of annual progress on each section as shall enable the Company to complete and equip the same and each of them, in running order, on or before the first day of May, 1891, by which date the Company hereby agree to

Period for completion.

complete and equip the said sections in conformity with this contract, unless prevented by the act of God, the Queen's enemies, intestine disturbances, epidemics, floods, or other causes beyond the control of the Company. And in case of the interruption or obstruction of the work of construction from any of the said causes, the time fixed for the completion of the railway shall be extended for a corresponding period.

As to portion of central section made by Government.

5. The Company shall pay to the Government the cost, according to the contract, of the portion of railway, 100 miles in length, extending from the city of Winnipeg westward, up to the time at which the work was taken out of the hands of the contractor and the expenses since incurred by the Government in the work of construction, but shall have the right to assume the said work at any time and complete the same paying the cost of construction as aforesaid, so far as the same shall the have been incurred by the Government.

Government to construct portions now under contract within periods fixed by contract.

6. Unless prevented by the act of God, the Queen's enemies, intestine disturbances, epidemics, floods or other causes beyond the control of the Government, the Government shall cause to be completed the said Lake Superior section by the dates fixed by the existing contracts for the construction thereof; and shall also cause to be completed the portion of the said Western section now under contract, namely, from Kamloops to Yale, within the period fixed by the contracts therefor, namely, by the thirtieth day of June, 1885; and shall also cause to be completed, on or before the first day of May, 1891, the remaining portion of the said Western section, lying between Yale and Port Moody, which shall be constructed of equally good quality in every respect with the standard hereby created for the portion hereby contracted for. And the said Lake Superior section, and the portions of the said Western section now under contract, shall be completed as nearly as practicable according to the specifications and conditions of the contracts therefor, except in so far as the same have been modified by the Government prior to this contract.

Completed railway to be property of company.

7. The railway constructed under the terms hereof shall be the property of the Company: and pending the completion of the Eastern and central sections, the Government shall transfer to the Company the possession and right to work and run the several portions of the Canadian Pacific Railway already constructed or as the same shall be completed. And upon

Transfer of portions constructed by the Government.

the completion of the Eastern and Central sections, the Government shall convey to the Company with a suitable number of station buildings and with water service (but without equipment), those portions of the Canadian Pacific Railway constructed or to be constructed by the Government which shall then be completed; and upon completion of the remainder of the portion of railway to be constructed by the Government, that portion shall also be conveyed to the Company; and the Canadian Pacific Railway shall become and be thereafter the absolute property of the Company. And the Company shall thereafter and for ever efficiently maintain, work and run the Canadian Pacific Railway.

Company to operate the railway forever.

8. Upon the reception from the Government of the possession of each of the respective portions of the Canadian Pacific Railway, the Company shall equip the same in conformity with the standard herein established for the equipment of the sections hereby contracted for, and shall thereafter maintain and efficiently operate the same.

Company to equip portions transferred to them.

9. In consideration of the premises, the Government agree to grant to the Company a subsidy in money of $25,000,000 and in land of 25,000,000 acres, for which subsidies the construction of the Canadian Pacific Railway shall be completed and the same shall be equipped, maintained and operated—the said subsidies respectively to be paid and granted as the work of construction shall proceed, in manner and upon the conditions following, that is to say:—

Subsidy in money and land.

(a) The said subsidy in money is hereby divided and appropriated as follows, namely:—

Apportionment of money.

CENTRAL SECTION
Assumed at 1,350 miles:—

1st—900 miles at $10,000 per mile	$9,000,000
2nd—450 miles at 13,333 per mile	6,000,000
	15,000,000

EASTERN SECTION
Assumed at 650 miles, subsidy equal to

$15,384.61 per mile	10,000,000
	$25,000,000

And the said subsidy in land is hereby divided and appropriated as follows, subject to the reserve hereinafter provided for:—

And of land.

CENTRAL SECTION

1st—900 miles at 12,500 acres per mile	$11,250,000
2nd—450 miles at 16,666.66 per mile	7,500,000
	18,750,000

EASTERN SECTION
Assumed at 650 miles, subsidy equal to

9,615.35 acres per mile	6,250,000
	25,000,000

(b) Upon the construction of any portion of the railway hereby contracted for, not less than 20 miles in length, and the completion thereof so as to admit of the running of regular trains thereon, together with such equipment thereof as shall be required for the traffic thereon, the Government shall pay and grant to the Company the money and land subsidies applicable thereto, according to the division and appropriation thereof made as hereinbefore

When to be paid or granted.

Option of company to take terminable bonds.

provided; the Company having the option of receiving in lieu of cash terminable bonds of the Government bearing such rate of interest, for such period and nominal amount as may be arranged, and which may be equivalent according to actuarial calculation to the corresponding cash payment—the government allowing four percent interest on moneys deposited with them.

Provision as to materials for construction delivered by company in advance.

(c) If at any time the Company shall cause to be delivered on or near the line of the said railway, at a place satisfactory to the Government, steel rails and fastenings to be used in the requirements for such construction, the Government, on the requisition of the Company, shall, upon such terms and conditions as shall be determined by the Government, advance thereon three-fourths of the value thereof at the place of delivery. And a proportion of the amount so advanced shall be deducted, according to such terms and conditions, from the subsidy to be thereafter paid, upon the settlement for each section of 20 miles of railway—which proportion shall correspond with the proportion of such rails and fastenings which have been used in the construction of such sections.

Option of the company during a certain time to substitute payment of interest on certain bonds instead of issuing land grant bonds.

(d) Until the first day of January, 1882, the Company shall have the option, instead of issuing land grant bonds as hereinafter provided, of substituting the payment by the Government of the interest (or part of the interest) on bonds of the Company mortgaging the railway and the lands to be granted by the Government, running over such term of years as may be approved by the Governor in Council, in lieu of the cash subsidy hereby agreed to be granted to the Company or any part hereof—such payments of interest to be equivalent according to actuarial calculation to the corresponding cash payment, the Government allowing four per cent interest on moneys deposited with them; and the coupons representing the interest on such bonds shall be guaranteed by the Government to the extent of such equivalent. And the proceeds of the sale of such bonds to the extent of not more than $25,000,000 shall be deposited with the Government, and

Deposit of proceeds of sale of such bonds.

the balance of such proceeds shall be placed elsewhere by the Company, to the satisfaction and under the exclusive control of the Government; failing which last condition the bonds in excess of those sold shall remain in the hands of the Government. And from time to time as the work proceeds, the Government shall pay over to the Company: firstly, out of the amount so to be placed—and, after the expenditure of that amount, out of the

Payments to company out of such deposits.

amount deposited with the Government—sums of money bearing the same proportion to the mileage cash subsidy hereby agreed upon, which the net proceeds of such sale (if the whole of such bonds are sold upon the issue thereof, or, if such bonds be not all then sold, the net proceeds of the issue, calculated at the rate at which the sale of part of them shall have been made), shall bear to the sum of $25,000,000. But if only a portion of the bond issue be sold, the amount earned by the Company according to the

Payment by delivery of bonds.

proportion aforesaid shall be paid to the Company, partly out of the bonds in the hands of the Government, and partly out of the cash deposited with the Government, in similar proportions to the amount of such bonds sold and remaining unsold respectively; and the Company shall receive the bonds so paid, as cash, at the rate at which the said partial sale thereof shall have been made. And the Government will receive and hold such sum of

Sinking fund.

money towards the creation of a sinking fund for the redemption of such bonds, and upon such terms and conditions as shall be agreed upon between the Government and the Company.

Alteration in apportionment of money grant in such case.

(e) If the Company avail themselves of the option granted by clause 'd', the sum of $2,000 per mile for the first eight hundred miles of the Central section shall be deducted pro rata from the amount payable to the Company in respect of the said eight hundred miles, and shall be appropriated to increase the mileage cash subsidy appropriated to the remainder of the said Central section.

Grant of land required for railway purposes.

10. In further consideration of the premises, the Government shall also grant to the Company the lands required for the road-bed of the railway, and for its stations, station grounds, workshops, dock ground and water frontage at the termini on navigable waters, buildings, yards and other appurtenances required for the convenient and effectual construction and working of the railway, in so far as such land shall be vested in the Government. And the Government

Admission of certain materials free of duty.

shall also permit the admission free of duty of all steel rails, fish plates and other fastenings, spikes, bolts and nuts, wire, timber and all material for bridges to be used in the original construction of the railway, and of a telegraph line in connection therewith, and all telegraphic apparatus required for the first equipment of such telegraph line; and will convey to the Company, at cost price,

Sale of certain materials to company by Government

with interest, all rails and fastenings bought in or since the year 1879, and other materials for construction in the possession of or purchased by the Government, at a valuation—such rails, fastenings and materials not being required by it for the construction of the said Lake Superior and Western sections.

Provision respecting land grant.

11. The grant of land, hereby agreed to be made to the Company, shall be so made in alternate sections of 640 acres each, extending back 24 miles deep, on each side of the railway, from Winnipeg to Jasper House, in so far as such lands shall be vested in the Government—the Company receiving the sections bearing uneven numbers. But should any of such sections consist in a material degree of land not fairly fit for settlement, the Company shall not be

Case of deficiency of land on line of railway provided for.

obliged to receive them as part of such grant; and the deficiency thereby caused and any further deficiency which may arise from the insufficient quantity of land along the said portion of railway, to complete the said 25,000,000 acres, or from the prevalence of lakes and water stretches in the sections granted (which lakes and water stretches shall not be computed in the acreage of such sections), shall be made up from other portions in the tract known as the fertile belt, that is to say, the land lying between parallels 49 and 57 degrees of north latitude, or elsewhere at the option of the Company, by the grant therein of similar alternate sections extending back 24 miles deep on each side of any branch line or lines of railway to be located by the Company, and to be shown on a map or plan thereof deposited with the Minister of

Selection by company in such case, with consent of Government.

Railways; or of any common front line or lines agreed upon between the Government and the Company—the conditions hereinbefore stated as to lands not fairly fit for settlement to be applicable to such additional grants. And the Company may, with the consent of the Government, select in the NorthWest Territories any tract or tracts of land not taken up as a means of supplying or partially supplying such deficiency. But such grants shall be made only from lands remaining vested in the Government.

12. The Government shall extinguish the Indian title affecting the lands herein appropriated, and to be hereafter granted in aid of the railway.

Location of the railway between certain terminal points.

13. The Company shall have the right, subject to the approval of the Governor in Council, to lay out and locate the line of the railway hereby contracted for, as they may see fit, preserving the following terminal points, namely: from Callander station to the point of junction with the Lake Superior section; and from Selkirk to the junction with the Western section at Kamloops by way of the Yellow Head Pass.

*Power to con-
struct branches.*

14. The Company shall have the right from time to time to lay out, construct, equip, maintain and work branch lines of railway from any point or points along their main line of railway, to any point or points within the territory of the Dominion. Provided always, that before commencing any branch they shall first deposit a map and plan of such branch in the Department of Railways. And the Government shall grant to the Company the lands required for the road-bed of such branches and for the stations, station grounds, buildings, workshops, yards and other appurtenances requisite for the efficient construction and working of such branches, in so far as such lands are vested in the Government.

*Lands necessary
for the same.*

*Restrictions as
to competing
lines for a limited
period.*

15. For twenty years from the date hereof, no line of railway shall be authorized by the Dominion Parliament to be constructed South of the Canadian Pacific Railway, from any point at or near the Canadian Pacific Railway, except such line as shall run South West or to the Westward of South West; nor to within fifteen miles of Latitude 49. And in the establishment of any new Province in the North-West Territories, provision shall be made for continuing such prohibition after such establishment until the expiration of the same period.

*Exemption from
taxation in N.W.
Territories.*

16. The Canadian Pacific Railway, and all stations and station grounds, workshops, buildings, yards and other property, rolling stock and appurtenances required and used for the construction and working thereof, and the capital stock of the Company, shall be forever free from taxation by the Dominion, or by any Province hereafter to be established, or by any Municipal Corporation therein; and the lands of the Company, in the North-West Territories, until they are either sold or occupied, shall also be free from such taxation for 20 years after the grant thereof from the Crown.

Land grant.

*Their nature
and conditions
of issue by the
company.*

*Deposit with
Government: for
what purposes
and on what
conditions.*

17. The Company shall be authorized by their act of incorporation to issue bonds, secured upon the land granted and to be granted to the Company, containing provisions for the use of such bonds in the acquisition of lands, and such other conditions as the Company shall see fit—such issue to be for $25,000,000. And should the Company make such issue of land grant bonds, then they shall deposit them in the hands of the Government; and the Government shall retain and hold one-fifth of such bonds as security for the due performance of the present contract in respect of the maintenance and continuous working of the railway by the Company, as herein agreed, for ten years after the completion thereof, and the remaining $20,000,000 of such bonds shall be dealt with as hereinafter provided. And as to the said one-fifth of the said bonds, so long as no default shall occur in the maintenance and working of the said Canadian Pacific Railway, the Government shall not present or demand payment of the coupons of such bonds, nor require payment of any interest thereon. And if any of such bonds, so to be retained by the Government, shall be paid off in the manner to be provided for the extinction of the whole issue thereof, the Government shall hold the amount received in payment thereof as security for the same purposes as the bonds so paid off, paying interest thereon at four per cent per annum so long as default is not made by the Company in the performance of the conditions hereof. And at the end of the said period of ten years from the completion of the said railway, if no default shall then have occurred in such maintenance and working thereof, the said bonds, or if any of them shall then have been paid off, the remainder of said bonds and the money received from those paid off, with accrued interest, shall be delivered back by the Government to the Company with all the coupons attached to

*If the company
make no default
in operating
railway.*

In case of such default.

such bonds. But if such default should occur, the Government may thereafter require payment of interest on the bonds so held and shall not be obliged to continue to pay interest on the money representing bonds paid off; and while the Government shall retain the right to hold the said portion of the said land grant bonds, other securities satisfactory to the Government may be substituted for them by the Company, by agreement with the Government.

Provision if such bonds are sold faster than lands are earned by the Company, and deposit on interest with Government, and payments by Government to Company.

18. If the Company shall find it necessary or expedient to sell the remaining $20,000,000 of the land grant bonds or a larger portion thereof than in the proportion of one dollar for each acre of land then earned by the Company, they shall be allowed to do so, but the proceeds thereof, over and above the amount to which the Company shall be entitled as herein provided, shall be deposited with the Government. And the Government shall pay interest upon such deposit half-yearly, at the rate of four per cent per annum, and shall pay over the amount of such deposit to the Company from time to time, as the work proceeds, in the same proportions, and at the same times and upon the same conditions as the land grant—that is to say: the Company shall be entitled to receive from the Government out of the proceeds of the said land grant bonds, the same number of dollars as the number of acres of the land subsidy which shall then have been earned by them, less one-fifth thereof, that is to say, if the bonds are sold at par, but

Lands to be granted subject to such bonds.

if they are sold at less than par, then a deduction shall be made therefrom corresponding to the discount at which such bonds are sold. And such land grant shall be conveyed to them by the Government, subject to the charge created as security for the said land grant bonds, and shall remain subject to such charge till relieved thereof in such manner as shall be provided for at the time of the issue of such bonds.

Company to pay certain expenses.

19. The Company shall pay any expenses which shall be incurred by the Government in carrying out the provisions of the last two preceding clauses of this contract.

If land bonds are not issued, one-fifth of land to be retained as security.

20. If the Company should not issue such land grant bonds, then the Government shall retain from out of each grant to be made from time to time, every fifth section of the lands hereby agreed to be granted, such lands to be so retained as security for the purposes and for the length of time, mentioned in section eighteen hereof. And such lands may be sold in such manner and at such prices as shall be agreed upon between

How to be disposed of.

the Government and the company; and in that case the price thereof shall be paid to, and held by the Government for the same period, and for the same purposes as the land itself, the Government paying four per cent per annum interest thereon. And other securities satisfactory to the Government may be substituted for such lands or money by agreement with the Government.

Company to be incorporated as by Schedule A.

21. The Company to be incorporated, with sufficient powers to enable them to carry out the foregoing contract, and this contract shall only be binding in the event of an Act of incorporation being granted to the Company in the form hereto appended as Schedule A.

Railway Act to apply.

22. The Railway Act of 1879, in so far as the provisions of the same are applicable to the undertaking referred to in this contract, and in so far as they are not inconsistent herewith or inconsistent with or contrary to the

Exceptions

provisions of the Act of incorporation to be granted to the Company, shall apply to the Canadian Pacific Railway.

In witness whereof the parties hereto have executed these presents at the City of Ottawa, this twenty-first day of October, 1880.

(Signed)   CHARLES TUPPER,
Minister of Railways and Canals.
GEO. STEPHEN
DUNCAN MCINTYRE
J. S. KENNEDY,
R. B. ANGUS,
J. J. HILL,

Per pro. Geo. Stephen.
MORTON, ROSE & CO.
KOHN, REINACH & CO.

By P. Du P. Grenfell.

Signed in presence of F. Braun,
and seal of the Department
hereto affixed by SIR CHARLES
TUPPER, in presence of

(Signed) F. BRAUN.

SCHEDULE A, REFERRED TO IN THE FOREGOING CONTRACT

INCORPORATION

Certain persons incorporated.

1. George Stephen, of Montreal, in Canada, Esquire; Duncan McIntyre, of Montreal, aforesaid, Merchant; John S. Kennedy, of New York, in the State of New York, Banker; the firm of Morton, Rose and Company, of London, in England, Merchants; the firm of Kohn, Reinach and Company, of Paris, in France, Bankers; Richard B. Angus, and James J. Hill, both of St. Paul, in the State of Minnesota, Esquires; with all such other persons and corporations as shall become shareholders in the company hereby incorporated, shall be and they are hereby constituted a body corporate and politic, by the name of the "Canadian Pacific Railway Company."

Corporate name.

Capital stock and shares.

Paid up shares.

2. The capital stock of the Company shall be twenty-five million dollars, divided into shares of one hundred dollars each—which shares shall be transferable in such manner and upon such conditions as shall be provided by the by-laws of the Company: and such shares, or any part thereof may be granted and issued as paid-up shares for value bona fide received by the Company, either in money at par or at such price and upon such conditions as the Board of Directors may fix, or as part of the consideration of any contract made by the Company.

Substitution of Company as contractors; and when.

Effect of such substitution

3. As soon as five million dollars of the stock of the Company have been subscribed, and thirty per centum thereof, paid up, and upon the deposit with the Minister of Finance of the Dominion of one million dollars in money or in securities approved by the Governor in Council, for the purpose and upon the conditions in the foregoing contract provided the said contract shall become and be transferred to the Company, without the execution of any deed or instrument in that behalf; and the Company shall, thereupon, become and be vested with all the right of the contractors named in the said contract, and shall be subject to, and liable for, all their duties and

obligations, to the same extent and in the same manner as if the said contract had been executed by the said Company instead of by the said contractors; and thereupon the said contractors, as individuals, shall cease to have any right or interest in the said contract, and shall not be subject to any liability or responsibility under the terms thereof otherwise than as members of the corporation hereby created. And upon the performance of the said conditions respecting the subscription of stock, the partial payment thereof, and the deposit of one million dollars to the satisfaction of the Governor in council, the publication by the Secretary of State in the *Canada Gazette*, of a notice that the transfer of the contract to the Company has been effected and completed shall be conclusive proof of the fact. And the Company shall cause to be paid up, on or before the first day of May next, a further installment of twenty per centum upon the said first subscription of five million dollars, of which call thirty days' notice by circular mailed to each shareholder shall be sufficient. And the Company shall call in, and cause to be paid up, on or before the 31st day of December, 1882, the remainder of the said first subscription of five million dollars.

Notice in the Canada Gazette

Further install-ment to be paid up.

And rest of $5,000,000.

4. All the franchises and powers necessary or useful to the Company to enable them to carry out, perform, enforce, use and avail themselves of, every condition, stipulation, obligation, duty, right, remedy, privilege, and advantage agreed upon, contained or described in the said contract are hereby conferred upon the Company. And the enactment of the special provisions hereinafter contained shall not be held to impair or derogate from the generality of the franchises and powers so hereby conferred upon them.

Necessary fran-chises and powers granted.

Proviso.

DIRECTORS

5. The said George Stephen, Duncan McIntyre, John S. Kennedy, Richard B. Angus, James J. Hill, Henry Stafford Northcote, of London, aforesaid, Esquires, Pascoe Du P. Grenfell, of London, aforesaid, Merchant, Charles Day Rose of London, aforesaid, Merchant, and Byron J. Reinach, of Paris, aforesaid, Bankers are hereby constituted the first directors of the Company, with power to add to their number, but so that the directors shall not in all exceed fifteen in number; and the majority of the directors, of whom the President shall be one, shall be British subjects. And the Board of Directors so constituted shall have all the powers hereby conferred upon the directors of the Company, and they shall hold office until the first annual meeting of the shareholders of the Company.

First directors of the Company.

Number limited.

6. Each of the directors of the Company, hereby appointed, or hereafter appointed or elected, shall hold at least two hundred and fifty shares of the stock of the Company. But the number of directors to be hereafter elected by the shareholders shall be such, not exceeding fifteen, as shall be fixed by by-law, and subject to the same conditions as the directors appointed by, or under the authority of the last preceding section; the number thereof may be hereafter altered from time to time in like manner. The votes for their election shall be by ballot.

Qualification of directors.

Alteration of num-ber by by-law.

Ballot.

7. A majority of the directors shall form a quorum of the board; and until otherwise provided by by-law, directors may vote and act by proxy—such proxy to be held by a director only; but no director shall hold more than two proxies, and no meeting of directors shall be competent to transact business unless at least three directors are present thereat in person, the remaining number of directors required to form a quorum being represented by proxies.

Quorum.

Proviso.

Three must be present.

Executive com-
mittee.

President to be
one.

Chief place of
business.

Other places.

How to be
notified.

Service of process
thereat.

And if Company
fail to appoint
places.

8. The Board of Directors may appoint, from out of their number, an executive Committee, composed of at least three directors, for the transaction of the ordinary business of the Company, with such powers and duties as shall be fixed by the by-laws; and the President shall be ex officio a member of such committee.

9. The chief place of business of the Company shall be at the City of Montreal, but the Company may, from time to time, by by-law, appoint and fix other places within or beyond the limits of Canada at which the business of the company may be transacted, and at which the directors or shareholders may meet, when called as shall be determined by the by-laws. And the company shall appoint and fix by by-law, at least one place in each Province or Territory through which the railway shall pass, where service of process may be made upon the Company, in respect of any cause of action arising within such Province or Territory, and may afterwards, from time to time, change such place by by-law. And a copy of any by-law fixing or changing any such place, duly authenticated as herein provided, shall be deposited by the Company in the office, at the seat of Government of the Province or Territory to which such by-law shall apply, of the clerk or prothonotary of the highest, or one of the highest, courts of civil jurisdiction of such Province or Territory. And if any cause of action shall arise against the Company within any Province or Territory, and any writ or process be issued against the Company thereon out of any court in such Province or Territory, service of such process may be validly made upon the Company at the place within such Province or Territory so appointed and fixed; but if the company fail to appoint and fix such place, or to deposit, as hereinbefore provided, the by-law made in that behalf, any such process may be validly served upon the Company, at any of the stations of the said railway within such Province or Territory.

SHAREHOLDERS

First and other
annual meetings.

Notice.

Special general
meetings; notice.

Place.

Provision if a
meeting is neces-
sary before notice
as aforesaid can
be given.

10. The first annual meeting of the shareholders of the Company, for the appointment of directors, shall be held on the second Wednesday in May, one thousand eight hundred and eighty two, at the principal office of the Company in Montreal; and the annual general meeting of shareholders, for the election of directors and the transaction of business generally, shall be held on the same day in each year thereafter at the same place unless otherwise provided by the by laws. And notice of each of such meetings shall be given by the publication thereof in the Canada Gazette for four weeks, and by such further means as shall, from time to time, be directed by the laws.

11. Special general meetings of the shareholders may be convened in such manner as shall be provided by the by-laws and except as hereinafter provided, notice of such meetings shall be given in the same manner as notices of annual general meetings, the purpose for which such meeting is called being mentioned in then notices thereof; and except as hereinafter provided, all such meetings shall be held at the chief place of business of the Company.

12. If at any time before the first annual meeting of the shareholders of the Company, it should become expedient that a meeting of the directors of the Company, or a special general meeting of the shareholders of the Company, should be held, before such meeting can be called, before such meeting can conveniently be called, and notice thereof given in the manner provided by this Act, or by the by-laws, or before by-laws in that behalf have been passed, and at a place other than at the chief place of business of the Company in

Montreal before the enactment of a by-law authorizing the holding of such meeting elsewhere; it shall be lawful for the President or for any three of the directors of the Company to call special meetings either of directors or

Notice in such case.

of shareholders, or of both, to be held at the City of London, in England, at times and places respectively, to be stated in the notices to be given of such meetings respectively. And notices of such meetings may be validly given by a circular mailed to the ordinary address of each director or shareholder

Meetings always valid if all shareholders or their proxies are present.

as the case may be, in time to enable him to attend such meeting, stating in general terms the purpose of the intended meeting. And in the case of a meeting of shareholders, the proceedings of such meeting shall be held to be valid and sufficient, and to be binding on the Company in all respects, if every shareholder of the Company be present thereat in person or by proxy, notwithstanding that notice of such meeting shall not have been given in the manner required by this Act.

Limitations as to votes and proxies.

13. No shareholder holding shares upon which any call is overdue and unpaid shall vote at any meeting of shareholders. And unless otherwise provided by the by-laws, the person holding the proxy of a shareholder shall be himself a shareholder.

And as to calls.

14. No call upon unpaid shares shall be made for more than twenty per centum upon the amount thereof.

RAILWAY AND TELEGRAPH LINE

Line and gauge of railway.

15. The Company may lay out, construct, acquire, equip, maintain and work a continuous line of railway, of the gauge of four feet eight and one-half inches; which railway shall extend from the terminus of the Canada Central Railway near Lake Nipissing, known as Callander Station, to Port Moody in the Province of British Columbia; and also a branch line of railway from some point on the main line of railway to Fort William on Thunder Bay; and also the existing branch line of railway from some point on the main

Commencement and completion.

line of railway to Fort William on Thunder Bay; and also the existing branch line of railway from Selkirk, in the Province of Manitoba, to Pembina in the said Province; and also other branches to be located by the Company from

Other branches.

time to time as provided by the said contract—the said branches to be of the gauge aforesaid; and the said main line of railway, and the said branch lines of railway, shall be commenced and completed as provided by the said contract; and together with such other branch lines as shall be hereafter constructed by the said Company, and any extension of the said main line of railway that shall hereafter be constructed or acquired by the Company, shall constitute the line

Name of Railway.

of railway hereinafter called THE CANADIAN PACIFIC RAILWAY.

16. The Company may construct, maintain and work a continuous telegraph line and telephone lines throughout and along the whole line of the Canadian

Company may construct lines of telegraph or telephone, and work them and collect tolls.

Pacific Railway, or any part thereof, and may also construct or acquire by purchase, lease or otherwise, any line or lines of telegraph connecting with the line so to be constructed along the line of the said railway, and may undertake the transmission of messages for the public by any such line or lines of telegraph or telephone, and collect tolls for so doing; or may lease such line or lines of telegraph or telephone, or any portion thereof; and if they think proper to undertake the transmission of messages for hire, they

Subject to Con. Stat. Can., c. 67, ss. 14, 15, 16.

shall be subject to the provisions of the fourteenth, fifteenth and sixteenth sections of chapter sixty-seven of the Consolidated Statutes of Canada. And they may use any improvement that may hereafter be invented (subject to the rights of patentees) for telegraphing or telephoning, and any other means of communication that may be deemed expedient by the Company at any time hereafter.

POWERS

Application of 42V., c. 9.

17. "*The Consolidated Railway act, 1879*", in so far as the provisions of the same are applicable to the undertaking authorized by this charger, and in so far as they are not inconsistent with or contrary to the provisions hereof, and save and except as hereinafter provided, is hereby incorporated herewith.

Exceptions as to such application.

18. As respects the said railway, the seventh section of "*The Consolidated Railway Act, 1879*", relating to POWERS, and the eighth section thereof relating to PLANS AND SURVEYS, shall be subject to the following provisions:—

As to lands of the Crown required.

a. The Company shall have the right to take, use and hold the beach and land below high-water mark, in any stream, lake, navigable water, gulf or sea, in so far as the same shall be vested in the Crown and shall not be required by the Crown, to such extent as shall be required by the Company for its railway and other works, and as shall be exhibited by a map or plan thereof deposited in the office of the Minister of Railways. But the provisions of this sub-section shall not apply to any beach or land lying East of Lake Nipissing except with the approval of the Governor in Council.

Plans and book of reference.

b. It shall be sufficient that the map or plan and book of reference for any portion of the line of the railway not being within any district or county for which there is a Clerk of the Peace, be deposited in the office of the Minister of Railways of Canada; and any omission, mis-statement or erroneous description of any lands therein may be corrected by the Company, with the consent of the Minister and certified by him; and the Company may then make the railway in accordance with such certified correction.

Deviations from line on plan.

c. The eleventh sub-section of the said eighth section of the Railway Act shall not apply to any portion of the railway passing over ungranted lands of the crown, or lands not within any surveyed township in any Province; and in such places, deviations not exceeding five miles from the line shown on the map or plan as aforesaid, deposited by the Company, shall be allowed, without any formal correction or certificate; and any further deviation that may be found expedient may be authorized by order of the Governor in

Deposit of plan of main line, etc.

Council and the Company may then make their railway in accordance with such authorized deviation.

d. The map or plan and book of reference of any part of the main line of the Canadian Pacific Railway made and deposited in accordance with this

And of branches.

section, after approval by the Governor in Council, and of any branch of such railway hereafter to be located by the said Company in respect of which the approval of the Governor in Council shall not be necessary, shall avail as if made and deposited as required by the said "*Consolidated Railway Act,*

Copies thereof.

1879", for all the purposes of the said act, and of this Act; and any copy of, or extract therefrom, certified by the said Minister or his deputy, shall be received as evidence in any court of law in Canada.

e. It shall be sufficient that a map or profile of any part of the completed

Registration thereof.

railway, which shall not lie within any county or district having a registry office, be filed in the office of the Minister of Railways.

Company may take materials from public lands and a greater extent for stations, and, etc., than allowed by 42 V., c.9.

19. It shall be lawful for the Company to take from any public lands adjacent to or near the line of the said railway, all stone, timber, gravel and other materials which may be necessary or useful for the construction of the railway; and also to lay out and appropriate to the use of the Company, a greater extent of lands, whether public or private, for stations, depots, workshops, buildings, sidetracks, wharves, harbours and roadway, and for establishing screens against snow, than the breadth and quantity mentioned in *"The Consolidated Railway Act, 1879,"*—such greater extent taken, in any case being allowed by the Government, and shown on the maps or plans deposited with the Minister of Railways.

Proviso.

20. The limit to the reduction of tolls by the Parliament of Canada provided for by the eleventh sub-section of the 17th section of *"The Consolidated Railway Act, 1879,"* respecting TOLLS, is hereby extended, so that such reduction may be to such an extent that such tolls when reduced shall not produce less than ten per cent. per annum profit on the capital actually expended in the construction of the railway, instead of not less than fifteen percent. per annum profit, as provided by the said sub-section; and so also that such reduction shall not be made unless the net income of the Company, ascertained as described in said sub-section, shall have exceeded ten per cent. per annum instead of fifteen percent. per annum as provided

Reduction by Governor in Council extended in like manner.

by the said sub-section. And the exercise by the Governor in Council of the power of reducing the tolls of the Company as provided by the tenth sub-section of said section seventeen is hereby limited to the same extent with relation to the profit of the Company, and to its net revenue, as that to which the power of Parliament to reduce tolls is limited by said sub-section eleven as hereby amended.

Restriction as to transfers of stock.

21. The first and second sub-sections of section 22 of *"The Consolidated Railway Act, 1879,"* shall not apply to the Canadian Pacific Railway Company; and it is hereby enacted that the transfer of shares in the undertaking shall be made only upon the books of the Company in person or by an attorney, and shall not be valid unless so made; and the form and mode of transfer shall be such as shall be, from time to time, regulated by the by-laws of the Company. And

Advances on, by Company forbidden.

the funds of the company shall not be used in any advance upon the security of any of the shares or stock of the Company.

Transfer or transmission to non-shareholders subject to vote of directors until completion of contract.

22. The third and fourth sub-sections of *"The Consolidated Railway Act, 1879,"* shall be subject to the following provisions, namely—that if before the completion of the railway and works under the said contract, any transfer should purport to be made of any stock or share in the Company, or any transmission of any share should be effected under the provisions of said sub-section four, to a person not already a shareholder in the company, and if in the opinion of the board it should not be expedient that the person (not being already a shareholder) to whom such transfer or transmission shall be made or effected should be accepted as a shareholder, the directors may by resolution veto such transfer or transmission; and thereafter, and until after the completion of the said railway and works under the said contract, such person shall not be, or be recognized as a shareholder in the Company; and the original shareholder, or his estate, as the case may be, shall remain

Proviso: as to transfer by a firm to a partner.

subject to all the obligations of a shareholder in the Company, with all the rights conferred upon a shareholder under this Act. But any firm holding paid-up shares in the Company may transfer the whole or any of such shares to any partner in such firm having already an interest as such partner in such shares, without being subject to such veto. And in the event of such veto being exercised, a note shall be taken of the transfer or transmission so vetoed in order that it may be recorded in the books of the Company after the completion of the railway and works as aforesaid; but until such completion, the transfer or transmission so vetoed shall not confer any rights, nor have any effect of any nature or kind whatever as respects the Company.

Certain other provisions of 42 V., c.9, not to apply.

23. Sub-section sixteen of section nineteen, relating to PRESIDENT AND DIRECTORS, THEIR ELECTION AND DUTIES; sub-section two of section twenty-four relating to by-laws, NOTICES, etc., sub-sections five and six of section twenty-eight, relating to GENERAL PROVISIONS, and section ninety-seven, relating to RAILWAY FUND, of "*The Consolidated Railway Act, 1879*", shall not, nor shall any of them apply to the Canadian Pacific Railway or to the Company hereby incorporated.

Company to afford reasonable facilities to and receive the like from certain other railway companies.

24. The said Company shall afford all reasonable facilities to the Ontario and Pacific Junction Railway Company, when their railway shall be completed to a point of junction with the Canadian Pacific Railway, and to the Canada Central Railway Company, for the receiving, forwarding and delivering of traffic upon and from the railways of the said Companies, respectively, and for the return of carriages, trucks and other vehicles; and no one of the said Companies shall give or continue any preference or advantage to, or in favour of either of the others, or of any particular description of traffic, in any respect whatsoever; nor shall any one of the said Companies subject any other thereof, or any particular description of traffic, to any prejudice or disadvantage in any respect whatsoever; and any one of the said Companies which shall have any terminus or station near any terminus or station of either of the others, shall afford all reasonable facilities for receiving and forwarding all the traffic arriving by either of the others, without any unreasonable delay, and without any preference or advantage, or prejudice or disadvantage, and so that no obstruction may be offered in the using of such railway as a continuous line of communication,

As to rates of carriage of traffic in such cases.

and so that all reasonable accommodation may, at all times, by the means aforesaid, be mutually afforded by and to the said several railway companies; and the said Canadian Pacific Railway Company shall receive and carry all freight and passenger traffic shipped to or from any point on the railway of either of the said above-named railway companies passing over the Canadian Pacific Railway or any part thereof, at the same mileage rate and subject to the same charges for similar services, without granting or allowing

Reservation as to purchasers of land, and immigrants.

any preference or advantage to the traffic coming from or going upon one of such railways over such traffic coming from or going upon the other of them, reserving, however, to the said Canadian Pacific Railway Company the right of making special rates for purchasers of land or for immigrants or intending immigrants, which special rate shall not govern or affect the rates of passenger traffic as between the said Company and the said two above-named Companies or either of them. And any agreement made between

Contrary agreements void.

any two of the said Companies contrary to the foregoing provisions, shall be unlawful, null and void.

Company may purchase or acquire by lease or otherwise certain other railways or amalgamate with them.

25. The Company, under the authority of a special general meeting of the shareholders thereof, and as an extension of the railway hereby authorized to be constructed, may purchase or acquire by lease or otherwise, and hold and operate, the Canada Central Railway, or may amalgamate therewith, and may purchase or acquire by lease or otherwise and hold and operate a line or lines of railway from the City of Ottawa to any point at navigable water on the Atlantic seaboard or to any intermediate point, or may acquire running

powers over all railway now constructed between Ottawa and any such point or intermediate point. And the Company may purchase or acquire any such railway, subject to such existing mortgages, charges or liens thereon as shall be agreed upon, and shall possess with regard to any lines or railway so purchased, or acquired, and becoming the property of the Company, the same powers as to the issue of bonds thereon, or on any of them, to an amount not exceeding twenty thousand dollars per mile, and as to the security for such bonds, as are conferred upon the Company by the twenty-eighth section hereof, in respect of bonds to be issued upon the Canadian Pacific Railway. But such issue of bonds shall not affect the right of any holder of mortgages or other charges already existing upon any line of railway so purchased or acquired; and the amount of bonds hereby authorized to be issued upon such line of railway shall be diminished by the amount of such existing mortgages or charges thereon.

And borrow to a limited amount on bonds in consequence.

Not to affect prior mortgages.

26. The Company shall have power and authority to erect and maintain docks, dockyards, wharves, slips and piers at any point on or in connection with the said Canadian Pacific Railway, and at all the termini thereof on navigable water, for the convenience and accommodation of vessels and elevators; and also to acquire and work elevators, and to acquire, own, hold, charter, work and run steam and other vessels for cargo and passengers upon any navigable water, which the Canadian Pacific Railway may reach or connect with.

Company may have docks, etc., and run vessels on any navigable water their railway touches.

BY-LAWS

By laws may provide for certain purposes.

27. The by-laws of the Company may provide for the remuneration of the president and directors of the Company, and of any executive committee of such directors; and for the transfer of stock and shares; the registration and inscription of stock, shares and bonds, and the transfer of registered bonds; and the payment of dividends and interest at any place or places within or beyond the limits of Canada; and for all other matters required by the said contract or by this Act to be regulated by by-laws: but the by-laws, unless they are approved by such meeting.

Must be confirmed at next general meeting.

BONDS

Amount of bonds limited.

28. The Company, under the authority of a special general meeting of the shareholders called for the purpose, may issue mortgage bonds to the extent of ten thousand dollars per mile of the Canadian Pacific Railway for the purposes of the undertaking authorized by the present Act; which issue shall constitute a first mortgage and privilege upon the said railway, constructed or acquired, and upon its property, real and personal, acquired and to be thereafter acquired, including rolling stock and plant and upon its tolls and revenues (after deduction from such tolls and revenues of working expenses), and upon the franchises of the Company; the whole as shall be declared and described as so mortgaged in any deed of mortgage as hereinafter provided. Provided always, however, that if the Company shall have issued or shall intend to issue land grant bonds under the provisions of the thirtieth section thereof, the lands granted and to be granted by the Government to the Company may be excluded from the operation of such mortgage and privilege: and provided also that such mortgage and privilege shall not attach upon any

Mortgages for securing the same on all the property of the company.

Proviso: in case land grant bonds have been issued under sec. 30.

property which the Company are hereby, or by the said contract, authorized to acquire or receive from the Government of Canada until the same shall have been conveyed by the Government to the company, but shall attach upon such property, if so declared in such deed, as soon as the same shall be conveyed to the Company. And such mortgage and privilege may be evidenced by a deed or deeds of mortgage executed by the Company, with the authority of its shareholders expressed by a resolution passed at such general meeting; and any such deed may contain such description of the property mortgaged by such deed, and such conditions respecting the payment of the bonds secured thereby and of the interest thereon, and the remedies which shall be enjoyed by the holders of such bonds or by any trustee or trustees for them in default of such payment, and the enforcement of such remedies, and may provide for such forfeitures and penalties, in default of such payment, as may be approved by such meeting; and may also contain, with the approval aforesaid, authority to the trustee or trustees, upon such default, as one of such remedies, to take possession of the railway and property mortgaged, and to hold and run the same for the benefit of the bondholders thereof for a time to be limited by such deed, or to sell the said railway and property, after such delay, and upon such terms and conditions as may be stated in such deed: and with like approval any such deed may contain provisions to the effect that upon such default and upon such other conditions as shall be described in such deed, the right of voting possessed by the shareholders of the Company, and by the holders of preferred stock therein, or by either of them, shall cease, and holders, or to them and to the holders of the whole or of any part of the preferred stock of the Company, as shall be declared by such deed: and such deed may also provide for the conditional or absolute cancellation after such sale of any or all of the shares so deprived of voting power, or of any or all of the preferred stock of the Company, or both; and may also, either directly by its terms, or indirectly by reference to the by laws of the Company, provide for the mode of enforcing and exercising the powers and authority to be conferred and defined by each deed, under the provisions hereof. And such deed, and the provisions thereof made under the authority hereof, and such other provisions thereof as shall purport (with like approval) to grant such further and other powers and privileges to such trustees or trustees and to such bondholders, as are not contrary to law or to the provisions of this Act, shall be valid and binding. But if any change in the ownership or possession of the said railway and property shall, at any time, take place under the provisions hereof, or of any such deed, or in any other manner, the said railway and property shall continue to be held and operated under the provisions hereof, and of "The Consolidated Railway Act, 1879," as hereby modified. And if the Company does not avail itself of the power of issuing bonds secured upon the land grant along as hereinafter provided, the issue of bonds hereby authorized may be increased to any amount not exceeding twenty thousand dollars per mile of the said Canadian Pacific Railway.

29. If any bond issue be made by the Company under the last preceding section before the said railway is completed according to the said contract, a proportion of the proceeds of such bonds, or a proportion of such bonds if they be not sold, corresponding to the proportion of the work contracted for then remaining incomplete, shall be received by the Government, and shall be held, dealt with, and from time to time paid over by the Government to the Company upon the same conditions, in the same manner and according to the same proportions as the proceeds of the bonds, the issue of which is

Evidence of mortgage and what conditions the bonds may contain.

Right of voting may, in each case, be transferred to bondholders.

Cancellation of shares deprived of voting power.

Enforcing conditions.

Further provisions under mortgage deed.

Increase of borrowing power if no land grant bonds are issued.

Provision if such bonds are issued before completion of railway.

Provision as to issue of land grant mortgage bonds

contemplated by subsection d of Clause 9 of the said contract, and by the thirty-first section hereof.

30. The Company may also issue mortgage bonds to the extent of twenty-five million dollars upon the lands granted in aid of the said railway and of the undertaking authorized by this Act; such issue to be made only upon similar authority to that required by this Act for the issue of bonds upon the railway; and when so made such bonds shall constitute a first mortgage upon such lands, and shall attach upon them when they shall be granted, if they are not actually granted at the time of the issue of such bonds. And such mortgage may be evidenced by a deed or deeds of mortgage to be executed under like authority to the deed securing the issue of bonds on the railway; and such deed or deeds under like authority may contain similar conditions, and may confer upon the trustee or trustees named thereunder, and upon the holders of the bonds secured thereby, remedies, authority, power and privileges, and may provide for forfeitures and penalties, similar to those which may be inserted and provided for under the provisions of this Act in any deed securing the issue of bonds on the railway, together with such other provisions and conditions, not inconsistent with law or with this Act as shall be so authorized. And such bonds may be styled Land Grant Bonds, and they and the proceeds thereof shall be dealt with in the manner provided in the said contract.

Evidence of mortgage and conditions.

Name of and how dealt with.

Issue of bonds in place of land grant bonds under agree-ment with Government.

31. The Company may, in the place and stead of the said land grant bonds, issue bonds, under the twenty-eighth section hereof, to such amount as they shall agree with the Government to issue, with the interest guaranteed by the government as provided for in the said contract; such bonds to constitute a mortgage upon the property of the Company and its franchises acquired and to be thereafter acquired—including the main line of the Canadian Pacific Railway, and the branches thereof hereinbefore described, with the plant and rolling-stock thereof acquired and to be thereafter acquired, but exclusive of such other branches thereof and of such personal property as shall be excluded by the deed of mortgage to be executed as security for such issue. And the provisions of the said twenty-eighth section shall apply to such issue of bonds, and to the security which may be given for the payment thereof, and they and the proceeds thereof shall be dealt with as hereby and by the said contract provided.

To include franchise as well as property of company.

Section 28 to apply.

Facilities for issue of mortgage bonds as to seal and signatures.

32. It shall not be necessary to affix the seal of the Company to any mortgage bond issued under the authority of this Act; and every such bond issued without such seal shall have the same force and effect, and be held, treated and dealt with by all courts of law and of equity as if it were sealed with the seal of the Company. And if it is provided by the mortgage deed executed to secure the issue of any bonds that any of the signatures to such bonds or to the coupons thereto appended may be engraved, stamped or lithographed thereon, such engraved, stamped or lithographed signatures shall be valid and binding on the Company.

"Working expenses" de-fined.

33. The phrase "working expenses" shall mean and include all expenses of maintenance of the railway, and of the stations, buildings, works and conveniences belonging thereto, and of the rolling and other stock and moveable plant used in the working thereof, and also all such tolls, rents or annual sums as may be paid in respect of the hire of engines, carriages or wagons let to the Company; also all rent, charges or interest on the purchase money of lands belonging to the Company, purchased but not paid for, or not fully paid for; and also all expenses of, and incidental to, working the railway and the traffic hereon, including stores and consumable articles; also rates, taxes, insurance and compensation

for accidents or losses; also all salaries and wages of persons employed in and about the working of the railway and traffic, and all office and management expenses, including directors' fees, agency, legal and other like expenses.

Currency on which bonds may be issued.

34. The bonds authorized by this Act to be issued upon the railway or upon the lands to be granted to the Company, or both, may be so issued in whole or in part in the denomination of dollars, pounds sterling, or francs, or in any or all of them, and the coupons may be for payment in denominations similar to those of the bond to which they are attached. And the whole or any of such bonds may be pledged, negotiated or sold upon such conditions and at such price as the Board of Directors shall from time to time determine. And provision may be made by the by-laws of the Company, that after the issue of any bond, the same may be surrendered to the Company by the holder thereof, and the Company, may, in exchange therefor, issue to such holder inscribed stock of the Company—which inscribed stock may be registered or inscribed at the chief place of business of the Company or elsewhere, in such manner, with such rights, liens, privileges and preferences, at such place, and upon such conditions, as shall be provided by the by-laws of the Company.

Price and conditions of sale.

May be exchanged for inscribed stock, etc.

Bonds need not be registered.

35. It shall not be necessary, in order to preserve the priority, lien, charge, mortgage or privilege, purporting to appertain to or be created by any bond issued or mortgage deed executed under the provisions of this Act, that such bond or deed should be enregistered in any manner, or in any place whatever. But every such mortgage deed shall be deposited in the office of the Secretary of State—of which deposit notice shall be given in the Canada Gazette. And in like manner any agreement entered into by the Company, under section thirty-six of this Act, shall also be deposited in the said office. And a copy of any such mortgage deed, or agreement, certified to be a true copy by the Secretary of State or his deputy, shall be received as *prima facie* evidence of the original in all courts of justice, without proof of the signatures or seal upon such original.

Mortgage deed, how deposited.

And agreements under s.36. Certified copies.

Agreement with bondholders, etc., for restricting issues.

36. If, at any time, any agreement be made by the Company with any persons intending to become bondholders of the Company, or be contained in any mortgage deed executed under the authority of this Act, restricting the issue of bonds by the Company, under the powers conferred by this Act, or defining or limiting the mode of exercising such powers, the Company, after the deposit thereof with the Secretary of State as hereinbefore provided, shall not act upon such powers otherwise than as defined, restricted and limited by such agreement. And no bond thereafter issued by the Company, and no order, resolution or proceeding thereafter made, passed or had by the Company, or by the Board of Directors, contrary to the terms of such agreement, shall be valid or effectual.

Effect thereof.

Company may issue guaranteed or preferred stock to a limited amount.

37. The Company may, from time to time, issue guaranteed or preferred stock, at such price, to such amount, not exceeding ten thousand dollars per mile, and upon such conditions as to the preferences and privileges appertaining thereto, or to different issues or classes thereof, and otherwise, as shall be authorized by the majority in value of the shareholders present in person or represented by proxy at any annual meeting or at any special general meeting thereof called for the purpose—notice of the intention to propose such issue at such meeting being given in the notice calling such meeting. But the guarantee or preference accorded to such stock shall not interfere with the lien, mortgage and privilege attaching to bonds issued under the authority of this Act. And the holders of such preferred stock shall have such power of voting at meetings of shareholders as shall be conferred upon them by the by-laws of the Company.

Not to affect privileges of bondholders.

Voting.

EXECUTION OF AGREEMENTS

Contracts, bills, etc., by its agents to bind the company. Proof thereof.

38. Every contract, agreement, engagement, scrip certificate or bargain made, and every bill of exchange drawn, accepted or endorsed, and every promissory note and cheque made, drawn or endorsed on behalf of the Company, by any agent, officer or servant of the Company, in general accordance with his powers as such under the by-laws of the Company, shall be binding upon the Company: and in no case shall it be necessary to have the seal of the Company affixed to any such bill, note, cheque, contract, agreement,

Non-liability of such agents.

engagement, bargain or scrip certificate, or to prove that the same was made, drawn, accepted or endorsed, as the case may be, in pursuance of any by-law or special vote or order; nor shall the party so acting as agent, officer or

Proviso: as to notes.

servant of the Company be subjected individually to any liability whatsoever to any third party therefore: Provided always, that nothing in this act shall be construed to authorize the Company to issue any note payable to the bearer thereof, or any promissory note intended to be circulated as money, or as the note of a bank, or to engage in the business of banking or insurance.

GENERAL PROVISIONS

Reports to Government.

39. The Company shall, from time to time, furnish such reports of the progress of the work, with such details and plans of the work as the Government may require.

Publication of notices.

40. As respects places not within any Province, any notice required by "The Consolidated Railway act, 1879", to be given in the "Official Gazette" of a Province, may be given in the Canada Gazette.

Form of deeds, etc., to the Company.

41. Deeds and conveyances of lands to the Company for the purpose of this act (not being letters patent from the Crown), may, in so far as circumstances will admit, be in the form following, that is to say:—

"Know all men by these presents, that I, A.B., in consideration of _____ ____ paid to me by the Canadian Pacific Railway Company, the receipt whereof is hereby acknowledged, grant, bargain, sell and convey unto the

Form.

said The Canadian Pacific Railway Company, their successors and assigns, all that tract or parcel of land (describe the land) to have and to hold the said land and premises unto the said Company, their successors and assigns for ever.

"Witness my hand and seal, this _____ day of one thousand eight hundred and

"Signed, sealed and delivered in presence of
A.B. (L. S.)

"C.D.
"E.F.

Obligation of the grantor.

or in any other form to the like effect. And every deed made in accordance herewith shall be held and construed to impose upon the vendor executing the same the obligation of guaranteeing the Company and its assigns against all dower and claim for dower and against all hypothecs and mortgages and against all liens and charges whatsoever, and also that he has a good, valid and transferable title thereto.

From Archives Canadian Pacific Railway. Contact: Canadian Pacific Archives, Windsor Station, P.O. Box 6042, Station Centre Ville, Montreal, Quebec, Canada H3C 3E4.

Related Cultural Documents—I

MAJOR DATES IN THE EARLY HISTORY OF CANADA'S RAILWAY TO THE PACIFIC

1871, July 20—Crown Colony of British Columbia joins Canada on condition that a railway linking Eastern Canada with the Pacific Coast will be commenced within two years of union and completed within 10 years.

1871–1873—Charters granted to several groups to construct the Pacific railway, including one headed by Sir Hugh Allan. Disclosure of Allan's financial contributions to Conservative party during elections brings about "Pacific Scandal" and fall of government.

1875, June 1—Ceremonial first sod turned at what is now Thunder Bay, Ont., marking the start of construction by federal government on Lake Superior Section of the Pacific railway.

1880—Construction of Western Section of railway is started by federal government.

1880, Oct. 21—Government of Canada signs contract with George Stephen and Duncan McIntyre of Montreal, and others, to construct Eastern and Central Sections and otherwise complete and put Pacific railway into operation by 1891.

1881, Feb. 15—Royal assent given to the Act, 44 Victoria, Chapter 1, incorporating Canadian Pacific Railway Company and embodying 1880 contract.

1881, May 1—Canadian Pacific company begins operations by assuming train services on government-built rail lines in Lake Superior Section, radiating from Winnipeg. Company also begins construction on Eastern and Central Sections.

1881, June 9—Canada Central Railway Company amalgamated with Canadian Pacific Railway Company, providing access eastward as far as Ottawa.

1882, Jan. 1—William C. Van Horne hired as general manager to direct all construction and operations of the CPR.

1882, Spring—Canadian Pacific purchases Western Division of the Quebec, Montreal, Ottawa & Occidental Railway, giving company access by rail in the east as far as Montreal.

1883, August—Rails reach Calgary from east on the central Section, giving continuous rail line from Thunder Bay to Calgary.

1885, Spring—Militia moved by rail to quell Northwest Rebellion, demonstrating transcontinental railway's strategic value.

1885, May 18—Driving of last spike at Jack Fish Bay on Lake Superior completes Eastern Section, enabling through train service between Montreal and Winnipeg to be inaugurated on Nov. 1, 1885.

1885, Nov. 7—Driving of last spike in Eagle Pass, British Columbia, completes Central Section and whole transcontinental line, more than five years earlier than stipulated in charter.

1886, June 28—Through train service inaugurated between Montreal, Toronto and Pacific tidewater at Port Moody, B.C.

1887, May 23—Pacific terminus moved to Vancouver from Port Moody.

From Archives Canadian Pacific Railway. Contact: Canadian Pacific Archives, Windsor Station, P.O. Box 6042, Station Centre Ville, Montreal, Quebec, Canada H3C 3E4.

Related Cultural Documents—II

INTERNATIONAL PRIME MERIDIAN CONFERENCE, WASHINGTON, DC, 1884

Count Lewenhaupt, Delegate for Sweden, then proposed that the resolutions passed by the Conference should be formally recorded in a Final Act, stating the votes on each resolution that was adopted.

The Conference took a recess, in order to allow the Delegates to examine a draft of the Final Act.

After the recess the Final Act was unanimously adopted, as follows:

FINAL ACT.

The President of the United States of America, in pursuance of a special provision of Congress, having extended to the Governments of all nations in diplomatic relations with his own, an invitation to send Delegates to meet Delegates from the United States in the city of Washington on the first of October, 1884, for the purpose of discussing, and, if possible fixing upon a meridian proper to be employed as a common zero of longitude and standard of time-reckoning throughout the whole world, this International Meridian Conference assembled at the time and place designated; and, after careful and patient discussion, has passed the following resolutions:

I.

"That it is the opinion of this Congress that it is desirable to adopt a single prime meridian or all nations, in place of the multiplicity of initial meridians which now exist?"

This resolution was unanimously adopted.

II.

"That the Conference proposes to the Governments here represented the adoption of the meridian passing through the centre of the transit instrument at the Observatory of Greenwich as the initial meridian for longitude."

The above resolution was adopted by the following vote:
In the affirmative:

Austria-Hungary,	Mexico,
Chile,	Netherlands,
Colombia,	Paraguay,
Costa Rica,	Russia,
Germany,	Salvador,
Great Britain,	Spain,
Guatemala,	Sweden,
Hawaii,	Switzerland,
Italy,	Turkey,
Japan,	United States,
Liberia,	Venezuela.

In the negative:
San Domingo.

Abstaining from voting :
Brazil, France.

Ayes 22; noes,1; abstaining, 2.

III.
"That from this meridian longitude shall be counted in two directions up to 180 degrees, east longitude being plus and west longitude minus."
This resolution was adopted by the following vote:
In the affirmative:

Chile,	Liberia,
Colombia,	Mexico,
Costa Rica,	Russia,
Great Britain,	Paraguay,
Guatemala,	Salvador,
Hawaii,	United States,
Japan,	Venezuela.

In the negative:

Italy,	Sweden,
Netherlands,	Switzerland.
Spain,	

Abstaining from voting:

Austria-Hungary,	Germany,
Brazil,	San Domingo,
France,	Turkey.

Ayes, 14; noes, 5; abstaining, 6.

IV.

"That the Conference proposes the adoption of a universal day for all purposes for which it may be found convenient and which shall not interfere with the use of local or standard time where desirable."

This resolution was adopted by the following vote:

In the affirmative:

Austria-Hungary,	Mexico,
Brazil,	Netherlands,
Chile,	Paraguay,
Colombia,	Russia,
Costa Rica,	Salvador,
France,	Spain,
Great Britain,	Sweden,
Guatemala,	Switzerland,
Hawaii,	Turkey,
Italy,	United States,
Japan,	Venezuela.
Liberia,	

Abstaining from voting:

Germany,	San Domingo.

Ayes, 23; abstaining, 2.

V.

"That this universal day is to be a mean solar day is to begin for all the world at the moment of mean midnight of the initial meridian, coinciding with the beginning of the civil day and date of that meridian and is to be counted from zero up to twenty-four hours."

This resolution was adopted by the following vote:

In the affirmative:

Brazil,	Liberia,
Chile,	Mexico,
Colombia,	Paraguay,
Costa Rica,	Russia,
Great Britain,	Turkey,
Guatemala,	United States,
Hawaii,	Venezuela.
Japan,	

In the negative:

Austria-Hungary,	Spain.

Abstaining from voting:

France,	San Domingo,
Germany,	Sweden,
Italy,	Switzerland.
Netherlands,	

Ayes, 15; noes, 2; abstaining, 7.

VI.

"That the Conference expresses the hope that as soon as may be practicable the astronomical and nautical days will be arranged everywhere begin at midnight."

This resolution was carried without division.

VII.

"That the Conference expresses the hope that the technical studies designed to regulate and extend the application of the decimal system to the division of angular space and of time shall be resumed, so as to permit the extension of this application to all cases in which it presents real advantages."

The motion was adopted by the following vote:

In the affirmative:

Austria-Hungary,	Mexico,
Brazil,	Netherlands,
Chile,	Paraguay,
Colombia,	Russia,
Costa Rica,	San Domingo,
France,	Spain,
Great Britain,	Switzerland,
Hawaii,	Turkey,
Italy,	United States,
Japan,	Venezuela.
Liberia,	

Abstaining from voting:

Germany,	Sweden.
Guatemala,	

Ayes, 21; abstaining, 3.

Done at Washington the 22nd of October, 1884.
C. R. P. RODGERS
President.
R. Steachey, J. Jannsen, L. Cruls,
Secretaries.

The following resolution was then adopted unanimously:

"That a copy of resolutions passed by this Conference shall be communicated to the Government of the United States of America, at whose instance and within whose territory the Conference has been convened."

Mr. Rutherford, Delegate of the United States, then presented the following resolution:

"Resolved, That the Conference adjourn, to meet upon its call of the President, for the purpose of verifying the protocols."

This resolution was then unanimously carried, and the Conference adjourned at half past three, to meet upon the call of the President.

From http://wwp.greenwichmeantime.com/info/conference-finalact.htm.

A modern-day shot of the Eiffel Tower and Champs de Mars. Courtesy of Corbis.

20
The Eiffel Tower

France

DID YOU KNOW . . . ?

➤ There were more than 5,300 plans and drawings for the tower.
➤ The tower was built in two years, two months, and five days, from 1887 to 1889. It was an instant financial success.
➤ There were 18,000 components made by 100 ironworkers off-site, then assembled by 130 workers on-site.
➤ It measures 410 feet (125 meters) on each side, stands 1,024.5 feet (312.27 meters) tall, and weighs 9,500 tons.
➤ The tower sways only 4.5 inches at the top.
➤ For many years, the tower was the world's tallest structure.
➤ Not one fatality occurred during construction.
➤ Guy de Maupassant, Alexander Dumas, Emile Zola, and other luminaries signed a petition objecting strenuously to the tower.
➤ Eiffel also designed the iron framework inside the Statue of Liberty.

"We, the writers, painters, sculptors, architects and lovers of the beauty of Paris, do protest with all our vigor and all our indignation, in the name of French taste and endangered French art and history, against the useless and monstrous Eiffel Tower" (http://www.discoverfrance.net/france/paris/). Clearly, initial reaction to the tower was mixed, as evidenced by this quote from a petition presented to the government of the city of Paris. The petition was signed by—among others—Guy de Maupassant, Alexander Dumas, Emile Zola, Charles Gounod, and Paul Verlaine.

Paris's soaring, open-lattice, wrought-iron Eiffel Tower, originally built for the International Exposition of 1889 commemorating the centennial of the French Revolution, remains a universally recognized symbol of France, and indeed all Europe. Over 700 proposals had been submitted by architects, engineers, sculptors, and artists. One was selected unanimously, the design by Gustave Eiffel.

The tower became an instant icon, the site of many romantic moments as well as staggering feats of individual bravado. In 1923, the man who would become mayor of the district of Montmartre showed his derring-do by bicycling down the tower using its legs as a ramp. In 1954, a mountain climber scaled its height, and in 1984 two English men parachuted from the top.

HISTORY

The plan for the tower was submitted to the design competition by the civil engineer Gustave Eiffel (1832–1923), already well known for such works as the arched Gallery of Machines for the Paris Exhibition of 1867, the dome for the Nice Observatory, a harbor in Chile, a 541-foot (165-meter) arched bridge in Garabit, France, a preconstructed spanned bridge in China, and an iron bridge at Bordeaux (the construction of which involved the first use of compressed air to drive piles). Eiffel's viaduct over the Truyère, which stretched 1,850 feet (564 meters), with a central arch span of 541 feet (165 meters), constituted an engineering record: with a height of 400 feet (122 meters) over the river, it was for years the world's highest bridge.

While Eiffel receives all the credit for the tower, it must be noted that the original conception for the 1889 exposition tower came from two engineers at Eiffel's firm: Maurice Koechlin and Emile Nouguier. It took years of work by more than 50 engineers and designers to prepare the approximately 5,300 plans and drawings for the tower.

Reproduced here is the official agreement of January 8, 1887, which outlines Gustave Eiffel's construction and operation of the tower. In addition to Eiffel, it was signed by Commerce and Industry Minister Edouard Lockroy—who, as commissioner general of the exposition, organized the design competition— and by Eugène Poubelle, prefect of the Seine.

Once approval was given, the project proceeded at a rapid pace. Excavation commenced on January 26, 1887, and assembly of the metal structure on July 1. The tower's 18,000 component parts were made by more than 100 ironworkers at the workshops of Eiffel's company in the outskirts of Paris, and were assembled by more than 130 workers at the exposition site.

The exposition was scheduled to open on May 6, 1889. Contrary to the expectations of many observers, Eiffel easily fulfilled his commitment to complete the project on schedule, finishing on March 30, 1889. In a ceremony the following day, a small group of dignitaries accompanied Eiffel to the top, where he raised a huge French flag with the letters "R. F." (*République Française*) and was awarded the Legion of Honor.

The tower rises from a square base measuring 410 feet (125 meters) on each side to a height of 1,024.5 feet (312.27 meters) (even higher today because of the addition of broadcasting antennas). Until the completion in 1930 of New York City's Chrysler Building, the Eiffel Tower had been the tallest man-made structure in the world. Despite its immense height, the tower weighed only 9,500 tons, with the metal framework accounting for 7,300 tons. Because of its cross-braced latticed structure, wind had little impact on its stability.

The Eiffel Tower was an immediate success. Construction costs, said to be approximately eight million gold francs, were quickly covered by receipts earned from visitors. By the time the fair closed in early November 1889, two million people had visited the tower. By the end of that year, receipts totaled 5.9 million gold francs. As of 2002, the total number of visitors to the tower had exceeded 200 million.

CULTURAL CONTEXT

France is the country that coined the phrase *les grands travaux* (large-scale engineering works). Gustave Eiffel was in the authentic tradition of *les grands travaux*, and those who built them. Perhaps the world's greatest artist in the medium of iron, Eiffel also wrote a book entitled *L'Architecture Métalique*.

He attended the Ecole Centrale, a school for the arts and manufacturing. But if his school prepared him for art, he also participated in its manufacturing mission, as Eiffel continued to design projects using iron—railway viaducts with supports of iron, and a bridge over the Douro River in Portugal. He decided to open his own factory just outside Paris in the town of Levallois-Perret. This combination of artistry, experience with iron, his own manufacturing and production facilities, and extensive business-management experience enabled Eiffel to be one of the first macroengineers to complete his great work not only within the prescribed budget but ahead of the estimated schedule.

While France was the nexus of great engineering works, it may have been Eiffel's experiences in the United States that sparked his idea for the tower. In 1876, he saw a proposal by two Americans, Samuel Fessenden Clarke and Arthur M. Reeves, who had designed a circular iron-framed tower intended as an engineering monument and icon for the Centennial Exposition of 1876. Clarke and Reeves's design was included in the 100th anniversary of the American Revolution—a fact that may have influenced Eiffel as he considered a piece for the 100th anniversary of the French Revolution. Eiffel himself credited Clarke and Reeves as the source of his inspiration.

Eiffel's design took advantage of new materials with which architects and engineers were beginning to become familiar. Before this time, *les grands travaux* were large in the sense of being wide and long, and were often based on stone construction—canals, aqueducts, and bridges; they were not especially high, except of course for the famous medieval cathedrals. Eiffel's monument is a marvel of physics. He was a pioneer in the aerodynamics of high frames, using a mathematical formula to determine the exact curve of the structure's base that would withstand the force of the wind against it and transform that force into added structural support. It is noteworthy that the Eiffel Tower sways only 4.5 inches at the top. The exact math of Eiffel's discoveries can be found in his 1913 work *The Resistance of the Air*. Eiffel's physics paved the way for the modern skyscrapers erected in later years.

The tower's height is an important factor in its endurance. Although the contract set 20 years as the length of time the tower would remain in the Parisian park the Champs de Mars, when the contract expired, the tower's height made it the obvious choice for siting communication antennae. So it remained in the park—to send telegraph signals.

Eiffel saw the advantages of adding communication antennae to the tower. When the first radio signals were sent by Eugène Ducretet in 1898, it was Gustave Eiffel who approached the military in 1901 and suggested that the tower be incorporated into an infrastructure for long-distance radio communications. By 1903, radio signaling had made major progress: the military was sending messages from the tower to bases around Paris, and by 1904 they were sending to the French east coast. A permanent radio station was installed in the tower in 1906. In 1910, its antenna became part of the International Time Service. Ever an enthusiast for the modern and new, Eiffel was gratified when the first European public radio broadcast came from the tower in 1921, just two years before his death. It would have pleased Eiffel that in 1957 television signals were added; no doubt he would highly approve of the webcam that today allows people from all over the world to experience the Eiffel Tower via the Internet.

PLANNING

The Eiffel Tower was perhaps more carefully and meticulously planned than any macro project in history—and all the planning was done by Eiffel himself. In fact, its planning may have taken longer than its building. As a result of such fastidious preparation, the project was finished ahead of schedule. In just two years, two months, and five days, Eiffel and his team successfully accomplished this engineering feat with such perfection that the individual pieces were tooled to an accuracy of 1/10 of a millimeter.

It took tremendous planning to foresee that most of the parts would have to be forged, machined, and assembled off-site and then installed. Workers at Eiffel's factory at Levallois-Perret made the parts; many had previous experience working on Eiffel's viaducts.

The on-site work crews required special organization. Teams of four men were needed to install each rivet: one held the rivet in place, one heated it red-hot, a third made sure the head of the rivet was positioned exactly right, and the fourth hammered it into place. Eiffel had planned the process with such precision that the hot rivets could cool right in place, expand as they cooled, and thereby strengthen the structure by taking advantage of natural thermodynamic principles.

The Eiffel Tower was one of the most meticulously planned and best-managed macro construction projects in history.

BUILDING

The tower was constructed of iron and held together by 2.5 million rivets, all resting on a masonry base. The foundation was made of caissons filled with concrete and sunk into the ground; these were 50 feet long, 22 feet wide, and 7 feet deep (15.2 × 6.8 × 2.1 meters). The tower consisted of two platforms and a laboratory at 896 feet (273 kilometers) for Eiffel's use. All sections of the tower were prefabricated; seven million holes were drilled off-site, and remarkably, there were no significant difficulties with on-site assembly.

The precision of the planning shows up in the contract presented here. For instance, one curious but highly specific detail is the reference in article 2: "Widow Bourouet-Aubertot, owner of the hotel on the avenue de la Bourdonnaye, 10." The widow had approached the prefect of the Seine threatening to require demolition of the Eiffel Tower, which was in the midst of construction. The contract states, "The aforementioned widow has withdrawn a provisional execution of a judgment to intervene." That settled the matter, and in the process made the Widow Bourouet-Aubertot forever famous (or infamous). Her story is just one of many complaints from abutters, interested parties, affected parties, and others whose viewpoints Eiffel had to confront.

Article 3 of the contract is noteworthy for more substantive reasons. It addresses park landscaping, which would be disturbed by the tower construction. The contract stipulates that Eiffel will be responsible for any plantings moved, and "will support the costs of removal by the gardeners of the city the trees, bushes, and plants that must be displaced." Note that the city's gardeners would do the work, not a work crew of riveters.

Another concern in article 3 was that "Mr. Eiffel will not cause any changes to the hydrants, sewer drains, or water pipes situated in the Garden of the city." Here we learn of the exquisite city planning for Paris, which located a water source under the park. There are two advantages to such a placement. First, the water pipes could be used to assist in the irrigation of the plantings. Second, the pipes are located beneath a surface that is subject to little vibration except running children; hence there would be few disturbances to the critical infrastructure of the Parisian water system. Gardeners and students of Indian architecture might recall that the Taj Mahal used a similar design, locating water pipes under the gardens.

Les égouts (the sewers) of Paris remain a tourist highlight, with an entrance beneath one of the bridges crossing the Seine. The sewer was designed during the administration of Baron Georges Eugène Haussmann, Napoleon III's famous prefect of the Seine. Paris had been one of the unhealthiest cities in the world, assailed by repeated epidemics. Haussmann insisted on cutting wide avenues that provided fresh air, thereby opening up narrow streets that had remained unchanged since medieval times.

Another portion of the contract presented here provides specific guidelines for the use of the Eiffel Tower in case of war. In article 8, Eiffel is instructed, "In order to facilitate scientific or military purposes or use, Mr. Eiffel will reserve on each floor a special room which will remain free for the disposition of persons designated by the Minister of the General Commission." In a note of finesse, it is added that said minister will get 300 free admissions per month, and the admissions can use the elevators.

Article 13 requires that in times of war or a state of siege, the government will have the right to use the tower, perhaps for its vantage point as a lookout, but more likely for sending signals. The contract is meticulous in outlining how Eiffel will be repaid for time lost during military use, and provides that "the term of concession will be extended one year for every period of three months or fraction of three months during which the suspension occurs."

A discussion of the construction of the Eiffel Tower is not complete without consideration of the elevators. Never before had *ascenseurs* risen to such heights. The French company Roux, Combaluzier, and Lepape built the first elevators, which carried passengers to the first platform by means of hydraulics, utilizing a double-looped chain for extra safety. In 1897, those elevators were replaced by equipment from the French firm Fives-Lille; these lasted 90 years, until they were improved in 1987. But even the Fives-Lille company only had the technical know-how to bring the elevators to the first level.

How could visitors get further without climbing the endless stairs upward? This was a task for the world's best-known elevator man, Elisha Graves Otis. In 1853, Otis introduced the world's first safety elevator in Yonkers, New York. From that point on, buildings could rise beyond the limitations of stairs. In one of his greatest works, Otis designed elevator cabins as two-decked rooms mounted on sloping runners and pulled by a cable powered by a hydraulic piston.

The pièce de résistance was the vertical lift designed by Leon Edoux to bring visitors to the top of the tower. Passengers changed cars halfway up, as only one car could continue upward, counterbalanced by the other going down, in a design not unlike a water clock and similarly powered by water tanks that helped provide the hydraulics. Of course, Edoux's ingenious engineering did not work in the winter, when frozen water in the tanks made operations impossible until the spring thaw. Edoux's marvelous invention operated until 1983, when technology had advanced sufficiently to offer a replacement.

At his own risk—and his own profit—Eiffel was free to conduct the construction in any manner he chose. He also was given the right to fix the fee for admission to the tower: higher on weekdays than on weekends. In return for these allowances, and for the right to lease and collect rents from the shops and cafés associated with the tower, Eiffel was required to pay 1,000 francs to the Exposition Commission.

Eiffel was also legally bound to pay the city of Paris for rental of the land on which the tower was built—a nominal 100 francs per annum (article 12).

The contract for building the Eiffel Tower contains insurance provisions, a feature lacking in many other contracts of similar vintage. Eiffel was required to put aside one percent to cover potential expenses for sick or wounded workers. In addition, Eiffel was bound to set aside a reserve fund to deal with accidents. It should be noted that there were no fatalities during the course of the construction—a tribute to Eiffel's well-orchestrated planning. The one recorded death of a workman occurred off-site and off-duty.

Like his fellow engineer/entrepreneur John Roebling, who built the Brooklyn Bridge in the United States using steel ropes manufactured in his own factory in New Jersey, Gustave Eiffel built the Eiffel Tower six years later utilizing a similar management process: coordination of on-site work with ongoing construction being prepared in his own nearby factory. Both of these mechanical engineers were not just designers but also business owners and managers—Eiffel was Mr. Iron, and Roebling was Mr. Steel. Both built structures higher than had ever been done before.

The tower bears the names of 72 of France's greatest technical minds, 18 luminaries on each of the four sides of the base. For most French people, however, one name is mentioned even more frequently than Eiffel's: Monsieur Poubelle, the prefect of the Seine, who signed the contract featured here on behalf of the government. Poubelle is the inventor of the garbage pail, which is still used all over Paris and France. The ubiquitous receptacle is actually called a *poubelle*.

IMPORTANCE IN HISTORY

What distinguishes the Eiffel Tower is not just its beauty or symbolism. Like the Colossus of Rhodes, it was a technological marvel of its time, pushing the limits of existing engineering knowledge. It is a little-known fact that Eiffel helped build the Statue of Liberty in New York Harbor. In 1885, he worked with Frederic Bartholdi to create the wrought-iron pylon inside the statue.

Eiffel was duly recognized for his tremendous achievement. On the 200-franc banknote (the currency of France before the adoption of the euro in 2002), there was a depiction of Gustave Eiffel and, on the reverse side, the tower. It is a distinction afforded few engineering projects, and a sign of his place in French history. Even in France today, the top of the social hierarchy is not, as some might imagine, painters, designers, or aristocrats, but instead engineers.

While standing at the top of his completed monument, Gustave Eiffel received the Legion of Honor, inducting him into that society that has been given special distinction and national renown in France.

The Eiffel Tower has made such an impression on the world that it was especially honored in Shanghai, China, in a cultural exchange featuring the Oriental Pearl TV Tower, built for telecommunications, and one of the world's highest buildings (1,535 feet, or 468 meters).

Eiffel Tower machinery with a man beside a wheel that raises the elevator, during the Paris Exposition. Courtesy of the Library of Congress.

Eiffel continues the tradition of *les grands travaux* as a worthy successor to Pierre-Paul Riquet (the Canal des Deux Mers), Ferdinand de Lesseps (the Suez Canal), French-born Isambard Kingdom Brunel (RMS *Great Eastern* and the Thames Tunnel), and more recently Louis Armand (modernizer of the French railways).

FOR FURTHER REFERENCE

Books and Articles

Clark, Ronald W. *Works of Man*. New York: Viking, 1985.

Fletcher, Banister. *Sir Banister Fletcher's A History of Architecture*. Revised by J.C. Palmer. New York: Charles Scribner's Sons, 1975.

Harris, S. *The Tallest Tower: Eiffel and the Belle Époque*. Washington, DC: Regnery Gateway, 1989.

Hawkes, Nigel. *Structures: Man-Made Wonders of the World*. London: Marshall Editions, 1990.

Internet

For wide-ranging information on the Eiffel Tower, see http://www.tour-eiffel.fr/. This website offers information on the construction process (including rivets and elevators), Eiffel's advocacy of radio signals from the tower, and a comprehensive bibliography of books and other media.

For real-time webcam views of the Eiffel Tower, see: http://www.abcparislive.com.

For a 3D model of the tower and many images, see http://www.greatbuildings.com/buildings/Eiffel_Tower.html.

Film and Television

Modern Marvels: The Eiffel Tower. Videocassette. Directed by Larayne Decoeur, Noah Morowitz. New York: Modern Marvels, 1994.

Documents of Authorization

AGREEMENT CONCERNING THE EIFFEL TOWER

Paris, Imprimerie et Librairie Centrales Des Chemins De Fer, Imprimerie Chaix, Société Anonyme au Capital de Six Millions, Rue Bergèree, 20, 1889.

Agreement Concerning the Eiffel Tower

Between Mr. Edouard LOCKROY, Minister of Commerce and Industry, Commissioner General of the Universal Exposition of 1889, acting in the name of the State;

Mr. Eugène POUBELLE, Prefect of the Seine, acting in the name of the City of Paris, thus here is that which was authorized by the Municipal Council during deliberations from the 22 of October to the 28 of December 1886 and that which is fixed and limited by the deliberations, which remain enclosed as exhibited here:

As one party;

And Mr. EIFFEL, Engineer-Constructor, living in Leval-lois-Perret, rue Fouquet, number 42; acting under his own personal name;

As the other party.

These parties have made the following agreements.

Article 1

Mr. Eiffel promises to the Minister of Commerce and Industry, The Commissioner General of the Universal Exposition of 1889, to construct, in the role of entrepreneur, in the enclosure of the Exposition, in the Champ de Mars, a Tower made of iron of the height of 300 meters as part of the construction of the Exposition, for which the plans and estimate are here attached to the present agreement.

This Tower will be completed and put into use by the opening of the Exposition of 1889.

Article 2

The Tower will be erected in one part of the Champ de Mars, placed according to the disposition of the minister of Commerce and Industry by the City of Paris, as has resulted from the deliberations of the Municipal Council of the City of Paris during the dates from October 22 to December 28 1886.

It will occupy the site indicated in the plan here attached, seen in the deliberations of the Municipal Council.

It is explained here, for purposes of clarification, according to the terms of the agreement dated in Paris on the 28 of December 1880, registered, handed over by the State to the City of Paris that a Park for the Exposition will be established in the Champ de Mars (next to the Seine) so that it can now proceed and be required, by the authority of the City of Paris, to complete and finalize the aforesaid Park and maintaining it in good condition with this stipulation:

"In the case where the Universal Exposition will take place in the Champ de Mars, the free use of the park will be assured by the State, with the responsibility of restoring it to good condition after the Exposition.

"And the City may transfer power to the vendors to construct on the frontage of the two avenues of Suffren and of La Bourdonnaye, two zones of 40 meters each in breadth, including the areas reaching up to the avenues to the Park."

It is agreed that the ground plot, already having been allocated into several lots, will be awarded to various vendors, following the official proceedings of diverse dates; and that the aforesaid proceedings, under the rubric: Origin of Rights of Ownership, recalling the outcomes of the agreement of the 28 of December 1880 between the State and the City of Paris; which, under the rubric: Terms and Conditions, is stated in article 5:

"Legal Rights and Methods of Access.

"Each proprietor will have the right of egress to the Park in the Champ de Mars through the means of a grill gate cross-barred and set within the gate [*editor's note:* such as in a convent, e.g., a gate within a gate]; this will be established during the Exposition in the Champ de Mars, and the right of egress will be suspended and the grillgate will be closed in some manner during the Exposition.

"The successful bidders for the area around the façade of Avenue A and Avenue B will have the rights of opening and exiting through the pathways."

It is further agreed that the enunciations and stipulations which preceded are being restated: two of the claimants to property have maintained that the establishment of the Eiffel Tower in the Park of the Champ de Mars was contrary to their legal rights that had already been consented to by the City of Paris.

That one of these, the Widow Bourouet-Aubertot, owner of the hotel on the avenue de La Bourdonnaye, 10, has herself addressed to the Prefect of the Seine, on the 6 of November 1886, a previous memo concerning a suit which she intended to introduce before the Civil Tribunal of the Seine, for the purpose of forbidding the City of Paris to act or to allow the establishing of the Eiffel Tower in the Park of the Champ de Mars and to require the demolition of all the work which had begun in contempt of the suit. The aforementioned widow has withdrawn a provisional execution of a judgment to intervene.

That the City of Paris may not mortgage the Park in the Champ de Mars for profit through real estate sales, but has the legal right to exercise the power being impeded or hindered by the establishment of the Eiffel Tower, and that the vendors are without legal rights, either to oppose this establishment or to claim legal damages.

This having been said, Mr. Eiffel promises to do his utmost to support the damages which could result from the construction and the exploitation of the Tower, without the power to invoke any of the previous cases or provisions here in the guaranty of the City or of the State.

Article 3

Mr. Eiffel may not occupy temporarily any part of the Garden of the city, except for the placement of the Tower, without the authorization of the Prefect of the Seine. He will support the costs of removal, by the gardeners of the city of the trees, bushes, and plants that must be displaced. He will furthermore be responsible for all the damage that may result to the garden during the construction of the Tower, and for these he will reimburse the City.

Mr. Eiffel will not cause any changes in the hydrants, sewer drains, or water pipes situated in the Garden of the city, without the prior authorization of the Prefect of the Seine. The expenses resulting from any changes or authorized displacements will be the responsibility of Mr. Eiffel.

Article 4

The enterprise will be comprised of the foundation of the basement, the substructure of masonry, the complete metal framework, the construction and interior equipment for all the rooms and floors, as well as all the lightning rods and their accessories; but it will not include the fitting out of the grounds around the perimeter of the Tower, nor the transformation of the avenues, squares, and other fixtures as per the wish of the Director of the Exposition, which in no case will be the responsibility of Mr. Eiffel.

Article 5

The Tower will be constructed in conformity with the preliminary plan which was submitted to the Special Commission named by the Minister of

Commerce and Industry by order of the date of the 12 of May 1886, modified in the estimate and the design sketches here attached following the conclusions of the report of that Commission. The project will moreover be completed with regard to the electrical works in conformity to the conclusions in the report of that Commission dated the 24 of June 1886, which is attached to the present document.

Article 6

Pursuant to these conditions, Mr. Eiffel will be responsible for the definitive plans and the complete execution of the Tower, which will be under his responsibility. He will present the project in the execution of the foundation, which will be submitted for examination to the Commission specially instituted by the order of the 12 of May 1886. The project thus is ordered to be put into execution, with the modifications that are judged necessary through common accord in the course of execution. Mr. Eiffel will remain, during the execution of the work, under the direction of the Engineers of the Exposition and the Controller of the Commission specially instituted on the 12 of May 1886.

Mr. Eiffel is required to obtain approval for all the project details and especially the elevators that will be employed in the interior of the Tower, and he may undertake the establishment of the elevators only after the approval of the Minister of the General Commission.

The Tower may not be put into use until after it has been approved and accepted by the special Commission, even if the acceptance requires some amendments, whatever they may be, and these will be the responsibility of the constructor. It is understood that the modifications or rework recognized as necessary by the special Commission will not give rise to any additional expenses that will be the responsibility of the Exposition.

Article 7

For the price of the work, such as is evaluated in the estimate here-attached, it is agreed that the payment to Mr. Eiffel the sum of 1,500,000 Francs for the credits allowed by the Exposition, and the enjoyment and dividends resulting from the exploitation of the Tower during the year of the Exposition and during the twenty years that follow, from the date of the first of January of eighteen hundred and ninety, during the whole of which time the following conditions must be followed:

The sum of 1,500,000 Francs will be paid out as follows:

Firstly, 500,000 (five hundred thousand) Francs when the metal framework has reached the height of the floor of the first stage;

500,000 (five hundred thousand) Francs when the metal framework has reached the second stage;

500,000 (five hundred thousand) Francs when the work is complete and has been provisionally received and accepted for exploitation and use;

Secondly, during the total duration of the Exposition, Mr. Eiffel will exploit to his own profit and at his own risk and peril the aforesaid construction in a manner that will be judged as best conforming to his own interests, as much as can be gained from the elevators for the public and from the installation of restaurants, cafes, etc.

He will remain the master for the fixing of tariffs to be applied without exceeding the maximums indicated as follows:

Elevators to the top:	Weekdays:	Fr. 5
	Sundays and holidays:	Fr. 2
Elevators to the first stage:	Weekdays:	Fr. 2
	Sundays and holidays:	Fr. 1

These numbers, which are applicable to the elevators effective between the hours of 11am and 6pm, may be modified at the request of Mr. Eiffel if experience demonstrates the necessity, and if the Minister of the General Commission judges necessary. In all cases, as concerns the cafés, restaurants, and other related establishments in the construction of the Tower, the concessions made to these third parties by Mr. Eiffel must be approved by the Minister and must follow the rules applied to establishments of this nature in the rest of the Exposition.

Mr. Eiffel will be responsible to the State or to the City for the third parties which may be thus substituted. In any case the aforesaid substitutes will not have any legal right or redress, for any cause whatever it may be, against the State and the City, and the enjoyment and use of their concession will be subordinated during the duration of the exploitation to Mr. Eiffel himself or to the Society afterward provided. Mr. Eiffel will give these rightful claimants or their substitutes full knowledge of the present agreements and take full responsibility for the communication of said agreements.

Mr. Eiffel will, therefore, pay to the Administration of the Exposition, a sum of 1,000 (one thousand) Francs for the right of establishing the beneficial rents during the duration of the Exposition.

Article 8

In order to facilitate scientific or military purposes and use, Mr. Eiffel will reserve on each floor a special room, which will remain free for the disposition of persons designated by the Minister of the General Commission.

In addition, Mr. Eiffel will give to the disposition of the Minister of the General Commission, for the same purpose, a number of free admissions, which will not exceed 300 (three hundred) per month. These admissions will be given the free right of passage in the elevators or the stairs.

Article 9

The Administration of the Exposition will dispose, for such rightful usage as it wishes, all of the land surrounding the Tower which is not occupied by the four base columns which support it. The Administration will, in all cases, hold in reserve, all around these base column supports, the means of access necessary for the public to arrive easily at the elevators, the stairs, and any other parts of the Tower and for provisions of any and all nature, necessary for the use of these, for entry.

Article 10

During the whole duration of the Exposition, Mr. Eiffel will have free admission to the Exposition for himself and for the personnel who are involved in the maintenance and service of the Tower.

Article 11

After the Exposition and the replacement of the Park in the Champ de Mars, the City will then become the owner and proprietor of the Tower, with all the advantages and obligations relevant to this; but Mr. Eiffel in the completion of the fee for the work, will retain the enjoyment until the expiration of twenty years from the date of 1 January 1890, at which point this enjoyment and use shall return to the City of Paris. After these twenty years, the return of the Tower will be made in good condition of maintenance, and during that twenty years' time it will be required of Mr. Eiffel to make any special, required repairs.

Article 12

During the whole of his period of exploitation, Mr. Eiffel will remain in all regards concerning this exploitation responsible for the same conditions that will apply during the Exposition, except that he will pay to the City of Paris a consideration of 100 (one hundred) Francs per year beginning on 1 January 1890 for the land occupied by the Tower and that which is necessary for its exploitation and also that which the City may substitute for the Minister of the General Commission, for the enjoyment of floors especially fitted out for scientific experiments.

It is intended that the City of Paris will always maintain in the Tower an access in keeping with the needs of exploitation, notably access by elevator car.

Article 13

In the case of war or the declaration of a state of siege, the State will have the clear right to take the place, actively or passively, of Mr. Eiffel for the use of the Tower. An appraisal will be made of the value of the installations, and each side will have the right of comparison and of agreement with said value.

During the entire time Mr. Eiffel is thus deprived of the use of the Tower, the State will be completely responsible.

For compensation of this suspension of use, the term of the concession will be extended one year for every period of three months or fraction of three months during which the suspension occurs.

Article 14

Mr. Eiffel will, at every moment, have the legal right to form an Association or Company either for the construction, or for the exploitation, of the Tower, which Association or Company will cede back all or part of its rights and obligations. The said substitution will be approved by the State or the City, following the period of exploitation in which the Association will be constituted.

Article 15

In the case where, after the advice of the Commission specially instituted by decree of the Minister, on the date of 12 May 1886, Mr. Eiffel or his rightful claimants, will not perform due diligence necessary to assure the completed execution for the engagements of the date of the opening of the Exposition, also in the case if Mr. Eiffel or his rightful claimants does not manifest the wish to continue the work, or the wish can be implied as such by the cessation of the work for a period of twenty days in which the work is stopped and remains ineffective, the State has the right to cancel the present agreements and to continue the execution of work as is here decreed:

First. In the case of cancellation, the State may leave the work in the situation in which they found it; in this case, Mr. Eiffel will be owed, from the original allocation of 1,500,000 Francs, a sum proportionate to the completed work.

The State can also demolish the construction completed to date. In this case, the net proceeds of the resale of the materials (after the deduction of the cost of demolition and the return of the place to its former state) will revert to Mr. Eiffel or his rightful claimants, for all that exceeds the portion of the sum of 1,500,000 Francs already given to Mr. Eiffel or to his rightful claimants.

Second. In the case where the State opts to continue the work, they will carry out the administration, by the means which the State might judge necessary, at the expense of Mr. Eiffel, whose help and surveillance that may be necessary obtained by entrepreneurs chosen by the State, the ensuing Contract resulting will be in effect for the rest of the disposition.

Article 16

As a guaranty for the engagement undertaken by him, Mr. Eiffel will deposit, according to a deadline that will be fixed by the Ministry, as a bond of surety, the sum of 100,000 (one hundred thousand) Francs in legal deposits, for yearly income to the State registered or mixed, or rent to the State and value to the Treasury, to the bearer, conforming to Articles 5 and following of the decree of 18 November 1882.

This sum will be returned to Mr. Eiffel after the completion of the achievement and the definite acceptance of the Tower, except for 10,000 (ten thousand) Francs which will be retained until the expiration of the present agreement.

Article 17

The parties will remain compliant to the clauses and conditions general and relative to the enterprises of the Exposition, decreed by the Minister of the General Commission on 5 August 1886, in all those aforesaid clauses which are not contrary to the present agreement.

In all times, as concerns Article 17 relative to the safety of workers who may be sick or wounded, it is intended that a reserve of one percent will be apportioned from the allocation of 1,500,000 francs granted by the state.

It is further intended that the relief allowed by the Administration of the Exposition, by virtue of the dispositions here recalled, will be the responsibility of Mr. Eiffel for accidents of every nature which shall exceed the reserve of 15,000 (fifteen thousand) Francs.

Article 18

The present agreement will only be legal when accompanied by a prescribed registration fee of three (3) Francs.

The present agreements are done in triplicate originals in Paris, the 8 of January, 1887.

Seen and approved: Seen and approved:
Signed: E. LOCKROY   Signed: G. EIFFEL
Seen and approved:
Signed: POUBELLE

From Société Nouvelle d'Exploitation de la Tour Eiffel.

Paris: Imprimerie et Librairie Centrales des Chemin de Fer, Imprimerie Chaix, Société Anonyme au Capital de Six Millions, 1889: 1-12.

21

The Trans-Siberian Railway

Russia

DID YOU KNOW . . . ?

- The Trans-Siberian Railway was one of the largest railway-construction projects in history.
- Construction began in 1891 and was completed in 1904.
- It extends nearly 5,780 miles (9,300 kilometers) from Moscow to Vladivostok.
- An old Russian sleigh track was chosen as the best route for the railway across Russia.
- 90,000 laborers, many of them soldiers and convicts, worked on the railway.
- The total number of workers employed on the project was estimated at approximately 10,000 in 1891, and rose to nearly 90,000 by 1895.
- The project cost at least one trillion rubles.
- Lake Baikal, the world's deepest lake, contains 23 percent of the world's freshwater reserves.
- The word *czar* derives from the Latin word *Caesar.*

While other national rail projects, like the U.S. transcontinental railroad, the Canadian Pacific Railway, and the Canadian National Railway, had as their purpose improved internal national communication, the project in Russia was different. Of course, the czars saw such a link as another way to create added routes within their extensive empire. Even more important, however, was the possibility of connecting Russia with Europe, Asia, and ultimately North

The Trans-Siberian Express passes by Lake Baikal. © Wolfgang Kaehler / Corbis.

America. One could say that other national railways looked inward; Russia's looked outward.

The Trans-Siberian Railway was begun by Czar Alexander III and completed by his son, Czar Nicholas. Once started, it became one of the largest rail-construction projects in history. It took 12 years to build, was 4,690 miles (7,500 kilometers) long, and was perhaps one of the most difficult construction projects in terms of natural, financial, and human resources.

HISTORY

Like the United States and Canada, nineteenth-century Russia constructed a great railway spanning a continent. In each case the new railway tied together a vast and diverse nation, and upon completion served as an engine for further economic development. Russia's Trans-Siberian Railway was built after the U.S. railroad and was far longer, running from European Russia to the Pacific coast of eastern Asia. Construction began in 1891 and was finished in 1904 on a route that cut across Manchuria. A longer, roundabout route that lay entirely within Russian territory was completed in 1916. Today, the railway extends nearly 5,780 miles (9,300 kilometers) from Moscow to Vladivostok, and most of it is electrified.

The Great Siberian Way, as it was called in czarist days, became a necessity when Vladivostok developed into a major port not long after its founding in 1860. An overland transportation link from there to Moscow and

St. Petersburg facilitated the development of Russia's vast Siberian hinterlands, some of them relatively recently annexed, and was of immense strategic value to the Russian military. The only alternative to a rail connection was a sea journey that covered more than 12,000 miles (20,000 kilometers). Because of the enormous anticipated costs, however, some government officials resisted building a railway.

Czar Alexander III began the push that produced action. He noted on a report from the general-governor of Irkutsk in 1886: "I have read many reports of the governors-general of Siberia and must own with grief, and share that until now the Government has done scarcely anything toward satisfying the needs of this rich, but neglected country! It is time, high time!" (in the czar's own hand) (Dmitriev-Mamonov and Zdsiarski, 56). Expeditions were mounted to explore possible routes. Meanwhile, concern grew over the threat posed by rumors of communication- and transportation-infrastructure development in Manchuria by the Chinese and the British.

The pace of development picked up after the czar authorized the construction of the railway in the statement reprinted here, which he wrote on the margins of a report in July 1890. In February 1891 a ministerial committee approved a proposal to start work on the railway simultaneously from Miass in the west and Vladivostok in the east.

An official declaration of the decision to start the project was sent by Alexander to his son, Czarevitch Nicholas, heir to the throne, who was traveling in the Far East. Nicholas received the rescript (also produced here) when he arrived at Vladivostok in May for the formal ceremony launching railway construction.

CULTURAL CONTEXT

The idea of a connected railway, especially one extending into the vast reaches of Siberia, had circulated among the Russian intelligentsia and the court since 1857. But the costs were so staggering and the terrain so difficult that the project remained inactive. The economic prospects of the coal and iron mines in the Ural Mountains suggested that a rail scheme might pay for itself. In 1873, the Ural Railway Company was established, and the idea of a national railway again came into consideration.

When Czar Alexander III rose to power in 1881, he felt a passionate love for his nation clouded by a nagging worry that Siberia was too disconnected from the centers of power. Siberia lay within easy reach of Asia, where borders were threatened. If a railway could reach Siberia, the goals of unification, security, and prosperity might all be achieved at the same time. As with other rulers throughout history, the czar recognized the advantage of a centrally controlled transit route where subjects could traverse the land without fear. New towns along the route would prosper quickly thanks to the expanded commerce that would inevitably develop.

A switch operator poses on the Trans-Siberian Railway, near the town of Ust Katav on the Yuryuzan River, 1910. Courtesy of the Library of Congress.

In 1894, Nicholas ascended the Russian throne following the death of his father, Alexander III. At his first meeting after becoming czar, the new imperial ruler addressed his ministers: "Gentlemen! To have begun the construction of the railway line across Siberia is one of the greatest achievements of the glorious reign of my never-to-be-forgotten Father. The fulfillment of this essentially peaceful work, entrusted to me by my beloved Father, is my sacred duty and my sincere desire. With your assistance, I hope to complete the construction of a Siberian line and to have it done economically and, most important of all, quickly and solidly."

PLANNING

Railways already existed in European Russia; the routes boomed with business. In 1857, after much discussion and planning, a committee of ministers brought to Czar Alexander II a proposal for a rail route through Russia. He was intrigued by their plan, but the proposal stalled. However, during 1877–78, a mining railway was completed that linked two commercial centers. In 1880, a bridge honoring Czar Alexander II was completed that crossed the Volga River, resulting in a continuous rail line linking European Russia with Orenbúrg in the Ural Mountain basin, enabling passage from Europe to eastern Russia near central Asia.

In 1882, the czar ordered the minister of ways of communication to submit new ideas for communicating with Siberia. In 1884, Adjutant General Konstantin Nikolevich Possiét put forward a plan complete with technical and commercial details. The czar was ready to consider plans for a rail link, and in January 1885 discussions began. A year later, three missions were sent to scout possible routes. They returned three years later with their findings—stark evidence of the size of this immense region and the difficulty of penetrating the terrain.

In 1890, following a report from the acting minister concerning the strategic need for the Ussúri line, the czar wrote his famous directive, the first authorization presented here. Soon thereafter, Alexander sent a note (reproduced here) to his son Nicholas giving orders to "start building a continuous railroad across all Siberia." How would such a vast undertaking be financed? Alexander directed that the funds be taken from the Russian treasury.

By May 1891, Czarevitch Nicholas had organized his father's edict into a plan. In a religious ceremony dedicating the project, the czar's son laid the first stone in the right-of-way. In a later ceremony, he laid a silver plate he had commissioned with silversmiths in St. Petersburg. Construction on the Vladivostok terminus officially began, and the race was on to finish the plan so carefully and ceremonially laid.

BUILDING

It was extremely difficult to identify a route. Build in the north? Too cold. Build in the south? Too swampy. Build in the middle? Too many forests. In the end, the Russians took the path of least resistance and laid the rail along old sleigh routes—an approach similar to the "Princeton method" in which the famous university paved its sidewalks only after students tramped the most convenient routes among the buildings.

Much of the credit for successful completion of construction goes to Sergey I. Witte, minister of finance from 1892 to 1903, who shepherded the work forward. The building of the railroad was a colossal operation, carried out under extremely difficult conditions with such rudimentary equipment as axes, saws, wheelbarrows, and shovels. Track was laid through forests, across permafrost and swampland, and over mountain passes. Numerous rivers were bridged, among them the mile-wide (nearly two-kilometer) Amur (a new Amur bridge has since been built after the old one was dismantled in 1999). Another challenge was the Khor River, characterized by catastrophic high water in the summer, which reportedly rose as much as 30 feet (9 meters). More than a dozen tunnels were built, including the mile-long (nearly two kilometer) Tarmanchukan.

Vast stretches of territory were undeveloped. Timber and many kinds of equipment and other materials had to be brought from thousands of miles away, in some cases even from Poland, Britain, and the United States.

Even with labor conscription, it was almost impossible to recruit the required number of workers. Soldiers and convicts comprised a large portion of the labor force, but workers were also brought in from European Russia. Sign-ups were held in the big cities, and those who agreed to work were transported to the hinterlands. Even so, maintaining a reliable supply of laborers was a constant challenge. As many as 20,000 people worked just on the section at Amurskaya. The total number employed on the project was estimated at approximately 10,000 in 1891, and rose to nearly 90,000 by 1895.

One obstacle seemed insurmountable: Lake Baikal, a huge lake the size of Switzerland directly in the path of the railway construction. At 5,340 feet (1,620 meters) at its deepest point, the lake was such a barrier that a boat would be needed to cross its 400-mile (250-meter) length. Russian managers contacted colleagues in England who had designed a boat that could break through ice. The Russians bought the boat, but it proved to be too large to ship; the vessel was disassembled, transported in pieces, and rebuilt on-site (more than a few parts went missing and had to be improvised).

With the boat, the railway could offer transport across the lake. The tracks stopped right at the water's edge, where passengers boarded the ferry and were invited to sip hot tea or perhaps vodka until they could board another train at the far end of Lake Baikal. During winter, transport reverted to the time-tested and dependable horse-drawn sleighs, which skated across the ice-covered lake with ease. A rail route, complete with 200 bridges and 33 tunnels, was constructed along the southern side of Lake Baikal, and with its completion, seamless transport by train became a reality. Construction figures for 1903 give some sense of the job's scale. Amount of earth moved: 100 million cubic meters. Number of crossbeams supporting track (known as "sleepers") made and laid: 12 million. Weight of rails laid: 1 million tons. The cost of the entire project has been estimated at one trillion rubles (http://www.waytorussia.net/TransSiberian/Intro.html).

By July 1907, there was talk of rebuilding the Siberian railway and recommendations for adding a second track. The Russo-Japanese war had caused huge backlogs along the railway. In 1914, 330,000 tons of coal sat in Archangel ready to be transported. In Vladivostok, steel, lead, trucks, and tools valued at $1 billion were laid up near the harbor. Warehouses were so full of cargo to be moved that some supplies had to be stored outdoors with tarps strapped across the piles to protect against wind and snow. The cargo sat for two years waiting for transport.

Then the czarist government fell. When the United States officially recognized the new government, it gave a $100 million loan to repair and revitalize the railway. An even more valuable gift was sent: John F. Stevens, engineer. He began an immediate study, which pronounced the tracks in good condition and worthy of further investment. America sent workers to make repairs, and in 1917 Russia and America entered into a cooperative agreement: 300 Americans from leading U.S. railway companies came to Russia. The association was called the Russian Railway Service Corps, directed by George Emerson. Each American who served in the corps was given a U.S. Army commission even though the project was not military.

By March 1925 the railway was running once again, and it remains in operation today.

IMPORTANCE IN HISTORY

In 2002, the Russian Ministry held a conference at which it was decided that this venerable railway route was still key to world commerce. More than 250,000 containers were transported via the Trans-Siberian Railway in 1989, with profits of $2 billion. Commercial activity in Asia was continuing to expand in 2002, so the goal was to increase container shipments to one million per year via this central transitway connecting Asia to Europe. (*Pravda*, http://english. pravda.ru/comp/2002/06/03/29609.html)

"Uniting Continents—A Global Task," a phrase used by Vadim Popov, was the headline of a proposal by Vadim's son, Dmitri Alekseevich Popov, a former professor of the Academy of Forestry and head of the academy's Ground Transportation Department. Popov espoused a vision that would connect Asia and America via the Bering Strait with the building of a railway along the rim of the Arctic Circle. As a forest expert, the elder Popov saw opportunities for transporting lumber from the great Siberian forest down the three rivers streaming from that region—the Lena, Ob, and Enisey Rivers—and then transporting the lumber on to Europe and America by train. Just as Alexander III had opened his son Nicholas's eyes to the vast possibilities for Russia if a trans-Siberian railway were built, now another father and son extended the vision. A 1995 letter from the Railway Ministry to the younger Popov states, "The most interesting proposal is the one that would construct the railroad in the area of the North Pole arch, since this version includes both transport communication and the possibility of more complete development of nearby territories using river, sea, and automobile transport" (http://segodnya.spb.rus.net/5-6-00/eng/19_e.htm).

There are other proponents of such a plan. Wallace Hickel, twice governor of Alaska, has been a consistent advocate for a link across the Bering Strait. He and his associate Mead Treadwell made several journeys to Russia to discuss this and other commercial and cultural connections. In 1998, a conference in Russia on "Development of the Arctic Transportation System in the 21st Century" revealed the extent of the mineral resources available in the Arctic that are ripe for development but as yet have not found suitable transport. The same strategic rationale applied to the development of the early sections of the Trans-Siberian Railway, the success of which was due to the mines that supplied profitable cargo.

The Trans-Siberian could also be extended to link Japan and Korea. A paper by Svetlana Kuzmichenko, a Vladivostok BISNIS (Business Information Service for the Newly Independent States) representative, discussed two proposals by the Russian Ministry of Railways to extend the Trans-Siberian in two directions: to Japan via Sakhalin Island for the Tokyo–Moscow–London line, and to South Korea via North Korea for the Seoul–Moscow–London line. These extensions would make possible huge amounts of cargo shipping. Today $600 billion (90 million tons) of cargo travels between East Asia and Europe, one-third in containers. Most cargo moves from Japan, South Korea, Singapore, and China to Europe. Kuzmichenko posits that if the Trans-Siberian were extended, receipts would be worth at least $100 million per year to Russia. In October 2000, Russia launched a study of projects to link the Trans-Siberian Railway to Seoul and Tokyo and has already made major investments in security for the cargo system and faster transport, with more than $3 billion already invested.

Technical difficulties caused by incompatibility of national railways must be surmounted. Russian rail width is different from Korean rails, and both are different from their European counterparts. In Europe, the solution might be to unload and continue on by truck, but other possibilities could be considered. In Japan, linkages could be made via bridge or tunnel. Indeed, a tunnel was begun in 1950, and it could be reactivated with more updated technology.

As a new generation of maglev tube trains becomes prototyped and tested, the Trans-Siberian Railway will look forward to transporting both freight and passengers at thousands of miles per hour. Such high-speed trains require long stretches of uninterrupted terrain to reach high velocity and reap the benefits. The Trans-Siberian route is particularly suited to maglev, and could eventually make it feasible to travel from Moscow to Beijing in a few hours.

Large-scale engineering projects now within range of technical feasibility suggest that heads of state must regard macroengineering research and management as one of their primary tasks, for only the authority of a head of state or government can reduce the extensive delays and inordinate expense of obtaining the numerous authorizations needed for construction. If a head of state champions a project, the lengthy permissions process is often greatly shortened.

FOR FURTHER REFERENCE

Books and Articles

Dmitriev-Mámonov, Aleksandr Ippolitovich, and Anton Feliksovich Zdziárski, eds. *Guide to the Great Siberian Railway with 2 Photographs, 360 Photo-gravurs, 4 Maps of Siberia and 3 Plans of Towns.* Revised by John Marshall. Translated by L. Kukol-Yasnopolsky. St. Petersburg: Artistic Printing Society, 1900. Reprinted by David & Charles Reprints. Newton Abbot: David & Charles, 1971.

Internet

For a link to an original *Scientific American* article on the Trans-Siberian Railway, published in August 26, 1899, with photos, see http://www.travelhistory.org/siberia/.
For a time line of the Russian czars, see http://killeenroos.com/4/Czar1800.htm.

For another view of the building process by the 27th Infantry Regimental Historical Society, see http://www.kolchak.org/History/Siberia/Trans-Siberian%20Railroad.htm.

For a review of the Lake Baikal spur, with pictures and maps, see http://baikal.irkutsk.org/railway.htm.

For Svetlana Kuzmichenko's paper, and a Web site that explores strategies for doing business in Russia and Eurasia, including extending the Trans-Siberian Railway to Japan and Korea, see http://www.bisnis.doc.gov/bisnis/bisdoc/010921rail.htm/s

To book passage on the Trans-Siberian, see "Practice Advice, Tips, and Secrets about the Trans-Siberian Railway" at http://www.waytorussia.net/TransSiberian/Advice.html.

Documents of Authorization—I

Czar Alexander's written instruction:

Necessary to proceed at once to the construction of the line.

From A. I. Dmitriev-Mamonov and A. F. Zdsiarski, eds., *Guide to the Great Siberian Railway with 2 Photographs, 360 Photo-gravurs, 4 Maps of Siberia and 3 Plans of Towns,* translated by L. Kukol-Yasnopolsky, revised by John Marshall (St. Petersburg: Artistic Printing Society, 1900), 58.

Documents of Authorization—II

An imperial rescript sent by Alexander on March 17, 1891 to his son, Czarevitch Nicholas:

Your Imperial Highness! Having given the order to start building a continuous line of railway across Siberia, which is to unite the rich Siberian provinces with the railway system of the Interior, I entrust to you to declare My will, upon your entering the Russian dominion after your inspection of the foreign countries of the East. At the same time, I desire you to lay the first stone at Vladivostok for the construction of Ussúri line, forming part of the Siberian Railway, which is to be carried at the cost of the state and under direction of the government. Your participation in the achievement of this work will be a testimony to My ardent desire to facilitate the communications between Siberia and the other countries of the empire, and to manifest My extreme anxiety to secure the peaceful prosperity of this country.

I remain, your sincerely loving Alexander

From A. I. Dmitriev-Mamonov and A. F. Zdsiarski, eds., *Guide to the Great Siberian Railway with 2 Photographs, 360 Photo-gravures, 4 Maps of Siberia and 3 Plans of Towns,* translated by L. Kukol-Yasnopolsky, revised by John Marshall (St. Petersburg: Artistic Printing Society, 1900), 62.

Related Cultural Documents

Spoken by Czarevitch Nicholas, at a ceremony celebrating the beginning of construction of the railway:

Gentlemen! To have begun the construction of the railway line across Siberia is one of the greatest achievements of the glorious reign of my never-to-be-forgotten

Father. The fulfillment of this essentially peaceful work, entrusted to me by my beloved Father, is my sacred duty and my sincere desire. With your assistance, I hope to complete the construction of a Siberian line and to have it done economically and, most important of all, quickly and solidly.

From A. I. Dmitriev-Mamonov and A. F. Zdsiarski, eds., *Guide to the Great Siberian Railway with 2 Photographs, 360 Photo-gravures, 4 Maps of Siberia and 3 Plans of Towns,* translated by L. Kukol-Yasnopolsky, revised by John Marshall (St. Petersburg: Artistic Printing Society, 1900), 65.

22
The Panama Canal

Panama

DID YOU KNOW . . . ?

➤ Ferdinand de Lesseps originated both the Suez and Panama Canals.

➤ Construction began in 1880 but halted in 1889. It began again in 1904, and the canal opened in 1914.

➤ The canal extends 40 miles (64.4 kilometers), connecting the Atlantic and Pacific Oceans and takes just 8–12 hours to transit.

➤ It shortened the distance between New York and San Francisco by nearly 7,900 miles (12,714 kilometers).

➤ 40,000 laborers died of yellow fever and 20,000 more in construction accidents

➤ It includes six locks (constructed in four years) and 12 chambers.

➤ The project cost $336 million.

➤ A postage stamp decided the discussion over where to locate the canal.

➤ Medical knowledge acquired during this project contributed to the eradication of yellow fever.

The Panama Canal not only split an isthmus but deconstructed a country while overcoming two major obstacles: land and politics. The long, rocky strip of Central America connected two enormous continents, and caused ships from East Coast ports to sail thousands of miles around South America to reach Pacific ports—a costly journey in money, months, and miles. A shortcut across the isthmus would telescope the distance between New York and San Francisco

Cargo ships moored in the Panama Canal. Courtesy of Corbis.

by nearly 7,900 miles (12,714 kilometers). The successful outcome launched a new country and a major new world trade route.

The Panama Canal is an excellent example of the key need for a champion at the highest level in order for a macroengineering project to succeed optimally. In such a case as this, the best intentions of a group of experienced individuals would have had little effect without a political leader who could make it all possible. That individual was Theodore Roosevelt. He became so involved in the Panama Canal that he was photographed running one of the Bucyrus shovels, looking for all the world as if he were building the canal himself—and some would say he did. Roosevelt not only authorized and launched it, but stayed closely involved in its management, going to the limits of his authority as president. Kublai Khan of China would approve; Louis XIV of France would agree. Presidents of the future could take note. Should there be a special macroengineering seminar for heads of state?

HISTORY

Just as the Suez Canal enabled mariners to avoid the arduous circumnavigation of Africa via the Cape of Good Hope, so too, as early as the sixteenth century, there was talk of a waterway that would cut across the Isthmus of Panama to shorten the journey around South America via Cape Horn. Columbus had searched in vain for a passage through the land that would lead him to the Indies, where treasures awaited; during the period 1862–67, when France tried to impose its rule on Mexico, Emperor Napoleon III pondered the idea of building a canal across the southern portion of Mexico, the Isthmus of Tehuantepec.

It was a onetime French diplomat, Ferdinand de Lesseps, builder of the Suez Canal, who began the mammoth macroengineering project that would become the Panama Canal. With his success at Suez behind him, some years later a French company headed by de Lesseps began work on a canal across Panama. He intended to join the Atlantic and Pacific at sea level by building a canal along the Chagres and Rio Grande rivers. But Panama posed quite different challenges from Suez. Besides mountains, which made digging a flat ditch extraordinarily difficult, engineers and laborers working in Panama daily dealt with rain forests and heavy flooding. Panama's swamps, mosquitoes, and poor drainage systems brewed fatal health hazards, and after two years of work, up to 40,000 men had died of yellow fever.

It was not until the early twentieth century that the canal finally became a reality after the United States—motivated by potential military benefit as well as commercial navigation advantages—exercised a little power diplomacy to obtain approval from local authorities and then committed American dollars and personnel to the job.

Creating and building the canal was a complex process, and several key documents in that process are reproduced here. In the first, the Clayton-Bulwer Treaty with Great Britain (Document I), the U.S. government officially declared its interest in bisecting Central America with a "ship canal." In the treaty, both nations agreed to work together to encourage the construction of "practicable communications, whether by canal or railway, across the isthmus."

In the next few years, an American firm built the Panama Railroad in the general area where the canal would eventually be located. At that time, Panama was a republic of Colombia, but in 1878 the Colombian government was persuaded to grant a contract for canal construction to Lieutenant Lucien N. B. Wyse, a representative of Ferdinand de Lesseps's company. Wyse had conducted exploratory surveys in the area, and the agreement became known as the Wyse Concession; it is reproduced here as Document II.

Soon after, de Lesseps pushed forward a plan for a sea-level canal that would tunnel through the Continental Divide. De Lesseps was already in his 70s by that time, and the company and the canal were quickly plagued by troubles, mostly financial. Several times, de Lesseps was forced to go back to his countrymen to garner funds, often as loans and once as a lottery. Work began in 1880, but in a few years it ground to a halt, for the French company found itself unable to overcome multiple challenges. Of no help was de Lesseps's insistence on a sea-level canal, as he had built at Suez, rather than a lock canal, despite the fact that the latter proved to be cheaper and more feasible according to reports at the time. Panama's volatile environment, relentless disease epidemics, and money and labor shortages contributed to the failure. By the end of the decade, de Lesseps's career as a manager of engineering was finished after revelations of swindle, blackmail, and excessive spending. Although he never served his sentence, he and his son were arrested, and his career ended. By 1888, the project was deemed a failure. A sum of $287 million had already been spent, but only 19 miles (31 kilometers) of canal were dug.

U.S. interest in a canal was once again sharpened by the Spanish-American War of 1898. A Nicaraguan route enjoyed considerable support, but in 1902 Congress enacted the Spooner Act (provided here as Document III), which endorsed the Panama option favored by President Theodore Roosevelt. The measure authorized acquisition of a newly reorganized French company's property and interests, and empowered Roosevelt to negotiate with Colombia for the necessary concessions. When Colombia would not ratify a proposed treaty, Roosevelt deployed U.S. troops and warships to support Panamanian separatists, who subsequently declared independence in November 1903. The Hay-Bunau-Varilla Treaty with Panama (reproduced here as Document IV) was quickly drawn up, granting the United States virtually sovereign control over a strip of Panamanian territory 10 miles (16 kilometers) wide within which the canal was to be built.

In 1912 Congress enacted the Panama Canal Act (reproduced here as Document V), which provided for the operation and defense of the canal and for the administration of the Canal Zone. Construction got underway in 1904; the waterway opened for traffic in 1914. Toward the end of the twentieth century, as the canal's military importance declined and Panamanians increasingly voiced displeasure that prime territory in the center of their country remained under foreign rule, the Canal Zone and the canal itself were gradually transferred to Panamanian control.

CULTURAL CONTEXT

Geography can control destiny. The isthmus that connects North and South America had long been eyed as a possible cut-through by explorers. In the sixteenth century, Europeans considered a canal in the region. U.S. President Grant studied the possibility and sent scouting teams to study the logistics. President McKinley was about to sign a bill in 1901 to authorize funds, but just before he was to put pen to official paper, he was assassinated.

There was no question that a transit route would be enormously profitable. Even the tiny railway that had been built across the isthmus made $7 million in its first six years. But that job held portent of things to come: the project claimed 6,000 lives. It took the vision of Ferdinand de Lesseps to get the job started. His success with the Suez Canal seemed reason enough to attempt such a massive undertaking as the Panama Canal, but he estimated this project would require 12 years of labor and cost $132 million. He was wrong, in many ways.

PLANNING

When Ferdinand de Lesseps first considered embarking on construction of the Panama Canal after his success in Suez, he purchased press coverage in France on behalf of the Compagnie Universelle du Canal Interoceanique. Newspaper accounts there pumped up the importance of the undertaking, hoping to excite public interest in the stock offering about to be launched. Unfortunately, the public did not respond, and de Lesseps raised only eight percent of what he needed. He should have stopped there. He ignored the advice of his son, Charles, who warned that Panama would be vastly different from Suez, where they had intimate knowledge of the terrain and close contacts in high places.

Undeterred by financial hurdles, perhaps hoping to make Panama his crowning achievement, the senior de Lesseps made several crucial errors. He visited a site in Panama at the wrong time—the dry season. Thus he had no idea what was in store when his crews began construction during the rainy season, were attacked by mosquitoes and endangered by mudslides, and endured every kind of tropical woe. In another blunder, de Lesseps decided the way to conquer the Chagres River was to dam it—something he determined on a site visit when observing the water at the low-water mark. Little did he suspect that it could become a raging torrent.

De Lesseps began work in 1881; by 1888 he had lost $287 million and 20,000 lives. That was after digging only 19 miles (31 kilometers). So many laborers perished that a chilling decision was made: fire people if they looked sick, so the company would not become responsible for their medical care. Things could not get worse for de Lesseps and his French investors. De Lesseps liquidated the company in 1889. However, an appraisal of the company's belongings—including equipment, maps, and the value of the land already excavated—was very high, and in 1894 the newly reorganized company, Compagnie Nouvelle du Canal de Panama, was opened. But so were court cases against Ferdinand and his son, Charles, who eventually left the country and hid in London. The French press, which had supported de Lesseps in 1880, turned against him, and his career ended in ignominy.

Legal documents have a life of their own, and the lease on the land held by the French company was still in effect until 1903. However, France had determined it could not possibly complete the canal, so a search began for a buyer. That offered a way for America to get in on the opportunity. The Army Corps of Engineers spent nearly three years on reconnaissance and planning. The U.S. Congress attempted in 1889 to charter a construction company in nearby Nicaragua—the Maritime Canal Company—which was financed by J. P. Morgan, but even his wealth was battered in the stock panic of 1893. Another attempt in 1901 halted when President McKinley, who was about to authorize a canal project, was assassinated.

When Theodore Roosevelt became president, he was now in a position to spur action on the canal. He was all too aware of the drawbacks of a lengthy sea passage, amply illustrated by the battleship *Oregon*, which tried to intervene on America's behalf during the war with Spain. Ocean passage for the ship had taken so long that by the time the *Oregon* reached the site of the battle, the war was almost over. In an effort to acquire France's lease before it expired,

Roosevelt engaged the help of two advisors who proved to be public-relations geniuses. William Nelson Cromwell and Phillipe Bunau-Varilla negotiated the lease price down to $40 million—a 60 percent reduction from the original price of $100 million.

If good public relations failed for de Lesseps, press coverage proved extremely helpful to Cromwell and Bunau-Varilla, who spent their own money to buy space in newspapers and magazines, printed special pamphlets, and conducted intense behind-the-scenes lobbying. Bunau-Varilla went on a publicity tour giving public lectures about why Panama should be chosen, and soon a few members of the U.S. Congress began touting Panama, where the work had already been sited and begun, as a preferable location. However, a single factor sealed the eventual decision: Nicaragua had volcanoes. Was it safe to build a canal near a volcano? Just before the congressional vote that would have authorized the location in Nicaragua, Bunau-Varilla managed to find 50 stamps issued by Nicaragua proudly depicting the natural wonder of the massive volcano, which just happened to be 20 miles from the proposed Nicaraguan site of the canal. In a brilliantly persuasive stroke, Bunau-Varilla sent each senator one of the volcano stamps, delivered on the morning of the vote. Then, a Panama supporter, Senator John Spooner, proposed an amendment that authorized purchase of the canal lease but changed the site to Panama. What became known as the Spooner Act sailed through Congress and was signed by President Roosevelt. The Panama Canal is an excellent example of the power of strong, well-focused public relations, combined with a high-level champion, to change the course of a macroengineering project—and history.

The final planning stage came with another round of public relations. The Spooner Act allocated $40 million to purchase the New Panama Canal Company and authorized talks with Colombia, which at the time owned the canal site in Panama. A proposed treaty presented terms to Colombia for a 100-year lease of the isthmus via a down payment of $10 million followed by $250,000 per year—perhaps one of the most expensive price-per-acre real estate deals in history. But the government of Colombia kept changing, and negotiations remained in flux. Panamanian residents favored the sale, but Colombia signaled its opposition. In a bid to resolve the impasse, and now holding the "big stick" for which he became so famous, President Roosevelt arranged a military confrontation. Assuming the role of commander-in-chief, Roosevelt sent the battleship *Nashville* to do what the *Oregon* had been unable to do earlier, and with some intimidation, preplanning, and quick action, a war was staged to help Panama declare its independence from Colombia; America quickly recognized the new country. With the *Nashville* looming ominously on the horizon, Colombia dared not challenge the coup; it reluctantly agreed to the terms of the Hay-Bunau-Varilla Treaty and signed a canal agreement with the new Panamanian government. Bunau-Varilla was installed as ambassador to Panama. His wife designed and sewed a new flag for Panama, just as Betsy Ross had done for the United States many years before. Not surprisingly, a new constitution was ready for consideration, drafted by American interests in confident anticipation of eventual success.

While the money paid informally to Colombian soldiers to encourage their surrender had been modest ($50 each), the big money went to Panama: $10 million from the United States for a long-term lease of the canal site and the rights to build. Funds had even been allocated to pay the poor French peasants who had banked their hopes on the failed de Lesseps, but somehow that $40 million disappeared. On November 3, 1903, the new nation of Panama was born, with a handsome trust fund assuring its growth and prosperity.

BUILDING

Lease in hand, America now faced the task of building the canal. The first engineer in charge, John Findlay Wallace, did no better than de Lesseps. In an alarming setback, Wallace returned home. America had spent $128 million with little to show but failure. Wallace was replaced by the second chief engineer, John Stevens, who was cut from the same cloth as Theodore Roosevelt—a big burly fellow who saw opportunity and solutions where others saw danger and failure. Stevens was a proven commodity, having had considerable success as a railroad builder, and he developed an innovative design that proved to be far more workable than de Lesseps's original insistence on a sea-level canal.

A shot looking south, showing construction of the center wall, 1910. Courtesy of the Library of Congress.

The terrain, Stevens posited, might be more easily penetrated with a combination of lakes and locks. He had to work behind the scenes in Congress to change the design, but finally received permission to dam up the unmanageable Chagres River, which created Gatún Lake, the largest man-made lake at the time. Then ships could be brought out of the lake through a chain of locks. This idea proved successful, and provided not only a solution to the Chagres River problem but also a source of irrigation and drinking water. Ironically, if de Lesseps had listened to French engineer Godin de Lépinay, who attended the meeting to propose a lock-based canal when de Lesseps insisted on a sea-level design, the history of the canal could have been entirely different.

Besides conquering terrain, Stevens also helped vanquish the rampant diseases that had completely undermined earlier progress. Stevens brought in Dr. William Gorgas, a public-health expert who had wiped out yellow fever in Cuba by eradicating mosquitoes. Before this time, precautions had been taken against armies of ants that swarmed over the laborers as they dug and even infested the hospital erected by the French at Ancon. Not understanding the nature of the problem, the hospital had situated potted plants around the premises to create a more pleasant environment for recovering workers, and the staff placed a container of water in each pot to keep out the ants. Of course, mosquitoes breed in standing water, and the hospital windows had no screens to repel them. As the recovering patients slept, they became reinfected with the disease that had put them in sick bay in the first place! Gorgas understood this cycle, and he oversaw a complete revamping of Panama's landscape. Swamps were drained, plumbing was installed, and roads were paved and ditches removed so they would not fill with water during the rainy season, thus inviting mosquitoes. Pesticides scented the air. Soon the new country became more attractive and hospitable; new towns began to flourish, churches were built, and schools were erected and staffed. Now that the army had a clean environment in which to live and a successful building plan could be implemented, Gorgas could leave.

Swashbuckler Stevens was a starter, not a finisher. He promised Roosevelt he would work on the Panama Canal only until it was clear whether it would succeed or fail. On February 12, 1907, an exhausted but satisfied Stevens turned over the job to army engineer Lieutenant Colonel Thomas Goethals, who oversaw the difficult Culebra Cut and eventually completed the job. In May 1913, the enormous Bucyrus steam shovels broke through, and the Chagres River waters poured into Gatún Lake. By September the troops were ready for a trial run, taking a small tugboat named *Gatún* across the lake and through the locks.

Just one year later, on August 15, 1914, the Panama Canal officially opened. It stretches 40 miles from the Atlantic Ocean at Limon Bay at Colon to the Pacific Ocean at the Bay of Panama at Balboa. Its six locks are made of concrete and include 12 chambers. The locks are 110 feet (69 meters) by 1,100 feet (688 meters) and were constructed in four years. Seventy holes in the chamber walls allow water to flood into the locks. Gigantic floating gates, some as high as 82 feet (52 meters) and weighing 745 tons, guard the chamber doors. Pressure

A work crew toiling on the Panama Canal. Courtesy of the Library of Congress.

from the water prevents the closed gates from admitting water. The canal is powered by electricity generated by a waterfall at Gatún. In total, $336 million was spent. It takes about eight hours for a ship sailing southward from Limón to Balboa to pass through the six locks and the Continental Divide.

In a poignant note, the completion of this project, which had been so driven by drama and heavy public relations, was relegated to the back pages of the newspapers by copy editors who instead devoted headlines to coverage of World War I.

IMPORTANCE IN HISTORY

While the canal's complex geopolitical history is intriguing, the project's organizational and engineering achievements are breathtaking. They could not have been accomplished without the public-health feat of overcoming the deadly mosquito-borne diseases that had stymied the French. Although the

U.S. part of the project saw more than 5,600 people killed by disease and accidents, this figure was just a fraction of the number estimated to have died in the French construction effort. Tens of thousands of laborers, many from various Caribbean islands, worked on the American project, which produced the biggest earthen dam (Gatún Dam) that had ever been built and subsequently resulted in what was then the largest artificial lake and the largest locks ever made. Today the canal stretches over 50 miles (80 kilometers) between the Pacific Ocean deep water on the west and the Caribbean Sea deep water on the east side, and it rises as high as 85 feet (26 meters) above sea level. At bottom, the channel ranges in width from 500 to 1,000 feet (150 to 300 meters).

In late 1999, just before transfer of the canal to Panamanian control was completed, the waterway became the first project to be honored by the American Society of Civil Engineers as a Civil Engineering Monument of the Millennium. The central role of the U.S. Army Corps of Engineers in the building of the Panama Canal illustrates the considerable, if intermittent, role of armies in the construction and maintenance of basic infrastructure. The ancient Roman army built the extensive network of roads that girdled the Mediterranean. By far the largest civil project entrusted to the Army Corps of Engineers was the Tennessee-Tombigbee Waterway, constructed from 1972 to 1985, involving the displacement of more earth than was moved to construct the Panama Canal at an overall cost of close to $2 billion. In the 1980s, China did not hesitate to mobilize five million soldiers to join the battle against the disastrous Yellow River floods. When Canada's Red River flooded in 1987, the city of Winnipeg, Manitoba, was saved from inundation with the assistance of Canadian army regiments that were mobilized to strengthen the dikes with thousands of sandbags.

What does the future hold for the Panama Canal? As of 2000, 13,000 ships had passed through, and the number is expected to double by 2050. Can the environment hold up? According to the Panama Canal Authority, the land surrounding the Panama Canal zone is becoming deforested by developers and showing signs of erosion. With insufficient rainfall it is sometimes difficult to fill Gatún Lake and the secondary reservoir, Madden. The Panama Canal authority warns that by 2010, the situation could become dire.

Global warming is assailing the environment. Some may see an option in the negatives of global warming, especially those who favor the development of an Arctic Circle Northwest Passage route that would be 4,700 nautical miles shorter than Panama for ships from Asia and Europe. But warming also creates rogue icebergs; who can forget the *Titanic*?

A design by MIT professor Ernst G. Frankel, a 2004 member of an advisory board of the Panama Canal Company, contains his vision of an above-ground canal built by constructing a road and walls on both sides that would be closed in and flooded, and might be a solution for Panama or even Nicaragua. There continues to be support for building a second, smaller canal alongside the main one to accommodate smaller ships and therefore reduce the burden on the current waterway.

The Panama Canal Authority has assumed the role of proactive preserver of the environment by participating in a benchmark program to replant 1.4 million acres (about 7% of Panama's land mass), which would preserve and protect its six forests, recapture rainfall, and increase active water storage by 600 percent.

FOR FURTHER REFERENCE

Books and Articles

Cunningham, Michelle. *Mexico and the Foreign Policy of Napoleon III*. Basingstoke, England, and New York: Palgrave, 2001.

McCullough, David. *The Path between the Seas: The Creation of the Panama Canal*. New York: Simon and Schuster, 1977.

Michel, Prince of Greece. *The Empress of Farewells: The Story of Charlotte, Empress of Mexico*. Translated by Vincent Aurora. New York: Atlantic Monthly Press, 2002.

Morris, Edmund. *Theodore Rex*. New York: Random House, 2001.

Newhouse, Elizabeth, ed. *The Builders: Marvels of Engineering*. Washington, DC: National Geographic Society, 1992.

Sandström, Gösta E. *Man the Builder*. New York: McGraw-Hill, 1970.

Stine, Jeffrey K. *Mixing the Waters: Environment, Politics, and the Building of the Tennessee-Tombigbee Waterway*. Akron, OH: University of Akron Press, 1993.

Internet

To access all the many legal documents, see the Panama Canal Web site: http://www.pancanal.com/eng/legal/law/index.html.

For coverage of the American role, see http://www.pbs.org/wgbh/amex/tr/panama.html.

For a summary by Tyler Jones, including details of the vagaries of financial and engineering leadership and bibliography, see http://www.june29.com/Tyler/nonfiction/pan2.html.

For the future of the canal and the activities of the Panama Canal Authority, see http://www.riverdeep.net/current/2000/12/120800_panama.jhtml.

Documents of Authorization—I

GREAT BRITAIN

APRIL 19, 1850

CLAYTON-BULWER TREATY

The United States of America and Her Britannic Majesty, being desirous of consolidating the relations of amity which so happily subsist between them, by setting forth and fixing in a Convention their views and intentions with reference to any means of communication by Ship Canal, which may be constructed between the Atlantic and Pacific Oceans, by the way of the River San Juan de Nicaragua and either or both of the Lakes of Nicaragua or Managua, to any port or place on the Pacific Ocean,—The President of the United States has conferred full powers on John M. Clayton, Secretary of State of the United States; and Her Britannic Majesty on the Right Honourable Sir Henry Lytton Bulwer, a Member of Her Majesty's Most Honourable Privy Council, Knight Commander of the Most Honourable Order of the Bath, and Envoy Extraordinary and Minister Plenipotentiary of Her Britannic Majesty to the United States, for the aforesaid purpose; and the said Plenipotentiaries having exchanged their full powers, which were found to be in proper form have agreed to the following articles:

Article I

The Governments of the United States and Great Britain hereby declare, that neither the one nor the other will ever obtain or maintain for itself any exclusive control over the said Ship Canal; agreeing, that neither will ever erect or maintain any fortifications commanding the same, or in the vicinity thereof, or occupy, or fortify, or colonize, or assume, or exercise any dominion over Nicaragua, Costa Rica, the Mosquito coast, or any part of Central America; nor will either make use of any protection which either affords or may afford, or any alliance which either has or may have to or with any state or people, for the purpose of erecting or maintaining any such fortifications, or of occupying, fortifying, or colonizing Nicaragua, Costa Rica, the Mosquito coast, or any part of Central America, or of assuming or exercising dominion over the same; nor will the United States or Great Britain take advantage of any intimacy, or use any alliance, connection, or influence that either may possess with any state or government through whose territory the said canal may pass, for the purpose of acquiring or holding, directly or indirectly, for the citizens or subjects of the one, any rights or advantages in regard to commerce or navigation through the said canal which shall not be offered on the same terms to the citizens or subjects of the other.

Article II

Vessels of the United States or Great Britain traversing the said canal shall, in case of war between the contracting parties, be exempted from blockade, detention, or capture by either of the belligerents; and this provision shall extend to such a distance from the two ends of the said canal as may hereafter be found expedient to establish.

Article III

In order to secure the construction of the said canal, the contracting parties engage that if any such canal shall be undertaken upon fair and equitable terms by any parties having the authority of the local government or governments through whose territory the same may pass, then the persons employed in making the said canal, and their property used, or to be used, for that object, shall be protected, from the commencement of the said canal to its completion, by the Governments of the United States and Great Britain, from unjust detention, confiscation, seizure, or any violence whatsoever.

Article IV

The contracting parties will use whatever influence they respectively exercise with any state, states, or governments possessing or claiming to possess any jurisdiction or right over the territory which the said canal shall traverse, or which shall be near the waters applicable thereto, in order to induce such states or governments to facilitate the construction of the said canal by every means in their power. And, furthermore, the United States and Great Britain agree to use their good offices, wherever or however it may be most expedient, in order to procure the establishment of two free ports, one at each end of the said canal.

Article V

The contracting parties further engage, that when the said canal shall have been completed, they will protect it from interruption, seizure, or unjust confiscation, and that they will guarantee the neutrality thereof, so that the said canal may forever be open and free, and the capital invested therein secure. Nevertheless, the Governments of the United States and Great Britain, in according their protection to the construction of the said canal, and guaranteeing its neutrality and security when completed, always understand that this protection and guarantee are granted conditionally, and may be withdrawn by both governments, or either government, if both governments, or either government, should deem that the persons or company undertaking or managing the same adopt or establish such regulations concerning the traffic thereupon as are contrary to the spirit and intention of this convention, either by making unfair discrimination in favor of the commerce of one of the contracting parties over the commerce of the other, or by imposing oppressive exactions or unreasonable tolls upon the passengers, vessels, goods, wares, merchandise, or other articles. Neither party, however, shall withdraw the aforesaid protection and guarantee without first giving six months' notice to the other.

Article VI

The contracting parties in this convention engage to invite every state with which both or either have friendly intercourse to enter into stipulations with them similar to those which they have entered into with each other, to the end that all other states may share in the honor and advantage of having contributed to a work of such general interest and importance as the canal herein contemplated. And the contracting parties likewise agree that each shall enter into treaty stipulations with such of the Central American states as they may deem advisable, for the purpose of more effectually carrying out the great design of this convention, namely, that of constructing and maintaining the said canal as a ship communication between the two oceans for the benefit of mankind, on equal terms to all, and of protecting the same; and they also agree, that the good offices of either shall be employed, when requested by the other, in aiding and assisting any differences that arise as to right or property over the territory through which the said canal shall pass between the states or governments of Central America, and such differences should in any way impede or obstruct the execution of the said canal, the Governments of the United States and Great Britain will use their good offices to settle such differences in the manner best suited to promote the interests of the said canal, and to strengthen the bonds of friendship and alliance which exist between the contracting parties.

Article VII

It being desirable that no time should be unnecessarily lost in commencing and constructing the said canal, the Governments of the United States and Great Britain determine to give their support and encouragement to such persons or company as may first offer to commence the same, with the necessary capital, the consent of the local authorities, and on such principles as accord with the spirit and intention of this convention; and, if any persons or company should already have, with any state through which the proposed ship canal may pass, a contract for the construction of such canal as that specified in this convention, to the stipulations of which contract neither of the contracting parties in this convention have any just cause to object, and the said persons or company shall moreover have made preparations, and expended time, money, and trouble, on the faith of such contract, it is hereby agreed that such persons or company shall have a priority of claim over every other person, persons, or company to the protection of the Governments of the United States and Great Britain, and be allowed a year from the date of the exchange of the ratifications of this convention for concluding their arrangements, and presenting evidence of sufficient capital subscribed to accomplish the contemplated undertaking; it being understood that if, at the expiration of the aforesaid period, such persons or company be not able to commence and carry out the proposed enterprise, then the Governments of the United States and Great Britain shall be free to afford their protection to any other persons or company that shall be prepared to commence and proceed with the construction of the canal in question.

Article VIII

The Governments of the United States and Great Britain having not only desired, in entering into this convention, to accomplish a particular object, but also to establish a general principle, they hereby agree to extend their protection, by treaty stipulations, to any other practicable communications, whether by canal or railway, across the isthmus which connects North and South America, and especially to the inter-oceanic communications, should the same prove to be practicable, whether by canal or railway, which are not proposed to be established by the way of Tehuantepec or Panama. In granting, however, their joint protection to any such canals or railways as are by this article specified, it is always understood by the United States and Great Britain that the parties constructing or owning the same shall impose no other charges or conditions of traffic thereupon, than the aforesaid governments shall approve of, as just and equitable; and, that the same canals, or railways, being open to the citizens and subjects of the United States and Great Britain on equal terms, shall, also, be open on like terms to the citizens and subjects of every other State which is willing to grant thereto, such protection as the United States and Great Britain engage to afford.

Article IX

The ratifications of this Convention shall be exchanged at Washington, within six months from this day, or sooner, if possible.

In faith whereof, we, the respective Plenipotentiaries, have signed this Convention, and have hereunto affixed our seals.

Done, at Washington, the nineteenth day of April, Anno Domini, one thousand eight hundred and fifty.

/s/ JOHN M. CLAYTON
/s/ HENRY LYTTON BULWER.

Documents of Authorization—II

WYSE CONCESSION

MARCH 20, 1878

CONTRACT FOR THE CONSTRUCTION OF AN INTER-OCEANIC CANAL ACROSS COLOMBIAN TERRITORY.

Eustorgio Salgar, Secretary of the Interior and of Foreign Relations of the United States of Colombia, duly authorized of the one part, and of the other part Lucien N. B. Wyse, chief of the Isthmus Scientific Surveying Expedition in 1876, 1877, and 1878, member and delegate of the board of directors of the International Inter-oceanic Canal Association, presided by General Etienne Turr, in conformity with powers bestowed at Paris, from the 27th to the 29th of October, 1877, have celebrated the following contract:

Article I. The Government of the United States of Colombia grants to Mr. Lucien N. B. Wyse, who accepts it in the name of the civil International Inter-oceanic Canal Association, represented by their board of directors, the exclusive privilege for the construction across its territory and for the operating of a canal between the Atlantic and Pacific oceans. Said canal may be constructed without restrictive stipulations of any kind.

This concession is made under the following conditions:

First. The duration of the privilege shall be for ninety-nine years from the day on which the canal shall be wholly or partially opened to public service, or when the grantees or their representatives commence to collect the dues on transit and navigation.

Second. From the date of approbation by the Colombian Congress for the opening of the inter-oceanic canal, the Government of the Republic can not construct, nor concede to any company or individual, under any consideration whatever, the right to construct another canal across Colombian territory which shall communicate the two oceans. Should the grantees wish to construct a railroad as an auxiliary to the canal, the Government (with the exception of existing rights) can not grant to any other company or individual the right to build another inter-oceanic railroad, nor do so itself, during the time allowed for the construction and use of the canal.

Third. The necessary studies of the ground and the route for the line of the canal shall be made at the expense of the grantees, by an international commission of individuals and competent engineers, in which two Colombian engineers shall take part. The commission shall determine the general route of the canal and report to the Colombian Government directly, or to its diplomatic agents in the United States or Europe, upon the results obtained, at the latest in 1881, unless unavoidable circumstances, clearly proven, should prevent their so doing. The report shall comprise in duplicate the scientific labors performed and an estimate of the projected work.

Fourth. The grantees shall then have a period of two years to organize a universal joint stock company, which shall take charge of the enterprise, and of the construction of the canal. This term shall commence from the date mentioned in the preceding paragraph.

Fifth. The canal shall be finished and placed at the public service within the subsequent twelve years after the formation of the company which will undertake its construction, but the executive power is authorized to grant a further maximum term of six years in the case of encountering superhuman obstacles beyond the power of the company, and if after one-third of the canal is built, the company should acknowledge the impossibility of concluding the work in the said twelve years.

Sixth. The canal shall have the length, depth, and all other conditions requisite in order that sailing vessels and steamships measuring up to 140 meters long, 16 meters in width, and 8 meters in draft shall, with lowered topmast, be able to pass the canal.

Seventh. All public lands which may be required for the route of the canal, the ports, stations, wharves, moorings, warehouses, and in general for the construction and service of the canal, as well as for the railway, should it be convenient to build it, shall be ceded gratis to the grantees.

Eighth. These unoccupied public lands shall revert to the Government of the Republic with the railroad and canal at the termination of this privilege. There is also granted for the use of the canal a belt of land 200 meters wide on each side of its banks throughout all the distance which it may run; but the owners of lands on its banks shall have free access to the canal and its ports, as well as to the right of use of any roads which the grantees may open there; and this without paying any dues to the company.

Ninth. If the lands through which the canal shall pass or upon which the railroad may be built, should, in whole or in part, be private property, the grantee shall have the right to demand their expropriation by the Government according to all the legal formalities in such cases. The indemnity which shall be made to the landowners, and which shall be based on their actual value, shall be at the expense of the company. The grantees shall enjoy in this case, and in those of the temporary occupation of private property, all the rights and privileges which the existing legislation confers.

Tenth. The grantees may establish and operate at their cost the telegraph lines which they may consider useful as auxiliaries in the building and management of the canal.

Eleventh. It is, however, stipulated and agreed that if, before the payment of the security determined upon in Article II, the Colombian Government should receive any formal proposal, sufficiently guaranteed in the opinion of the said Government, to construct the canal in less time and under more advantageous conditions for the United States of Colombia, said proposal shall be communicated to the grantees or their representatives that they may be substituted therein, in which case they shall be preferred; but if they do not accept such substitution the Colombian Government, in the new contract which they may celebrate, shall exact, besides the guarantee mentioned in Article II, the sum of $300,000 in coin, which shall be given as indemnity to the grantees.

Article II. Within the term of twelve months from the date at which the international commission shall have presented the definite results of their studies, the grantees shall deposit in the bank or banks of London, to be designated by the national executive power, the sum of 750,000 francs, to the exclusion of all paper money, as security for the execution of the work. The receipt of said banks shall be a voucher for the fulfillment of said deposit. It is understood that if the grantees should lose that deposit by virtue of the stipulations contained in clauses 2 and 3 of Article XXII of the present contract, the sum referred to, with interest accrued, shall become in toto the property of the Colombian Government. After the conclusion of the canal, said sum, without interest, which latter will in this case belong to grantees, shall remain for benefit of the treasury, for the outlays which it may have incurred or may incur in the construction of buildings for the service of the public officers.

Article III. If the line of the canal to be constructed from sea to sea should pass to the west and to the north of the imaginary straight line which joins

Cape Tiburon with Garachiné Point, the grantees must enter into some amicable arrangement with the Panama Railroad Company, or pay an indemnity, which shall be established in accordance with the provisions of law 46, of August 16, 1867, "approving the contract celebrated on July 5, 1867, reformatory of the contract of April 15, 1850, for the construction of an iron railroad from one ocean to the other through the Isthmus of Panama."

In case the international commission should choose the Atrato, or some other stream already navigable, as one of the entrances to the canal, the ingress and egress by such stream, and the navigation of its waters, so long as it is not intended to cross the canal, shall be open to commerce and free from all imposts.

Article IV. Besides the lands granted in paragraphs 7 and 8 or Article I, there shall be awarded to the grantees, as an aid for the accomplishment of the work, and not otherwise, 500,000 hectares of public lands, with the mines they may comprise, in the localities which the company may select. This award shall be made directly by the national executive power. The public lands situated on the seacoast, on the borders of the canal or of the rivers, shall be divided in alternate lots between the Government and the company, forming areas of from one to two thousand hectares. The measurements for the allotment or locating shall be made at the expense of the grantees and with the intervention of Government commissioners. The public lands thus granted, with the mines they may hold, shall be awarded to the grantees as fast as the work of construction of the canal progresses, and in accordance with rules to be laid down by the executive power.

Within a belt of 2 myriameters on each side of the canal, and during five years after the termination of the work, the Government shall not have the right to grant other lands beyond the said lots until the company shall have called for the whole number of lots granted by this article.

Article V. The Government of the Republic hereby declares the ports at each end of the canal, and the water of the latter from sea to sea, to be neutral for all time; and, consequently, in case of war among other nations, the transit through the canal shall not be interrupted by such event, and the merchant vessels and individuals of all nations of the world may enter into said ports and travel on the canal without being molested or detained. In general, any vessel may pass freely without any discrimination, exclusion, or preference of nationalities or persons, on payment of the dues and the observance of the rules established by the company for the use of the canal and its dependencies. Exception is to be made of foreign troops, which shall not have the right to pass without permission from Congress, and of the vessels of nations which, being at war with the United States of Colombia, may not have obtained the right to pass through the canal at all times, by public treaties wherein is guaranteed the sovereignty of Colombia over the Isthmus of Panama and over the territory whereon the canal is to be cut, besides immunity and neutrality of the said canal, its ports, bays, and dependencies and the adjacent seas.

Article VI. The United States of Colombia reserve to themselves the right to pass their vessels, troops, ammunitions of war at all times and without paying any dues whatever. The passage of the canal is strictly closed to war vessels of nations at war, and which may not have acquired, by public treaty with the Colombian Government, the right to pass by the canal at all times.

Article VII. The grantees will enjoy the right during the whole time of the privilege to use the ports at the termini of the canal, as well as at intermediate points, for the anchorage and repair of ships and the loading, depositing, trans-shipping, or landing of merchandise. The ports of the canal shall be open and free to the commerce of all nations, and no import duties shall be exacted except on merchandise destined to be introduced for the consumption of the rest of the Republic. The said ports shall therefore be open to importations from the commencement of the work, and the custom-houses and the revenue service which the Government may deem convenient for the collection of duties on merchandise destined for other portions of the Republic shall be established, in order to prevent introduction of smuggled goods.

Article VIII. The executive power shall dictate, for the protection of the financial interests of the Republic, the regulations conducive to the prevention of smuggling, and shall have the power to station, at the cost of the nation, the number of men which they may deem necessary for that service.

Out of the indispensable officials for that service, ten shall be paid by the company, and their salaries shall not exceed those enjoyed by employees of the same rank in the Baranquilla customhouse.

The company shall carry gratis through the canal, or on the auxiliary railway, the men destined for the service of the nation, for the service of the State through whose territory the canal may pass, or for the service of the police, with the object of guarding against foreign enemies, or for the preservation of public order, and shall also transport gratis the baggage of such men, their war materials, armament, and clothing which they may need for the service assigned to them.

The subsistence of the public force which may be deemed necessary for the safety of the inter-oceanic transit shall likewise be at the expense of the company.

Article IX. The grantee shall have the right to introduce, free of import or other duties of whatever class, all the instruments, machinery, tools, fixtures, provisions, clothing for laborers which they may need during all the time allowed to them for the construction and use of the canal. The ships carrying cargoes for the use of the enterprise shall enjoy free entry into whatever point shall afford them easy access to the line of the canal.

Article X. No taxes, either national, municipal, of the State, or of any other class, shall be levied upon the canal, the ships that navigate it, the tugs and vessels at the service of the grantees, their warehouses, workshops, and offices, factories of whatever class, storehouses, wharves, machinery, or other works or property of whatever character belonging to them, and which they may need for the service of the canal and its dependencies during the time conceded for its construction and operation. The grantees shall also have the right to take from unoccupied lands the materials of any kind which they may require without paying any compensation for the same.

Article XI. The passengers, money, precious metals, merchandise, and articles and effects of all kinds which may be transported over the canal, shall also be exempt from all duties, national, municipal, transit, and others. The same exemption is extended to all articles and merchandise for interior or exterior commerce which may remain in deposit, according to the conditions which may be stipulated, with the company in the storehouses and stations belonging to them.

Article XII. Ships desiring to cross the canal shall present at the port of the terminus of the canal at which they may arrive their respective registers and other sailing papers, prescribed by the laws and public treaties, so that the vessels may navigate without interruption. Vessels not having said papers, or which should refuse to present them, may be detained and proceeded against according to law.

Article XIII. The Government allows the immigration and free access to the lands and shops of the grantees of all the employees and workingmen of whatever nationality, who may be contracted for the work, or who may come to engage themselves to work on the canal, on condition that such employees or laborers shall submit to the existing laws, and to the regulations established by the company. The Government promises them support and protection, and the enjoyment of their rights and guaranties, in conformity with the national constitution and laws during the time they may sojourn on the Colombian territory.

The national peons and laborers employed on the work of the canal shall be exempt from all requisition of military service, national as well as of the State.

Article XIV. In order to indemnify the grantees of the construction, maintenance, and working expenses incurred by them, they shall have, during the whole period of the privilege, the exclusive right to establish and collect for the passage of the canal and its ports, the dues for light-houses, anchorage, transit, navigation, repairs, pilotage, towing, hauling, storage, and of station according to the tariff which they may issue, and which they may modify at any time under the following express conditions:

First. They shall collect these dues, without any exceptional favor, from all vessels in like circumstances.

Second. The tariffs shall be published four months before their enforcement in the Diario Official of the Government, as well as in the capitals and the principal commercial ports of the countries interested.

Third. The principal navigation dues to be collected shall not exceed the sum of 10 francs for each cubic meter resulting from the multiplication of the principal dimensions of the submerged part of the ship in transit (length, breadth, and draft).

Fourth. The principal dimensions of the ship in transit—that is to say, the maximum exterior length and breadth at the water line, as well as the greatest draft—shall be the metrical dimension inserted in the official clearance papers, excepting any modifications supervening during the voyage. The ship's captains and the company's agents may demand a new measurement, which operations shall be carried out at the expense of the petitioner; and,

Fifth. The same measurement—that is to say, the number of cubic meters contained in the parallelopipedon circumscribing the submerged part of the ship—shall serve as a basis for the determination of the other accessory dues.

Article XV. By way of compensation for the rights and exemptions which are allowed to the grantees in this contract, the Government of the Republic shall be entitled to a share amounting to 5 per cent on all collections made by the company, by virtue of the dues which may be imposed in conforming with Article XIV, during the first twenty-five years after the opening of the canal to the public service. From the twenty-sixth up to the fiftieth year, inclusive, it shall be entitled to a share of 6 per cent; from the fifty-first to the seventy-fifth year, inclusive, it shall be entitled to a share of 6 per cent; from the fifty-first to the seventy-fifth to 7 per cent, and from the seventy-sixth to the termination of the privilege to 8 per cent. It is understood that these shares shall be reckoned, as has been said, on the gross income from all sources, without any deduction whatever for expenses, interest on shares or on loans or debts against the company. The Government of the Republic shall have the right to appoint a commissioner or agent, who shall intervene in the collections and examine the accounts, and the distribution or payment of the shares coming to the Government shall be made in due half-yearly installments. The product of the 5, 6, 7, and 8 per cent shall be distributed as follows:

Four-fifths of it shall go to the Government of the Republic and the remaining one-fifth to the government of the State through whose territory the canal may pass.

The company guarantees to the Government of Colombia that the share of the latter shall in no case be less than the sum of $250,000 a year, which is the same as that received as its share in the earnings of the Panama Railroad, so that if in any year the 5 per cent share should not reach said sum, it shall be completed out of the common funds of the company.

Article XVI. The grantees are authorized to require payment in advance of any charges which they may establish; nine-tenths of these charges shall be made payable in gold, and only the remaining one-tenth part shall be payable in silver of 25 grams, of a fineness of 900.

Article XVII. The ships which shall infringe upon the rules established by the company shall be subject to the payment of a fine which said company shall fix in its regulations, of which due notice shall be given to the public at the time of the issue of the tariff. Should they refuse to pay said fine, or furnish sufficient security, they may be detained and prosecuted according to the laws. The same proceedings may be observed for the damages they may have caused.

Article XVIII. If the opening of the canal shall be deemed financially possible, the grantees are authorized to form, under the immediate protection of the

Colombian Government, a universal joint-stock company, which shall undertake the execution of the work, taking charge of all financial transactions which may be needed. As this enterprise is essentially international and for public utility, it is understood that it shall always be kept free from political influences.

The company shall take the name of the Universal Inter-oceanic Canal Association; its residence shall be fixed in Bogota, New York, London, or Paris, as the grantees may choose; branch offices may be established wherever necessary. Its contracts, shares, bonds, and titles of its property shall never be subjected by the Government of Colombia to any charges for registry, emission, stamps, or any similar imposts upon the sale or transfer of these shares of bonds, as well as on the profits produced by these values.

Article XIX. The company is authorized to reserve as much as 10 per cent of the shares emitted, to form a fund of shares, to the benefit of the founders and promoters of the enterprise. Of the products of the concern, the company take, in the first place, what is necessary to cover all expenses of repairs, operations, and administration, and the share which belongs to the Government, as well as the sums necessary for the payment of the interest and the amortization of the bonds, and, if possible, the fixed interest or dividend of the shares; that which remains will be considered as net profit, out of which 80 per cent at least will be divided among the shareholders.

Article XX. The Colombian Government may appoint a special delegate in the board of the directors of the company whenever it may consider it useful to do so. This delegate shall enjoy the same advantages as are granted to the other directors by the by-laws of the company.

The grantees pledge themselves to appoint in the capital of the Union, near the national Government, a duly authorized agent for the purpose of clearing up all doubts and presenting any claims to which this contract may give rise. Reciprocally and in the same sense, the Government shall appoint an agent who shall reside in the principal establishment of the company situated on the line of the canal; and, according to the national constitution, the difficulties which may arise between the contracting parties shall be submitted to the decision of the federal supreme court.

Article XXI. The grantees, or those who in the future may succeed them in their rights, may transfer these rights to other capitalists or financial companies, but it is absolutely prohibited to cede or mortgage them under any consideration whatever to any nation or foreign government.

Article XXII. The grantees or their representatives, shall lose the right hereby acquired in the following cases:

First. If they do not deposit, on the terms agreed upon, the sum which by way of security must insure the execution of the work.

Second. If, in the first year of the twelve that are allowed for the construction of the canal, the works are not already commenced, in this case the company shall lose the sum deposited by way of security, together with the interest that may have accrued, all of which will remain for the benefit of the Republic.

Third. If, at the end of the second period fixed in paragraph 5 of Article I, the canal is not transitable, in this case also the company shall lose the sum deposited as security, which, with the interests accrued, shall remain for the benefit of the Republic.

Fourth. If they violate the prescriptions of Article XXI; and

Fifth. If the service of the canal should be interrupted for a longer period than six months without its being occasioned by the acts of God, etc.

In cases 2, 3, 4, 5, the federal supreme court shall have the right to decide whether the privilege has become annulled or not.

Article XXIII. In all cases of decisions of nullity the public lands mentioned in clauses 7 and 8 of Article I, and such lands as are not settled or inhabited from among those granted by Article IV, shall revert to the possession of the Republic in the condition they may be found in, and without any indemnity whatever, as well as the buildings, materials, works, and improvements which the grantees may possess along the canal and its accessories. The grantees shall only retain their capital, vessels, provisions, and in general all movable property.

Article XXIV. Five years previous to the expiration of the ninety-nine years of the privilege, the executive power shall appoint a commissioner to examine the condition of the canal and annexes, and, with the knowledge of the company or its agents on the Isthmus, to make an official report, describing in every detail the condition of the same and pointing out what repairs may be necessary. This report will serve to establish in what condition the canal and its dependencies shall be delivered to the National Government on the day of expiration of the privilege now granted.

Article XXV. The enterprise of the canal is reputed to be of public utility.

Article XXVI. This contract, which will serve as a substitute for the provisions of law 38 of May 26, 1876, and the clauses of the contract celebrated on the 28th of May of the same year, shall be submitted for the approval of the President of the Union and the definite acceptance by the Congress of the nation.

In witness whereof they sign the present in Bogota on the 20th March, 1878.

Eustorgio Salgar.
Lucien N. B. Wyse.
Bogota, March 23, 1878
Approved: Aquileo Parro,
President of the Union.
Eustorgio Salgar,
Secretary of the Interior and of Foreign Relations.

[Note from Lucien N.B. Wyse, wherein he declares he accepts all the modifications made by law 28 to the contract for the construction of the inter-oceanic canal.]

To the honorable Secretary of the Interior and Foreign Relations:

I have the honor to inform you that I accept each and all of the modifications introduced by Congress to the contract which I celebrated with Señor Eustorgio Salgar, your worthy predecessor in the department of the interior and foreign relations, for the construction of the inter-oceanic canal, which contract was approved by the executive power under date of March 23 last.

The modifications to which I have alluded are those recorded in law No. 28 of the 18th instant.

I hasten to lay this declaration before the Government of Colombia, so that it may be taken in consideration, in order that said law may be effective in all its parts.

Bogota, May 18, 1878.

Lucien N. B. Wyse

Chief of the International Scientific Commission for the Survey of the Isthmus, Member and Delegate from the Board of Directors of the Inter-oceanic Canal Association.

Documents of Authorization—III

LAWS OF THE UNITED STATES RELATING TO THE PANAMA CANAL

THE SPOONER ACT

APPROVED JUNE 28, 1902

An Act To provide for the construction of a canal connecting the waters of the Atlantic and Pacific oceans.

Be it enacted by the Senate and House of Representatives of the United States of America in Congress assembled, That the President of the United States is hereby authorized to acquire, for and on behalf of the United States, at a cost not exceeding forty millions of dollars, the rights, privileges, franchises, concessions, grants of land, right of way, unfinished work, plants, and other property, real, personal, and mixed, of every name and nature, owned by the New Panama Canal Company, of France, on the Isthmus of Panama, and all its maps, plans, drawings, records on the Isthmus of Panama and in Paris, including all the capital stock, not less however, than sixty-eight thousand eight hundred and sixty-three shares of the Panama Railroad Company, owned by or held for the use of said canal company, provided a satisfactory title to all of said property can be obtained.

Sec. 2. That the President is hereby authorized to acquire from the Republic of Colombia, for and on behalf of the United States, upon such terms as he may deem reasonable, perpetual control of a strip of land, the territory of the Republic of Colombia, not less than six miles in width, extending from the Caribbean Sea to the Pacific Ocean, and the right to use and dispose of the waters thereon, and to excavate, construct, and to perpetually maintain, operate, and protect thereon a canal, of such depth and capacity as will afford convenient passage of ships of the greatest tonnage and draft now in use, from the Caribbean Sea to the Pacific Ocean, which control shall include the right to perpetually maintain and operate the Panama Railroad, if the ownership thereof, or a controlling interest therein shall have been acquired by the United States, and also jurisdiction over said strip and the ports at the ends thereof to make such police and sanitary rules and regulations as shall be necessary to preserve order and preserve the public health thereon, and to establish such judicial tribunals as may be agreed upon thereon as may be necessary to enforce such rules and regulations.

The President may acquire such additional territory and rights from Colombia as in his judgment will facilitate the general purpose hereof.

Sec. 3. That when the President shall have arranged to secure a satisfactory title to the property of the New Panama Canal Company, as provided in section one hereof, and shall have obtained by treaty control of the necessary territory from the Republic of Colombia, as provided in section two hereof, he is authorized to pay for the property of the New Panama Canal Company forty millions of dollars and to the Republic of Colombia such sum as shall have been agreed upon, and a sum sufficient for both said purposes is hereby appropriated, out of any money in the Treasury not otherwise appropriated, to be paid on warrant or warrants drawn by the President.

The President shall then through the Isthmian Canal Commission hereinafter authorized cause to be excavated, constructed, and completed, utilizing to that end as far as practicable the work heretofore done by the New Panama Canal Company of France and its predecessor company, a ship canal from the Caribbean sea to the Pacific Ocean. Such canal shall be of sufficient capacity and depth as shall afford convenient passage for vessels of the largest tonnage and greatest draft now in use, and such as may be reasonably anticipated, and shall be supplied with all necessary locks and other appliances to meet the necessities of vessels passing through the same from ocean to ocean; and he shall also cause to be constructed such safe and commodious harbors at the termini of said canal, and make such provisions for defense as may be necessary for the safety and protection of said canal and harbors. That the President is authorized for purposes aforesaid to employ such persons as he may deem necessary, and to fix their compensation.

Sec. 4. That should the President be unable to obtain for the United States a satisfactory title to the property of the New Panama Canal Company and the control of the necessary territory of the Republic of Colombia and the rights mentioned in sections one and two of this Act, within a reasonable time and upon reasonable terms, then the President, having first obtained for the United States perpetual control by treaty of the necessary territory from Costa Rica and Nicaragua, upon terms which he may consider reasonable, for the construction, perpetual maintenance, operation, and protection of a canal connecting the Caribbean Sea with the Pacific Ocean by what is commonly known as the Nicaragua route, shall through the said Isthmian Canal Commission cause to be excavated and constructed a ship canal and waterway from a point on the shore of the Caribbean sea near Greytown, by way of Lake Nicaragua, to a point near Brito on the Pacific Ocean. Said canal shall be of sufficient capacity and depth to afford convenient passage for vessels of the largest tonnage and greatest draft now in use, and such as may be reasonably anticipated, and shall be supplied with all necessary locks and other appliances to meet the necessities of vessels passing through the same from ocean to ocean; and he shall also construct such safe and commodious harbors at the termini of said canal as shall be necessary for the safe and convenient use thereof, and shall make such provisions for defense as may be necessary for the safety and protection of said harbors and canal; and such sum or sums of money as may be agreed upon by such treaty as compensation to be paid to Nicaragua and Costa Rica for the concessions and rights hereunder provided to be acquired by the United States, are hereby appropriated, out of any money in the Treasury not otherwise appropriated, to be paid on warrant or warrants drawn by the President.

The President shall cause the said Isthmian Canal Commission to make such surveys as may be necessary for said canal and harbors to be made, and in making such surveys and in the construction of said canal may employ such persons as he may deem necessary and may fix their compensation.

In the excavation and construction of said canal the San Juan River and Lake Nicaragua, or such parts of each as may be made available, shall be used.

Sec. 5. That the sum of ten million dollars is hereby appropriated, out of any money in the Treasury not otherwise appropriated, toward the project herein contemplated by either route so selected.

And the President is hereby authorized to cause to be entered into such contract or contracts as may be deemed necessary for the proper excavation, construction, completion, and defense of said canal, harbors, and defenses, by the route finally determined upon under the provisions of this Act. Appropriations therefore shall from time to time be hereafter made, not to exceed in the aggregate the additional sum of one hundred and thirty-five millions of dollars should the Panama route be adopted, or one hundred and eighty millions of dollars should the Nicaragua route be adopted.

Sec. 6. That in any agreement with the Republic of Colombia, or with the States of Nicaragua and Costa Rica, the President is authorized to guarantee to said Republic or to said States the use of said canal and harbors, upon such terms as may be agreed upon, for all vessels owned by said States or by citizens thereof.

Sec. 7. That to enable the President to construct the canal and works appurtenant thereto as provided in this Act, there is hereby created the Isthmian Canal Commission, the same to be composed of seven members, who shall be nominated and appointed by the President, by and with the advice and consent of the Senate, and who shall serve until the completion of said canal unless sooner removed by the President, and one of whom shall be named as the chairman of said Commission. Of the seven members of said Commission at least four of them shall be persons learned and skilled in the science of engineering, and of the four at least one shall be an officer of the United States Army, and at least one other shall be an officer of the United States Navy, the said officers respectively being either upon the active or the retired list of the Army or of the Navy. Said commissioners shall each receive such compensation as the President shall prescribe until the same shall have been otherwise fixed by the Congress. In addition to the members of said Isthmian Canal Commission, the President is hereby authorized through said Commission to employ in said service any of the engineers of the United States Army at his discretion, and likewise to employ any engineers in civil life, at his discretion, and any other persons necessary for the proper and expeditious prosecution of said work. The compensation of all such engineers and other persons employed under this Act shall be fixed by said Commission, subject to the approval of the President. The official salary of any officer appointed or employed under this Act shall be deducted from the amount of salary or compensation provided by or which shall be fixed under the terms of this Act. Said Commission shall in all matters be subject to the direction and control of the President, and shall make to the President annually and at such other periods as may be required, either by law or by the order of the President, full and complete reports of all their actings and doings and of all moneys received and expended in the construction of said work and in the performance of their duties in connection therewith, which said reports shall be by the President transmitted to Congress. And the said Commission shall furthermore give to Congress, or either House of Congress, such information as may at any time be required either by act of Congress or by the order of either House of Congress. The President shall cause to be provided and assigned for the use of the Commission such offices as may, with the suitable equipment of the same, be necessary and proper, in his discretion, for the proper discharge of the duties thereof.

Sec. 8. That the Secretary of the Treasury is hereby authorized to borrow on the credit of the United States from time to time, as the proceeds may be required to defray expenditures authorized by this Act (such proceeds when received to be used only for the purpose of meeting such expenditures), the sum of one hundred and thirty million dollars, or so much thereof as may be necessary, and to prepare and issue therefor coupon or registered bonds of the United States in such form as he may prescribe, and in denominations of twenty dollars or some multiple of that sum, redeemable in gold coin at the pleasure of the United States after ten years from the date of their issue, and payable thirty years from such date, and bearing interest payable quarterly in gold coin at the rate of two per centum per annum; and the bonds herein authorized shall be exempt from all taxes or duties of the United States, as well as from taxation in any form by or under State, municipal, or local authority: Provided, That said bonds may be disposed of by the Secretary of the Treasury at not less than par, under such regulations as he may prescribe, giving to all citizens of the United States an equal opportunity to subscribe therefor, but no commissions shall be allowed or paid thereon; and a sum not exceeding one-tenth of one per centum of the amount of the bonds herein authorized is hereby appropriated, out of any money in the Treasury not otherwise appropriated, to pay the expense of preparing, advertising, and issuing the same.

Documents of Authorization—IV

PANAMA REPUBLIC
NOVEMBER 18, 1903
HAY-BUNAU-VARILLA TREATY

The United States of America and the Republic of Panama being desirous to insure the construction of a ship canal across the Isthmus of Panama to connect the Atlantic and the Pacific oceans, and the Congress of the United States of America having passed an act approved June 28, 1902, in furtherance of that object, by which the President of the United States is authorized to acquire within a reasonable time the control of the necessary territory of the Republic of Colombia, and the sovereignty of such territory being actually vested in the Republic of Panama, the high contracting parties have resolved for that purpose to conclude a convention and have accordingly appointed as their plenipotentiaries,—The President of the United States of America, John Hay, Secretary of State, and The Government of the Republic of Panama, Phillipe Bunau-Varilla, Envoy Extraordinary and Minister Plenipotentiary of the Republic of Panama, thereunto specially empowered by said government, who after communicating with each other their respective full powers, found to be in good and due form, have agreed upon and concluded the following articles:

Article I

The United States guarantees and will maintain the independence of the Republic of Panama.

Article II

The Republic of Panama grants to the United States in perpetuity the use, occupation and control of a zone of land and land under water for the construction, maintenance, operation, sanitation and protection of said Canal of the width of ten miles extending to the distance of five miles on each side of the center line of the route of the Canal to be constructed; the said zone beginning in the Caribbean Sea three marine miles from lean low water mark and extending to and across the Isthmus of Panama into the Pacific Ocean to a distance of three marine miles from mean low water mark with the proviso that the cities of Panama and Colon and the harbors adjacent to said cities, which are included within the boundaries of the zone above described, shall not be included within this grant. The republic of Panama further grants to the United States in perpetuity the use, occupation and control of any other lands and waters outside of the zone above described which may be necessary and convenient for the construction, maintenance, operation, sanitation and protection of the said Canal or of any auxiliary canals or other works necessary and convenient for the construction, maintenance, operation, sanitation and protection of the said enterprise.

The Republic of Panama further grants in like manner to the United States in perpetuity all islands within the limits of the zone above described and in addition thereto the group of small islands in the Bay of Panama, named Perico, Naos, Culebra and Flamenco.

Article III

The Republic of Panama grants to the United States all the rights, power and authority within the zone mentioned and described in Article II of this agreement and within the limits of all auxiliary lands and waters mentioned and described in said Article II which the United States would possess and exercise if it were the sovereign of the territory within which said lands and waters are located to the entire exclusion of the exercise of the Republic of Panama of any such sovereign rights, power or authority.

Article IV

As rights subsidiary to the above grants the Republic of Panama grants in perpetuity to the United States the right to use the rivers, streams, lakes and other bodies of water within its limits for navigation, the supply of water or water-power or other purposes, so far as the use of said rivers, streams, lakes and bodies of water and the waters thereof may be necessary and convenient for the construction, maintenance, operation, sanitation and protection of the said Canal.

Article V

The Republic of Panama grants to the United States in perpetuity a monopoly for the construction, maintenance and operation of any system of communication by means of canal or railroad across its territory between the Caribbean Sea and the Pacific Ocean.

Article VI

The grants herein contained shall in no manner invalidate the titles or rights of private land holders or owners of private property in the said zone or in or to any of the lands or waters granted to the United States by the provisions of any Article of this Treaty, nor shall they interfere with the rights of way over the public roads passing through the said zone or over any of the said lands or waters unless said rights of way or private rights shall conflict with rights herein granted to the United States in which case the rights of the United States shall be superior. All damages caused to the owners in this treaty or by reason of the operations of the United States, its agents or employees, of private lands or private property of any kind by reason of the grants contained or by reason of the construction, maintenance, operation, sanitation and protection of the said Canal or of the works of sanitation and protection herein provided for, shall be appraised and settled by a joint Commission appointed by the Governments of the United States and of the Republic of Panama, whose decisions as to such damages shall be final and whose awards as to such damages shall be paid solely by the United States. No part of the work on said Canal or the Panama Railroad or on any auxiliary works relating thereto and authorized by the terms of this treaty shall be prevented, delayed or impeded by or pending such proceedings to ascertain such damages. The appraisal of said private land and private property and the assessment of damages to them shall be based upon their value before the date of this convention.

Article VII

The Republic of Panama grants to the United States within the limits of the cities of Panama and Colon and their adjacent harbors and within the territory adjacent thereto the right to acquire by purchase or by the exercise of the right of eminent domain, any lands, buildings, water rights or other properties necessary and convenient for the construction, maintenance, operation and protection of the Canal and of any works of sanitation, such as the collection and disposition of sewage and the distribution of water in the said cities of Panama and Colon, which, in the discretion of the United States may be necessary and convenient for the construction, maintenance, operation, sanitation and protection of the said Canal and railroad. All such works of sanitation, collection and disposition of sewage and distribution of water in the cities of Panama and Colon shall be made at the expense of the United States, and the Government of the United States, its agents or nominees shall be authorized to impose and collect water rates and sewerage rates which shall be sufficient to provide for the payment of interest and the amortization of the principal of the cost of said works within a period of fifty years and upon the expiration of said term of fifty years the system of sewers and water works shall revert to and become the properties of the cities of Panama and Colon respectively, and the use of the water shall be free to the inhabitants of Panama and Colon, except to the extent that water rates may be necessary for the operation and maintenance of said system of sewers and waters.

The Republic of Panama agrees that the cities of Panama and Colon shall comply in perpetuity with the sanitary ordinances whether of a preventive or curative character prescribed by the United States and in case the Government of Panama is unable or fails in its duty to enforce this compliance by the cities of Panama and Colon with the sanitary ordinances of the United States the Republic of Panama grants to the United States the right and authority to enforce the same.

The same right and authority are granted to the United States for the maintenance of public order in the cities of Panama and Colon and the territories and harbors adjacent thereto in case the Republic of Panama should not be, in the judgment of the United States, able to maintain such order.

Article VIII

The Republic of Panama grants to the United States all rights which it now has or hereafter may acquire to the property of the New Panama Canal Company and the Panama Railroad Company as a result of the transfer of sovereignty from the Republic of Colombia to the Republic of Panama over the Isthmus of Panama and authorizes the New Panama Canal Company to seal and transfer to the United States its rights, privileges, properties, and concessions as well as the Panama Railroad and all the shares or part of the shares of that company; but the public lands situated outside of the zone described in Article II of this treaty now included in the concessions to both said enterprises and not required to the construction or operation of the Canal shall revert to the Republic of Panama except any property now owned by or in the possession of said companies within Panama or Colon or the ports or terminals thereof.

Article IX

The United States agrees that the ports at either entrance of the Canal and the waters thereof and the Republic of Panama agrees that the towns of Panama and Colon shall be free for all time so that there shall not be imposed or collected custom house tolls, tonnage, anchorage, lighthouse, wharf, pilot, or quarantine dues or any other charges or taxes of any kind upon any vessel using or passing through the Canal or belonging to or employed by the United States, directly or indirectly, in connection with the construction, maintenance, operation, sanitation and protection of the main Canal, or auxiliary works, or upon the cargo, officers, crew, or passengers of any such vessels, except such tolls and charges as may be imposed by the United States for the use of the Canal and other works, and except tolls and charges imposed by the Republic of Panama upon merchandise destined to be introduced for the consumption of the rest of the Republic of Panama, and upon vessels touching at the ports of Colon and Panama and which do not cross the Canal.

The Government of the Republic of Panama shall have the right to establish in such ports and in the towns of Panama and Colon such houses and guards as it may deem necessary to collect duties on importations destined to other portions of Panama and to prevent contraband trade. The United States shall have the right to make use of the towns and harbors of Panama and Colon as places of anchorage, and for making repairs, for loading, unloading, depositing, or transshipping cargoes either in transit or destined for the service of the Canal and for other works pertaining to the Canal.

Article X

The Republic of Panama agrees that there shall not be imposed any taxes, national, municipal, departmental, or of any other class upon the Canal, the railways and auxiliary works, tugs and other vessels employed in the service of the Canal, store houses, work shops, offices, quarters for laborers, factories of all kinds, warehouses, wharves, machinery and other works, property, and effects appertaining to the Canal or railroad and auxiliary works, or their officers or employees, situated within the cities of Panama and Colon, and that there shall not be imposed contributions or charges of a personal character of any kind upon officers, employees, laborers, and other individuals in the service of the Canal and railroad and auxiliary works.

Article XI

The United States agrees that the official dispatches of the Government of the Republic of Panama shall be transmitted over any telegraph and telephone lines established for canal purposes and used for public and private business at rates not higher than those required from officials in the service of the United States.

Article XII

The Government of the Republic of Panama shall permit the immigration and free access to the lands and workshops of the Canal and its auxiliary works of all employees and workmen of whatever nationality under contract to work upon or seeking employment upon or in any wise connected with the said Canal and its auxiliary works, with their respective families and all such persons shall be free and exempt from the military service of the Republic of Panama.

Article XIII

The United States may import at any time into the said zone and auxiliary lands, free of custom duties, imposts, taxes, or other charges, and without any restrictions, any and all vessels, dredges, engines, cars, machinery, tools, explosives, materials, supplies, and other articles necessary and convenient in the construction, maintenance, operation, sanitation and protection of the Canal and auxiliary works, and all provisions, medicines, clothing, supplies and other things necessary and convenient for the officers, employees, workmen and laborers in the service and employ of the United States and for their families. If any such articles are disposed of for use outside of the zone and auxiliary lands granted to the United States and within the territory of the Republic; they shall be subject to the same import or other duties as like articles imported under the laws of the Republic of Panama.

Article XIV

As the price or compensation for the rights, powers and privileges granted in this convention by the Republic of Panama to the United States, the Government of the United States agrees to pay to the Republic of Panama the sum of ten million dollars ($10,000,000) in gold coin of the United States on the exchange of the ratification of this convention and also an annual payment during the life of this convention of two hundred and fifty thousand dollars ($250,000) in like gold coin, beginning nine years after the date aforesaid.

The provisions of this Article shall be in addition to all other benefits assured to the Republic of Panama under this convention.

But no delay or difference of opinion under this Article or any other provisions of this treaty shall affect or interrupt the full operation and effect of this convention in all other respects.

Article XV

The joint commission referred to in Article VI shall be established as follows:

The President of the United States shall nominate two persons and the President of the Republic of Panama shall nominate two persons and they shall proceed to a decision; but in case of disagreement of the Commission (by reason of their being equally divided in conclusion) an umpire shall be appointed by the two Governments who shall render the decision. In the event of the death, absence, or incapacity of a Commission or Umpire, or of his omitting, declining or ceasing to act, his place shall be filled by the appointment of another person in the manner above indicated. All decisions by a majority of the Commission or by the umpire shall be final.

Article XVI

The two Governments shall make adequate provision by future agreement for the pursuit, capture, imprisonment, detention and delivery within said zone and auxiliary lands to the authorities of the Republic of Panama of persons charged with the commitment of crimes, felonies or misdemeanors without said zone and for the pursuit, capture, imprisonment detention and delivery without said zone to the authorities of the United States of persons charged with the commitment of crimes, felonies, and misdemeanors within said zone and auxiliary lands.

Article XVII

The Republic of Panama grants to the United States the use of all the ports of the Republic open to commerce as places of refuge for any vessels employed in the Canal enterprise, and for all vessels passing or bound to pass through the Canal which may be in distress and be driven to seek refuge in said ports. Such vessels shall be exempt from anchorage and tonnage dues on the part of the Republic of Panama.

Article XVIII

The Canal, when constructed, and the entrances thereto shall be neutral in perpetuity, and shall be opened upon the terms provided for by Section I of Article III, of and in conformity with all the stipulations of, the treaty entered into by the Governments of the United States and Great Britain on November 18, 1901.

Article XIX

The Government of the Republic of Panama shall have the right to transport over the Canal its vessels and its troops and munitions of war in such vessels at all times without paying charges of any kind. The exemption is to be extended to the auxiliary railway for the transportation of persons in the service of the Republic of Panama, or of the police force charged with the preservation of public order outside of said zone, as well as to their baggage, munitions of war and supplies.

Article XX

If by virtue of any existing treaty in relation to the territory of the Isthmus of Panama, whereof the obligations shall descend or be assumed by the Republic of Panama, there may be any privilege or concession in favor of the Government or the citizens and subjects of a third power relative to an inter-oceanic means of communication which in any of its terms may be incompatible with the terms of the present convention, the Republic of Panama agrees to cancel or modify such treaty in due form, for which purpose it shall give to the said third power the requisite notification within the term of four months from the date of the present convention, and in case the existing treaty contains no clause permitting its modifications or annulment, the Republic of Panama agrees to procure its modifications or annulment in such form that there shall not exist any conflict with the stipulations of the present convention.

Article XXI

The rights and privileges granted by the Republic of Panama to the United States in the preceding Articles are understood to be free of all anterior debts, liens, trusts, or liabilities, or concessions or privileges to other Governments, corporations, syndicates or individuals, and consequently, if there should arise any claims on account of the present concessions and privileges or otherwise, the claimants shall resort to the Government of the Republic of Panama and not to the United States for any indemnity or compromise which may be required.

Article XXII

The Republic of Panama renounces and grants to the United States the participation to which it might be entitled in the future earnings of the Canal under Article XV of the concessionary contract with Lucien N.B. Wyse now owned by the New Panama Canal Company and any and all other rights or claims of a pecuniary nature arising under or relating to said concession, or arising under or relating to the concessions to the Panama Railroad Company or any extension or modification thereof; and it likewise renounces, confirms and grants to the United States, now and hereafter, all the rights and property reserved in the said concessions which otherwise would belong to Panama at or before the expiration of the terms of ninety-nine years of the concessions granted to or held by the above mentioned party and companies, and all right, title and interest which it now has or may hereafter have, in and to the lands, canal, works, property and rights held by the said companies under said concessions or otherwise, and acquired or to be acquired by the United States from or through the New Panama Canal Company, including any property and rights which might or may in the future either by lapse of time, forfeiture or otherwise, revert to the Republic of Panama under any contracts or concessions, with said Wyse, the Universal Panama Canal Company, the Panama Railroad Company and the New Panama Canal Company.

The aforesaid rights and property shall be and are free and released from any present or reversionary interest in or claims of Panama and the title of the United States thereto upon consummation of the contemplated purchase by the United States from the New Panama Canal Company, shall be absolute, so far as concerns the Republic of Panama, excepting always the rights of the Republic specifically secured under this treaty.

Article XXIII

If it should become necessary at any time to employ armed forces for the safety or protection of the Canal, or of the ships that make use of the same, or the railways and auxiliary works, the United States shall have the right, at all times and in its discretion, to use its police and its land and naval forces or to establish fortifications for these purposes.

Article XXIV

No change either in the Government or in the laws and treaties of the Republic of Panama shall, without the consent of the United States, affect any right of the United States under the present convention, or under any treaty stipulation between the two countries that now exists or may hereafter exist touching the subject matter of this convention.

If the Republic of Panama shall hereafter enter as a constituent into any other Government or into any union or confederation of states, so as to merge her sovereignty or independence in such Government, union or confederation, the rights of the United States under this convention shall not be in any respect lessened or impaired.

Article XXV

For the better performance of the engagements of this convention and to the end of the efficient protection of the Canal and the preservation of its neutrality, the Government of the Republic of Panama will sell or lease to the United States lands adequate and necessary for naval or coaling stations on the Pacific coast and on the western Caribbean coast of the Republic at certain points to be agreed upon with the President of the United States.

This convention when signed by the Plenipotentiaries of the Contracting Parties shall be ratified by the respective Governments and the ratifications shall be exchanged at Washington at the earliest date possible.

In faith whereof the respective Plenipotentiaries have signed the present convention in duplicate and have hereunto affixed their respective seals.

Done at the City of Washington the 18th day of November in the year of our Lord nineteen hundred and three.

John Hay

P. Bunau-Varilla

Documents of Authorization—V

TEMPORARY GOVERNMENT

Approved April 28, 1904

An Act to provide for the temporary government of the Canal Zone at Panama, the protection of the canal works, and for other purposes.

Be it enacted by the Senate and House of Representatives of the United States of America in Congress assembled, That the President is hereby authorized, upon the acquisition of the property of the New Panama Canal Company and the payment to the Republic of Panama of the ten millions of dollars provided by article fourteen of the treaty between the United States and the Republic of Panama, the ratifications of which were exchanged on the twenty-sixth day of February, nineteen hundred and four, to be paid to the latter Government, to take possession of and occupy on behalf of the United States the zone of land and land under water of the width of ten miles, extending to the distance of five miles on each side of the center line of the route of the canal to be constructed thereon, which said zone begins in the Caribbean Sea three marine miles from mean low-water mark and extends to and across the Isthmus of Panama into the Pacific Ocean to the distance of three marine miles from mean low-water mark, and also of all islands within said zone, and in addition thereto the group of islands in the Bay of Panama named Perico, Naos, Culebra, and Flamenco, and, from time to time, of any lands and waters outside of said zone which may be necessary and convenient for the construction, maintenance, operation, sanitation, and protection of the said canal, or of any auxiliary canals or other works necessary and convenient for the construction, maintenance, operation, sanitation, and protection of said enterprise, the use, occupation, and control whereof were granted to the United States by article two of said treaty. The said zone is hereinafter referred to as "the Canal Zone". The payment of the ten millions of dollars provided by article fourteen of said treaty shall be made in lieu of the indefinite appropriation made in the third section of the Act of June twenty-eighth, nineteen hundred and two, and is hereby appropriated for said purpose.

Sec. 2. That until the expiration of the Fifty-eighth Congress, unless provision for the temporary government of the Canal Zone be sooner made by Congress, all the military, civil, and judicial powers as well as the power to make all rules and regulations necessary for the government of the Canal Zone and all the rights, powers, and authority granted by the terms of said treaty to the United States shall be vested in such person or persons and shall be exercised in such manner as the President shall direct for the government of said Zone and maintaining and protecting the inhabitants thereof in the free enjoyment of their liberty, property, and religion.

Documents of Authorization—VI

COLOMBIA-PANAMA REPUBLIC

JANUARY 9, 1909

ROOT-CORTES, ROOT-AROSEMENA, AND CORTES-AROSEMENA TREATIES

In an effort to compose the differences between the United States and Colombia arising out of the revolution at Panama, and the differences between Colombia and the new Republic of Panama, three treaties were signed at Washington on Jan. 9, 1909, between Secretary Root and Minister Cortes of Colombia; Secretary Root and Minister Arosemena of Panama; and Ministers Cortes and Arosemena. The treaties were of a tripartite nature, each depending on the others. They were ratified by the United States Senate on Feb. 24, 1909, but did not become operative by reason of the failure of Colombia to accept them.

The Root-Cortes treaty provided for peace and friendship between the United States and Colombia; granted freedom of passage through the Panama Canal to the troops and war ships of Colombia; exempted provisions, cattle, and other Colombia products from duty in the Canal Zone; granted free passage of Colombian mails in the zone; agreed to pay $250,000 annually to Colombia instead of Panama, from 1908 to 1917 inclusive, on condition that Colombia should recognize the independence of Panama and that Panama should be released from obligation to pay any of the public debt of Colombia; granted the use of ports in Colombia to American vessels; renounced all Colombia's rights in all canal contracts and concessions; and provided for a revision of the old treaty of 1846 (New Granada).

The Root-Arosemena treaty provided that the annual payment of $250,000 from the United States to Panama should begin four years instead of nine years from Nov. 18, 1903; consented to the assignment and transfer to Colombia of the first ten annual payments; provided for the delimitation of the cities of Panama and Colon and the adjacent harbors, and authorized the purchase by Panama of such portions of the waterworks in those cities as lie outside of the Canal Zone; provided for the arbitration of all questions arising out of the Hay-Bunau-Varilla treaty; and provided for reciprocal liberty of commerce and navigation, and granted to Panama most-favored-nation rights in the use of the Panama Canal facilities.

The Cortes-Arosemena treaty provided for the recognition of the independence of Panama by Colombia; provided for mutual and inviolable peace and friendship between Colombia and Panama; assigned to Colombia the first ten annual payments of $250,000 each payable by the United States to Panama under the Hay-Bunau-Varilla treaty; released Panama from the payment of any part of the public debt of Colombia; released each nation from all pecuniary claims held by the other on Nov. 3, 1903; confirmed the abandonment by Panama of all right and title to stock of the New Panama Canal Co.; granted reciprocal most-favored-nation rights to the citizens of each country, and established the status of citizens in the respective territories; provided that neither republic should extend its territory by force at the expense of the other; provided for the negotiation of additional treaties covering commerce, arbitration, and other relations; and provided for the establishment of the boundary line between Colombia and Panama.

Documents of Authorization—VII

THE PANAMA CANAL ACT

Approved August 24, 1912

An Act To provide for the opening, maintenance, protection, and operation of the Panama Canal, and the sanitation and government of the Canal Zone.

Be it enacted by the Senate and House of Representatives of the United States of America in Congress assembled, That the zone of land and land under water of the width of ten miles extending to the distance of five miles on each side of the center line of the route of the canal now being constructed thereon, which zone begins in the Caribbean sea three marine miles from mean low-water mark and extends to and across the Isthmus of Panama into the Pacific Ocean to the distance of three marine miles from mean low-water mark, excluding therefrom the cities of Panama and Colon and their adjacent harbors located within said zone, as excepted in the treaty with the Republic of Panama dated November eighteenth, nineteen hundred and three, but including all islands within said described zone, and in addition thereto the group of islands in the Bay of Panama named Perico, Naos, Culebra, and Flamenco, and any lands and waters outside of said limits above described which are necessary or convenient or from time to time may become necessary or convenient for the construction, maintenance, operation, sanitation, or protection of the said canal or of any auxiliary canals, lakes, or other works necessary or convenient for the construction, maintenance, operation, sanitation, or protection of the said canal, or of any auxiliary canals, lakes or other works necessary or convenient for the construction, maintenance, operation, sanitation, or protection of said canal, the use, occupancy, or control whereof were granted to the United States by the treaty between the United States and the Republic of Panama, the ratifications of which were exchanged on the twenty-sixth day of February, nineteen hundred and four, shall be known and designated as the Canal Zone, and the canal now being constructed thereon shall hereafter be known and designated as the Panama Canal. The President is authorized, by treaty with the Republic of Panama, to acquire any additional land or land under water not already granted, or which was excepted from the grant, that he deems necessary for the operation, maintenance, sanitation, or protection of the Panama Canal, and to exchange any land or land under water not deemed necessary for such purposes for the land or land under water which may be deemed necessary for such purposes, which additional land or land under water so acquired shall become part of the Canal Zone.

Sec. 2. That all laws, orders, regulations, and ordinances adopted and promulgated in the Canal Zone by order of the President for the government and sanitation of the Canal Zone and the construction of the Panama Canal are hereby ratified and confirmed as valid and binding until Congress shall otherwise provide. The existing courts established in the Canal Zone by Executive order are recognized and confirmed to continue in operation until the courts provided for in this Act shall be established.

Sec. 3. That the President is authorized to declare by Executive order that all land and land under water within the limits of the Canal Zone is necessary for the construction, maintenance, operation, sanitation, or protection of the Panama Canal, and to extinguish, by agreement when advisable, all claims and titles of adverse claimants and occupants. Upon failure to secure by agreement title to any such parcel of land or land under water the adverse claim or occupancy shall be disposed of and title thereto secured in the United States and compensation therefor fixed and paid in the manner provided in the aforesaid treaty with the Republic of Panama, or such modification of such Treaty as may hereafter be made.

Sec. 4. That when in the judgment of the President the construction of the Panama Canal shall be sufficiently advanced toward completion to render the further services of the Isthmian Canal Commission unnecessary the President is authorized by Executive order to discontinue the Isthmian Canal Commission, which, together with the present organization, shall then cease to exist; and the President is authorized thereafter to complete, govern, and operate the Panama Canal and govern the Canal Zone, or cause them to be completed, governed, and operated, through a governor of the Panama Canal and such other persons as he may deem competent to discharge the various duties connected with the completion, care, maintenance, sanitation, operation, government, and protection of the canal and Canal Zone. If any of the persons appointed or employed as aforesaid shall be persons in the military or naval service of the United States, the amount of the official salary paid to any such person shall be deducted from the amount of salary or compensation provided by or which shall be fixed under the terms of this Act. The governor of the Panama Canal shall be appointed by the President, by and with the advice and consent of the Senate, commissioned for a term of four years, and until his successor shall be appointed and qualified. He shall receive a salary of ten thousand dollars a year. All other persons necessary for the completion, care, management, maintenance, sanitation, government, operation and protection of the Panama Canal and Canal Zone shall be appointed by the President, or by his authority, removable at his pleasure, and the compensation for such persons shall be fixed by the President, or by his authority, until such time as Congress may by law regulate the same, but salaries or compensation fixed hereunder by the President shall in no instance exceed by more than twenty-five per centum the salary or compensation paid for the same or similar services to persons employed by the Government in continental United States. That upon the completion of the Panama Canal the President shall cause the same to be officially and formally opened for use and operation.

Before the completion of the canal, the Commission of Arts may make report to the President of their recommendation regarding the artistic character of the structures of the canal, such report to be transmitted to Congress.

Sec. 5. That the President is hereby authorized to prescribe and from time to time change the tolls that shall be levied by the Government of the United States for the use of the Panama Canal: Provided, That no tolls, when prescribed as above, shall be changed, unless six months' notice thereof shall have been given by the President by proclamation. No tolls shall be levied upon vessels engaged in the coastwise trade of the United States. That section forty-one hundred and thirty-two of the Revised Statutes is hereby amended to read as follows:

"Sec. 4132. Vessels built within the United States and belonging wholly to citizens thereof; and vessels which may be captured in war by citizens of the United States and lawfully condemned as prize, or which may be adjudged to be forfeited for a breach of the laws of the United States; and seagoing vessels, whether steam or sail, which have been certified by the Steamboat Inspection Service as safe to carry dry and perishable cargo, not more than five years old at the time they apply for Registry, wherever built, which are to engage only in trade with foreign countries or with the Philippine Islands and islands of Guam and Tutuila, being wholly owned by citizens of the United States or corporations organized and chartered under the laws of the United States or of any State thereof, the president and managing directors of which shall be citizens of the United States or corporations organized and chartered under the laws of the United States or of any State thereof, the President and managing directors of which shall be citizens of the United States, and no others, may be registered as directed in this title. Foreign-built vessels registered pursuant to this Act shall not engage in the coastwise trade: Provided, That a foreign-built yacht, pleasure boat, or vessel not used or intended to be used for trade admitted to American registry pursuant to this section shall not be exempt from the collection of ad valorem duty provided in section thirty-seven of the Act approved August fifth, nineteen hundred and nine, entitled 'An Act to provide revenue, equalize duties, and encourage the industries of the United States, and for other purposes.' That all materials of foreign production which may be necessary for the construction or repair of vessels built in the United States and all such materials necessary for the building or repair of their machinery and all articles necessary for the construction or repair of vessels built in the United States and all materials necessary for their outfit and equipment may be imported into the United States free of duty under such regulations as the Secretary of the Treasury may prescribe: Provided further, That such vessels so admitted under the provisions of this section may contract with the Postmaster General under the Act of March third, eighteen hundred and ninety-one, entitled 'An Act to provide for ocean mail service between the United States and foreign ports, and to promote commerce,' so long as such vessels shall in all respects comply with the provisions and requirements of said Act."

Tolls may be based upon gross or net registered tonnage, displacement tonnage, or otherwise, and may be based on one form of tonnage for warships and another for ships of commerce. The rate of tolls may be lower upon vessels in ballast than upon vessels carrying passengers or cargo. When based upon net registered tonnage for ships of commerce the tolls shall not exceed one dollar and twenty-five cents per net registered ton, nor be less, other than for vessels of the United States and its citizens, than the estimated proportionate cost of the actual maintenance and operation of the canal subject, however, to the provisions of article nineteen of the convention between the United States and the Republic of Panama, entered into November eighteenth, nineteen hundred and three. If the tolls shall not be based upon net registered tonnage, they shall not exceed the equivalent of one dollar and twenty-five cents per net registered ton as nearly as the same may be determined, nor be less than the equivalent of seventy-five cents per net registered ton. The toll for each passenger shall not be more than one dollar and fifty cents. The President is authorized to make and from time to time amend regulations governing the operation of the Panama Canal, and the passage and control of vessels through the same or any part thereof, including the locks and approaches thereto, and all rules and regulations affecting pilots and pilotage in the canal or the approaches thereto through the adjacent waters.

Such regulations shall provide for prompt adjustment by agreement and immediate payment of claims for damages which may arise from injury to vessels, cargo, or passengers from the passing of vessels through the locks under the control of those operating them under such rules and regulations. In case of disagreement suit may be brought in the district court of the Canal Zone against the governor of the Panama Canal. The hearing and disposition of such cases shall be expedited and the judgment shall be immediately paid out of moneys appropriated or allotted for canal operations.

The President shall provide a method for the determination and adjustment of all claims arising out of personal injuries to employees thereafter occurring while directly engaged in actual work in connection with the construction, maintenance, operation, or sanitation of the canal or of the Panama Railroad, or of any auxiliary canals, locks, or other works necessary and convenient for the construction, maintenance, operation, or sanitation of the canal, whether such injuries result in death or not, and prescribe a schedule of compensation therefor, and may revise and modify such method and schedule at any time; and such claims, to the extent they shall be allowed on such adjustment, if allowed at all, shall be paid out of the moneys hereafter appropriated for that purpose or out of the funds of the Panama Railroad Company, if said company was responsible for said injury, as the case may require. And after such method and schedule shall be provided by the President, the provisions of the Act entitled "An Act granting to certain employees of the United States the right to receive from it compensation for injuries sustained in the course of their employment," approved February twenty-fourth, nineteen hundred and eight, and of the Act entitled "An Act relating to injured employees on the Isthmian Canal," approved February twenty-fourth, nineteen hundred and nine, shall not apply to personal injuries thereafter received and claims for which are subject to determination and adjustment as provided in this section.

Sec. 6. That the President is authorized to cause to be erected, maintained, and operated, subject to the International Convention and the Act of Congress to regulate radio-communication, at suitable places along the Panama Canal and the coast adjacent to its two terminals, in connection with the operation of said canal, such wireless telegraphic installations as he may deem necessary for the operation, maintenance, sanitation, and protection of the said canal, and for other purposes. If it is found necessary to locate such installations upon territory of the Republic of Panama, the President is authorized to make such agreement with said Government as may be necessary, and also to provide for the acceptance and transmission, by said system, of all private and commercial messages, and those of the Government of Panama, on such terms and for such tolls as the President may prescribe: Provided, That the messages of the Government of the United States and the departments thereof, and the management of the Panama Canal, shall always be given precedence over all other messages. The President is also authorized, in his discretion, to enter into such operating agreements or leases with any private wireless company or companies as may best insure freedom from interference with the wireless telegraphic installations established by the United States. The President is also authorized to establish, maintain, and operate, through the Panama Railroad Company, or otherwise, dry docks, repair shops, yards, docks, wharves, warehouses, storehouses, and other necessary facilities and appurtenances, for the purpose of providing coal and other materials, labor, repairs, and supplies for vessels of the Government of the United States and, incidentally, for supplying such at reasonable prices to passing vessels, in accordance with appropriations hereby authorized to be made from time to time by Congress as a part of the maintenance and operation of the said canal. Moneys received from the conduct of said business may be expended and reinvested for such purposes without being covered into the Treasury of the United States; and such moneys are hereby appropriated for such purposes, but all deposits of such funds shall be subject to the provisions of existing law relating to the deposit of other public funds of the United States, and any net profits accruing from such business shall annually be covered into the Treasury of the United States. Monthly reports of such receipts and expenditures shall be made to the President by the persons in charge, and annual reports shall be made to the Congress.

Sec. 7. That the governor of the Panama Canal shall, in connection with the operation of such canal, have official control and jurisdiction over the Canal Zone and shall perform all duties in connection with the civil government of the Canal Zone, which is to be held, treated, and governed as an adjunct of such Panama Canal. Unless in this Act otherwise provided all existing laws of the Canal Zone referring to the civil governor or the civil administration of the Canal Zone shall be applicable to the governor of the Panama Canal, who shall perform all such executive and administrative duties required by existing law. The President is authorized to determine or cause to be determined what towns shall exist in the Canal Zone and subdivide and from time to time resubdivide said Canal Zone into subdivisions, to be designated by name or number, so that there shall be situated one town in each subdivision, and the boundaries of each subdivision shall be clearly defined. In each town there shall be a magistrate's court with exclusive original jurisdiction coextensive with the subdivision in which it is situated of all civil cases in which the principal sum claimed does not exceed three hundred dollars, and all criminal cases wherein the punishment that may be imposed shall not exceed a fine of one hundred dollars, or imprisonment not exceeding thirty days, or both, and all violations of police regulations and ordinances and all actions involving possession or title to personal property or the forcible entry and detainer of real estate. Such magistrates shall also hold preliminary investigations in charges of felony and offenses under section ten of this Act, and commit or bail in bailable cases to the district court. A sufficient number of magistrates and constables, who must be citizens of the United States, to conduct the business of such courts, shall be appointed by the governor of the Panama Canal for terms of four years and until their successors are appointed and qualified, and the compensation of such persons shall be fixed by the President, or by his authority, until such time as Congress may by law regulate the same. The rules governing said courts and prescribing the duties of said magistrates and constables, oaths and bonds, the times and places of holding such courts, the disposition of fines, costs, forfeitures, enforcements of judgments, providing for appeals therefrom to the district court, and the disposition, treatment, and pardon of convicts shall be established by order of the President. The governor of the Panama Canal shall appoint all notaries public, prescribe their powers and duties, their official seal, and the fees to be charged and collected by them.

Sec. 8. That there shall be in the Canal Zone one district court with two divisions, one including Balboa and the other including Cristobal; and one district judge of the said district, who shall hold his court in both divisions at such time as he may designate by order, at least once a month in each division. The rules of practice in such district court shall be prescribed or amended by order of the President. The said district court shall have original jurisdiction of all felony cases, of offenses arising under section ten of this Act, all causes in equity; admiralty and all cases at law involving principal sums exceeding three hundred dollars and all appeals from judgments rendered in magistrates' courts. The jurisdiction in admiralty herein conferred upon the district judge and the district court shall be the same that is exercised by the United States district judges and the United States district courts, and the procedure and practice shall also be the same. The district court or the judge thereof shall also have jurisdiction of all other matters and proceedings not herein provided for which are now within the jurisdiction of the Supreme Court of the Canal Zone, of the Circuit Court of the Canal Zone, the District Court of the Canal Zone, or the judges thereof. Said judge shall provide for the selection, summoning, serving, and compensation of jurors from among the citizens of the United States, to be subject to jury duty in either division of such district, and a jury shall be had in any criminal case or civil case at law originating in said court on the demand of either party. There shall be a district attorney and a marshal for said district. It shall be the duty of the district attorney to conduct all business, civil and criminal, for the Government, and to advise the governor of the Panama Canal on all legal questions touching the operation of the canal and the administration of civil affairs. It shall be the duty of the marshal to execute all process of the court, preserve order therein, and do all things incident to the office of marshal.

The district judge, the district attorney, and the marshal shall be appointed by the President, by and with the advice and consent of the Senate, for terms of four years each, and until their successors are appointed and qualified, and during their terms of office shall reside within the Canal Zone, and shall hold no other office nor serve on any official board or commission nor receive any emoluments except their salaries. The district judge shall receive the same salary paid the district judges of the United States, and shall appoint the clerk of said court, and may appoint one assistant when necessary, who shall receive salaries to be fixed by the President. The district judge shall be entitled to six weeks' leave of absence each year with pay. During this absence or during any period of disability or disqualification from sickness or otherwise to discharge his duties the same shall be temporarily performed by any circuit or district judge of the United States who may be designated by the President, and who, during such service, shall receive the additional mileage and per diem allowed by law to district judges of the United States when holding court away from their homes. The district attorney and the marshal shall be paid each a salary of five thousand dollars per annum.

Sec. 9. That the records of the existing courts and all causes, proceedings, and criminal prosecutions pending therein as shown by the dockets thereof, except as herein otherwise provided, shall immediately upon the organization of the courts created by this Act be transferred to such new courts having jurisdiction of like cases, be entered upon the dockets thereof, and proceed as if they had originally been brought therein, whereupon all the existing courts, except the supreme court of the Canal Zone, shall cease to exist. The President may continue the supreme court of the Canal Zone and retain the judges thereof in office for such time as to him may seem necessary to determine finally any causes and proceedings which may be pending therein. All laws of the Canal Zone imposing duties upon the clerks or ministerial officers of existing courts shall apply and impose such duties upon the clerks and ministerial officers of the new courts created by this Act having jurisdiction of like cases, matters and duties.

All existing laws in the Canal Zone governing practice and procedure in existing courts shall be applicable and adapted to the practice and procedure in the new courts.

The Circuit Court of Appeals of the Fifth Circuit of the United States shall have jurisdiction to review, revise, modify, reverse, or affirm the final judgments and decrees of the District Court of the Canal Zone and to render such judgments as in the opinion of the said appellate court should have been rendered by the trial court in all actions and proceedings in which the Constitution, or any statute, treaty, title, right, or privilege of the United States is involved and a right thereunder denied, and in cases in which the value in controversy exceeds one thousand dollars, to be ascertained by the oath of either party, or by other competent evidence, and also in criminal causes wherein the offense charged is punishable as a felony. And such appellate jurisdiction, subject to the right of review by or appeal to the Supreme Court of the United States as in other cases authorized by law, may be exercised by said circuit court of appeals in the same manner, under the same regulations, and by the same procedure as nearly as practicable as is done in reviewing the final judgments and decrees of the district courts of the United States.

Sec. 10. That after the Panama Canal shall have been completed and opened, for operation the governor of the Panama Canal shall have the right to make such rules and regulations, subject to the approval of the President, touching the right of any person to remain upon or pass over any part of the Canal Zone as may be necessary. Any person violating any of such rules or regulations shall be guilty of a misdemeanor, and on conviction in the District Court of the Canal Zone shall be punished by a fine not exceeding five hundred dollars or by imprisonment not exceeding a year, or both, in the discretion of the court. It shall be unlawful for any person, by any means or in any way, to injure or obstruct, or attempt to injure or obstruct, any part of the Panama Canal or the locks thereof or the approaches thereto. Any person violating this provision shall be guilty of a felony, and on conviction in the District Court of the Canal Zone shall be punished by a fine not exceeding ten thousand dollars or by imprisonment not exceeding twenty years, or both, in the discretion of the court. If this act shall cause the death of any person within a year and a day thereafter, the person so convicted shall be guilty of murder and shall be punished accordingly.

Sec. 11. That section five of the Act to regulate commerce, approved February fourth, eighteen hundred and eighty-seven, as heretofore amended, is hereby amended by adding thereto a new paragraph at the end thereof, as follows:

> "From and after the first day of July, nineteen hundred and fourteen, it shall be unlawful for any railroad company or other common carrier subject to the Act to regulate commerce to own, lease, operate, control, or have any interest whatsoever (by stock ownership or otherwise, either directly, indirectly, through any holding company, or by stockholders or directors in common, or in any other manner) in any common carrier by water operated through the Panama Canal or elsewhere with which said railroad or other carrier aforesaid does or may compete for traffic or any vessel carrying freight or passengers upon said water route or elsewhere with which said railroad or other carrier aforesaid does or may compete for traffic; and in case of the violation of this provision each day in which such violation continues shall be deemed a separate offense."

Jurisdiction is hereby conferred on the Interstate Commerce Commission to determine questions of fact as to the competition or possibility of competition, after full hearing, on the application for any railroad company or other carrier. Such application may be filed for the purpose of determining whether any existing service is in violation of this section and pray for an order permitting the continuance of any vessel or vessels already in operation, or for the purpose of asking an order to install new service not in conflict with the provisions of this paragraph. The commission may on its own motion or the application of any shipper institute proceedings to inquire into the operation of any vessel in use by any railroad or other carrier which has not applied to the commission and had the question of competition or the possibility of competition determined as herein provided. In all such cases the order of said commission shall be final.

If the Interstate Commerce Commission shall be of the opinion that any such existing specified service by water other than through the Panama Canal is being operated in the interest of the public and is of advantage to the convenience and commerce of the people, and that such extension will neither exclude, prevent, nor reduce competition on the route by water under consideration, the Interstate Commerce Commission may, by order, extend the time during which such service by water may continue to be operated beyond July first, nineteen hundred and fourteen. In every case of such extension the rates, schedules, and practices of such water carrier shall be filed with the Interstate Commerce Commission and shall be subject to the act to regulate commerce and all amendments thereto in the same manner and to the same extent as is the railroad or other common carrier controlling such water carrier or interested in any manner in its operation: Provided, Any application for extension under the terms of this provision filed with the Interstate Commerce Commission prior to July first, nineteen hundred and fourteen, but for any reason not heard and disposed of before said date, may be considered and granted thereafter.

No vessel permitted to engage in the coastwise or foreign trade of the United States shall be permitted to enter or pass through said Canal if such ship is owned, chartered, operated, or controlled by any person or company which is doing business in violation of the provisions of the Act of Congress approved July second, eighteen hundred and ninety, entitled "An Act to protect trade and commerce against unlawful restraints and monopolies," or the provisions of sections seventy-three to seventy-seven, both inclusive, of an Act approved August twenty-seventh, eighteen hundred and ninety-four, entitled "An Act to reduce taxation, to provide revenue for the Government, and for other purposes," or the provisions of any other Act of Congress amending or supplementing the said act of July second, eighteen hundred and ninety, commonly known as the Sherman Antitrust act, and amendments thereto, or said sections of the Act of August twenty-seventh, eighteen hundred and ninety-four. The question of fact may be determined by the judgment of any court of the United States of competent jurisdiction in any case pending before it to which the owners or operators of such ship are parties. Suit may be brought by any shipper or by the Attorney General of the United States.

That section six of said Act to regulate commerce, as heretofore amended, is hereby amended by adding a new paragraph at the end thereof, as follows:

"When property may be or is transported from point to point in the United States by rail and water through the Panama canal or otherwise, the transportation being by a common carrier or carriers, and not entirely within the limits of a single State, the Interstate Commerce Commission shall have jurisdiction of such transportation and of the carrier, both by rail and by water, which may or do engage in the same, in the following particulars, in addition to the jurisdiction given by the Act to regulate commerce, as amended June eighteenth, nineteen hundred and ten:

"(a) To establish physical connection between the lines of the rail carrier and the dock of the water carrier by directing the rail carrier to make suitable connection between its line and a track or tracks which have been constructed from the dock to the limits of its right of way, or by directing either or both the rail and water carrier, individually or in connection with one another, to construct and connect with the lines of the rail carrier a spur track or tracks to the dock. This provision shall only apply where such connection is reasonably practicable, can be made with safety to the public, and where the amount of business to be handled is sufficient to justify the outlay.

"The commission shall have full authority to determine the terms and conditions upon which these connecting tracks, when constructed, shall be operated, and it may, either in the construction or the operation of such tracks, determine what sum shall be paid to or by either carrier. The provisions of this paragraph shall extend to cases where the dock is owned by other parties than the carrier involved.

"(b) To establish through routes and maximum joint rates between and over such rail and water lines, and to determine all the terms and conditions under which such lines shall be operated in the handling of the traffic embraced.

"(c) To establish maximum proportional rates by rail to and from the ports to which the traffic is brought, or from which it is taken by the water carrier, and to determine to what traffic and in connection with what vessels and upon what terms and conditions such rates shall apply. By proportional rates are meant those which differ from the corresponding local rates to and from the port and which apply only to traffic which has been brought to the port or is carried from the port by a common carrier by water.

"(d) If any rail carrier subject to the Act to regulate commerce enters into arrangements with any water carrier operating from a port in the United States to a foreign country, through the Panama Canal or otherwise, for the handling of through business between interior points of the United States and such foreign country, the Interstate Commerce Commission may require such railway to enter into similar arrangements with any or all other lines of steamships operating from said port to the same foreign country."

The orders of the Interstate Commerce Commission relating to this section shall only be made upon formal complaint or in proceedings instituted by the commission of its own motion and after full hearing. The orders provided for in the two amendments to the Act to regulate commerce enacted in this section shall be served in the same manner and enforced by the same penalties and proceedings as are the orders of the commission made under the provisions of section fifteen of the Act to regulate commerce, as amended June eighteenth, nineteen hundred and ten, and they may be conditioned for the payment of any sum or the giving of security for the payment of any sum or the discharge of any obligation which may be required by the terms of said order.

Sec. 12. That all laws and treaties relating to the extradition of persons accused of crime in force in the United States, to the extent that they may not be in conflict with or superseded by any special treaty entered into between the United States and the Republic of Panama with respect to the Canal Zone, and all laws relating to the rendition of fugitives from justice as between the several States and Territories of the United States, shall extend to and be considered in force in the Canal Zone, and for such purposes and such purposes only the Canal Zone shall be considered and treated as an organized territory of the United States.

Sec. 13. That in time of war in which the United States shall be engaged, or when, in the opinion of the President, war is imminent, such officer of the Army as the President may designate shall, upon the order of the President, assume and have exclusive authority and jurisdiction over the operation of the Panama Canal and all its adjuncts, appendants, and appurtenances, including the entire control and government of the Canal Zone, and during a continuance of such condition the governor of the Panama Canal shall, in all respects and particulars as to the operation of such Panama Canal, and all duties, matters, and transactions affecting the Canal Zone, be subject to the order and direction of such officer of the Army.

Sec. 14. That this Act shall be known as, and referred to as, the Panama Canal Act, and the right to alter, amend, or repeal any or all of its provisions or to extend, modify, or annul any rule or regulation made under its authority is expressly reserved.

Documents of Authorization—VIII

PANAMA CANAL TOLL RATES

PROCLAMATION BY THE PRESIDENT, NOV. 13, 1912

I, William Howard Taft, President of the United States of America, by virtue of the power and authority vested in me by the act of Congress approved August twenty-fourth, nineteen hundred and twelve, to provide for the opening, maintenance, protection and operation of the Panama Canal and the sanitation and government of the Canal Zone, do hereby prescribe and proclaim the following rates of toll to be paid by vessels using the Panama Canal:

1. On merchant vessels carrying passengers or cargo, one dollar and twenty cents ($1.20) per net vessel ton—each one hundred (100) cubic feet—of actual earning capacity.
2. On vessels in ballast without passengers or cargo, forty (40) per cent less than the rate of tolls for vessels with passengers or cargo.
3. Upon naval vessels, other than transports, colliers, hospital ships, and supply ships, fifty (50) cents per displacement ton.
4. Upon army and navy transports, colliers, hospital ships, and supply ships, and one dollar and twenty cents ($1.20) per net ton, the vessels to be mea-

sured by the same rules as are employed in determining the net tonnage of merchant vessels.

The Secretary of War will prepare and prescribe such rules for the measurement of vessels and such regulations as may be necessary and proper to carry this proclamation into full force and effect.

In witness whereof I have hereunto set my hand and caused the seal of the United States to be affixed.

Done at the city of Washington this thirteenth day of November, in the year of our Lord one thousand nine hundred and twelve, and of the independence of the United States the one hundred and thirty-seventh.

[Seal] Wm. H. Taft
By the President:
P. C. Knox, Secretary of State

Documents of Authorization—IX

EXECUTIVE ORDER

January 27, 1914

CREATING A PERMANENT ORGANIZATION FOR THE PANAMA CANAL

Effective April 1, 1914

By virtue of the authority vested in me, I hereby enact the following order, creating a permanent organization for the Panama Canal, under the Act of Congress "To provide for the opening, maintenance, protection and operation of the Panama canal and the sanitation and government of the Canal Zone," approved August 24, 1912.

Section 1. The organization for the completion, maintenance, operation, government and sanitation of the Panama Canal and its adjuncts and the government of the Canal Zone shall consist of the following departments, offices and agencies, and such others as may be established by the Governor of the Panama Canal on the Isthmus or elsewhere with the approval of the President, all to be under the direction of the governor, subject to the supervision of the Secretary of War.

DEPARTMENT OF OPERATION AND MAINTENANCE—There shall be a Department of Operation and Maintenance under the immediate supervision and direction of the Governor of the Panama Canal. This Department shall be charged with the construction of the Canal and with its operation and maintenance when completed, including all matters relating to traffic of the Canal and its adjuncts, and the operation and maintenance of beacons, lights and lighthouses; the supervision of ports and waterways, including pilotage; the admeasuring and inspecting of vessels, including hulls and boilers; the operation and maintenance of the Panama Railroad upon the Isthmus, including telephone and telegraph systems; the operation of locks, coaling plants, shops, dry-docks and wharves; office engineering, including meteorology and hydrography; the construction of buildings and sanitary and municipal engineering, including the construction and maintenance of drainage ditches, streets, roads and bridges.

PURCHASING DEPARTMENT—There shall be a Purchasing Department under the supervision and direction of the Governor. This department shall be charged with the purchase of all supplies, machinery or necessary plant.

SUPPLY DEPARTMENT—There shall be a Supply Department, under the supervision and direction of the Chief Quartermaster. This department shall store and distribute all material and supplies for use of the Panama Canal and of its employees; and for other departments of the Government on the Isthmus and their employees; and for vessels of the United States and for other vessels, when required. The Supply Department shall operate commissaries, hotels and messes; shall be in charge of the maintenance of buildings, the assignment of quarters and the care of grounds; shall recruit and distribute unskilled labor; and shall have charge of the necessary animal transportation.

ACCOUNTING DEPARTMENT—There shall be an Accounting Department under the supervision and direction of the auditor, with an assistant in the United States. The duties of the department shall include all general bookkeeping, auditing and accounting, both for money and property, costkeeping, the examination of payrolls and vouchers, the inspection of time books and of money and property accounts, the preparation of statistical data, and the administrative examination of such accounts as are required to be submitted to the United States Treasury Department; and the collection, custody and disbursement of funds for the Panama Canal and the Canal Zone. These same duties shall be performed for the Panama Railroad Company on the Isthmus when not inconsistent with the charter and by-laws of that Company. The Department shall be charged with the handling of claims for compensation on account of personal injuries and of claims for damages to vessels. Within the limits fixed by law, the duties and financial responsibilities of the officers and employees charged with the receipt, custody, disbursement, auditing and accounting for funds and property shall be prescribed in regulations issued by the Governor, with the approval of the President. The Auditor shall maintain such a system of bookkeeping as will enable him to furnish at any time full, complete and correct information in regard to the status of appropriations made by Congress, the status of all other funds, and the amounts of net profits on all operations, which are to be covered into the Treasury as required by the Panama Canal Act.

HEALTH DEPARTMENT—There shall be a Health Department under the supervision and direction of the Chief Health Officer. This department shall be charged with all matters relating to maritime sanitation and quarantine in the ports and waters of the Canal Zone and in the harbors of the cities of Panama and Colon, and with land sanitation in the Canal Zone and sanitary matters in said cities in conformity with the Canal Treaty between the United States and the Republic of Panama and existing agreements between the two governments thereunder, and all matters relating to hospitals and charities.

EXECUTIVE SECRETARY—There shall be an Executive Secretary who, under the direction of the Governor of the Panama Canal, shall be charged with the supervision of all matters relating to the keeping of time of employees; to post offices, customs, taxes and excises, excepting the collection thereof; police and prisons; fire protection; land office; schools, clubs and law library; the custody of files and records; and the administration of estates of deceased and insane employees. He shall, in person or through one of his assistants, perform the duties of a Shipping Commissioner. He shall conduct all correspondence and communications between the authorities of the Canal Zone and the Government of the Republic of Panama and such other correspondence as may be given him in charge by the Governor. He shall have charge of the seal of the Government of the Canal Zone and shall attest such acts of the Government as are required by law to be performed and done under the seal.

The duties herein prescribed for the foregoing departments, offices and agencies will be assigned to divisions or bureaus thereunder by the Governor of the Panama Canal, as the necessities therefor arise. Each of the foregoing departments shall discharge such further duties as may be assigned to it from time to time by the Governor; and the Governor, with the approval of the President, may transfer from time to time specific duties from one department to another.

Section 2. The organization provided for in Section 1 shall be, in general, in accordance with the outline chart accompanying the memorandum of Jan. 27, 1914, entitled "Memorandum to accompany Executive Order of Jan. 27, 1914, providing for a permanent organization for the Panama Canal," and officers from certain departments shall be detailed in accordance with that memorandum.

Section 3. This order shall take effect from and after the 1st day of April, 1914, from which date the Isthmian Canal Commission, together with the present organization for the Panama Canal and the Canal Zone, shall cease to exist, in accordance with the terms of the above-mentioned Act of Congress.

The White House
WOODROW WILSON
January 27, 1914.

Documents of Authorization—X

REPEAL OF TOLLS EXEMPTION CLAUSE

Approved June 15, 1914

An Act To amend section five of "An Act to provide for the opening, maintenance, protection, and operation of the Panama Canal and the sanitation and government of the Canal Zone," approved August twenty-fourth, nineteen hundred and twelve.

Be it enacted by the Senate and House of Representatives of the United States of America in Congress assembled, That the second sentence in section five of the Act entitled "An Act to provide for the opening, maintenance, protection and operation of the Panama Canal, and the sanitation and government of the Canal

Zone," approved August twenty-fourth, nineteen hundred and twelve, which reads as follows: "No tolls shall be levied upon vessels engaged in the coastwise trade of the United States," be, and the same is hereby, repealed.

Sec. 2. That the third sentence of the third paragraph of said section of said Act be so amended as to read as follows: "When based upon net registered tonnage for ships of commerce the tolls shall not exceed $1.25 per net registered ton, nor be less than 75 cents per net registered ton, subject however, to the provisions of article nineteen of the convention between the United States and the Republic of Panama, entered into November eighteenth, nineteen hundred and three": Provided, That the passage of this Act shall not be construed or held as a waiver or relinquishment of any right the United States may have under the treaty with Great Britain, ratified the twenty-first of February, nineteen hundred and two, or the treaty with the Republic of Panama, ratified February twenty-sixth, nineteen hundred and four, or otherwise, to discriminate in favor of its vessels by exempting the vessels of the United States or its citizens from the payment of tolls for passage through said canal, or as in any way waiving, impairing, or affecting any right of the United States under said treaties, or otherwise, with respect to the sovereignty over or the ownership, control, and management of said canal and the regulation of the conditions or charges of traffic through the same.

Documents of Authorization—XI

ADMISSION OF FOREIGN-BUILT SHIPS

TO AMERICAN REGISTER

Approved August 18, 1914

An Act To provide for the admission of foreign-built ships to American registry for the foreign trade, and for other purposes.

Be it enacted by the Senate and House of Representatives of the United States of America in Congress assembled, That the words "not more than five years old at the time they apply for registry" in section five of the Act entitled "An act to provide for the opening, maintenance, protection, and operation of the Panama Canal and the sanitation and government of the Canal Zone," are hereby repealed.

Sec. 2. That the President of the United States is hereby authorized, whenever in his discretion the needs of foreign commerce may require, to suspend by order, so far and for such length of time as he may deem desirable, the provisions of law prescribing that all the watch officers of vessels of the United States registered for foreign trade shall be citizens of the United States.

Under like conditions, in like manner, and to like extent the President of the United States is also hereby authorized to suspend the provisions of the law requiring survey, inspection, and measurement by officers of the United States of foreign-built vessels admitted to American registry under this Act.

Sec. 3. This Act shall take effect immediately.

Automobiles drive into the Fort Washington Interchange of the Pennsylvania Turnpike, 2004. Courtesy of AP / Wide World Photos.

23

The Federal Highway System

United States

DID YOU KNOW . . . ?

- ➤ The highway system is a modern equivalent of the ancient Roman road network that girdled the Mediterranean.
- ➤ The first federal funds ($10,000) for a study of road conditions were given to the Department of Agriculture in 1893.
- ➤ In 1902, 23,000 cars and 17 million horses shared the roads.
- ➤ In 1904, there were only 154,000 miles (248,000 kilometers) of paved roads, versus 2 million miles (3 million kilometers) unpaved.
- ➤ The entire national highway system totals 160,093 miles (275,645 kilometers).
- ➤ The U.S. highway system was inspired by the German *Autobahn*.
- ➤ Interstates account for 24 percent of all travel on U.S. streets and roads.

The concept of a network of roads that would cover and unite America could be thought of as the modern-day equivalent of the roads of ancient Rome. Roman roads were built by the Roman army, and the justification for a comprehensive U.S. interstate road system was also military. The martial aspects were two: the need for a reliable system of evacuation routes, and the employment of soldiers returning from duty in World War II.

In 1900, there were only 8,000 cars in a country that would soon become a drive-through nation. By 1902, when the American Automobile Association (AAA) was founded, there were 23,000 cars and 17 million horses—and roads were designed with the majority users in mind. From the moment Henry Ford rolled his first Model T car off the production line at the Piquette Avenue Plant in Detroit, Michigan, on October 1, 1908, automobiles held a special interest for Americans. So it wasn't long before attention began to focus on the roads over which they would have to drive their "horseless carriages."

HISTORY

In 1892 a populist group called the Good Roads Movement lobbied Congress to get a bill through the Senate and House to study road conditions, but they were unsuccessful. In 1893, Congress passed an authorization regarding roads, giving $10,000 to the secretary of agriculture for the purpose. With funding, on October 3, 1893, the new Office of Road Inquiry was launched, headed by General Roy Stone, whose title was "Special Agent and Engineer for Road Inquiry." Inquiry it was, for the authorization specifically forbade any design of a road system; it called only for a study of current conditions, such study to be conducted by asking users their opinions about what was needed. The purpose? To assure that agricultural colleges had sufficient information on the transport of crops—for instance, cooling food during transit. General Stone took his mission with great seriousness; he and a single clerk managed to survey every state governor in the United States and its territories, all members of Congress, and all railroad presidents. They even compared rail rates for transporting goods to the cost of assembling the requisite materials for road building. In less than a year, in June 1894, Stone and his clerk had written and printed nine different bulletins on all the relevant subjects.

By 1904, the Office of Road Inquiry had become the Office of Public Roads, and by 1912 the results of the surveys began to make clear that the United States would soon have to put considerable funds and effort into improving the road system. By 1921, there were 387,000 miles of paved roads, most of them constructed using nineteenth-century technology.

Two Scottish engineers, John MacAdam and his colleague Thomas Telford, are responsible for most of the early American roads. These builders were best known for their insistence on proper drainage. They made the roads with crushed stone bound with gravel on a firm base of large stones. A camber, which made the road slightly convex, ensured that rainwater rapidly drained off the road and did not penetrate the foundations. This way of building roads later became known as the *Macadamized system*.

In 1909, Wayne County, Michigan, built the world's first mile of concrete highway. Road builders came from everywhere to learn how the concrete held up under the "heavy" traffic of that period. The success of this experiment speeded the development of modern automobile highways. It cost $13,537, including $1,000 in state aid. The road was replaced in 1922 by a broad thoroughfare. Another concrete highway, 24 miles long, was built in Pine Bluff, Arkansas.

As road-building technology advanced, regulations required sub-bases of gravel, crushed stone, or slag (a steel by-product) so heavy trucks could be accommodated. In 1919, Oregon came up with the idea of financing roads through a tax on gasoline. And in the 1930s the first toll road was built: the Pennsylvania Turnpike was constructed on a railroad right-of-way, denoting the evolution from one form of transit to another.

In the twentieth century, in an effort to satisfy the demand for better highways following the invention of the automobile, the U.S. government sponsored legislation authorizing the creation of a national highway system of unprecedented dimensions. Presented here are key legislative acts from 1916, 1921, and 1956.

The first step was passage in 1916 of the Federal-Aid Road Act, which provided for federal assistance to the states to construct "rural post roads." The measure defines specific appropriations through 1921, and places administration of the program under the Department of Agriculture, more specifically in the Bureau of Public Roads. The money was stipulated only for construction and improvement of roads; maintenance would be the responsibility of the states, and states were responsible for proposing other needed projects. In any case, roads built under the act were to be entirely toll-free.

The 1916 act was amended five years later to underscore the importance of a coordinated network of high-quality roads. The secretary of agriculture was directed to approve projects that would "expedite the completion of an adequate and connected system of highways, interstate in character." Federal aid was limited to seven percent of a state's highway miles (the roads to be considered for this amount were selected by the state); three-sevenths of the roads receiving federal aid had to be so-called primary or interstate highways, and the rest secondary or intercounty roads. Under the 1921 measure, the United States was required to pay no more than half the estimated cost of projects.

Following World War I, prosperity mushroomed. Many people were driving the Model T and popular songs began to romanticize the many possibilities of individualized transportation ("You can go as far as you like with me, in my merry Oldsmobile" [Edwards, 1905]). But the 1929 stock-market crash and the following Great Depression cast a cloud over the scene, and highway funds were diverted to pay for dole programs.

In 1934, the Hayden-Cartwright Act made two important changes. First, the act forbade the diversion of highway funds to nonhighway uses, and second, it allowed states to use 1.5 percent of the highway funds to develop plans and conduct surveys. Tremendous planning followed; for the first time, investigators drove over every mile of highway, making notations on its condition, including width and load factors. This was the first time that roads were categorized by three criteria: road mapping, traffic analysis, and economic yield. During the inventory taking, extensive mapping of the United States was also conducted, noting every farm, every house, every business, every plant, every school, every hospital—and the roads that served these places.

World War II sharpened awareness of the need for an enhanced system of highways in the United States. A 1944 measure (the Federal-Aid Highway Act of 1944) authorized establishment of a "National System of Interstate Highways." After the war, some states began building roads as part of the network, but design standards varied, and progress was slow, particularly since Congress failed to increase federal support for highways until 1952, when it authorized use of a (relatively) minuscule $25 million to match state funds on a 50-50 basis. By the time Dwight D. Eisenhower became President in 1953, a mere 6,000 miles (10,000 kilometers) had been constructed. Eisenhower pointed out that an interstate network could offer military as well as economic benefits, since it would facilitate the movement of military forces as well as a rapid evacuation of the population in the event of a nuclear attack—an argument that reflected the tensions of the cold war then being waged between the United States and the Soviet Union.

The landmark Federal-Aid Highway Act of 1956, which was accompanied by legislation that raised taxes on fuel and tires (the Highway Revenue Act of 1956), imposed new taxes on big trucks and other large vehicles, and created a Highway Trust Fund to help finance the interstate highways and other federally aided roads. The 1956 act set the federal portion of the cost of interstate system projects at a minimum of 90 percent, and required that the roads meet uniform standards set by the secretary of commerce (now responsible for highways). Roads, bridges, and tunnels for which tolls were charged were allowed to be part of the interstate system, but the use of federal funds in their construction was subject to severe restrictions. Congress expressed its desire that the interstate system be completed within 13 years. Although this target was not met, by 1986 only two percent of the network remained to be built.

The system of superhighways was a massive undertaking, which Eisenhower regarded as one of the biggest peacetime construction projects ever. It played a major role in enhancing the country's prosperity in the second half of the twentieth century. Not only did it generate many manufacturing and construction jobs, but it also raised the productivity level of the U.S. economy. In 1990, Congress renamed the highway network the Dwight D. Eisenhower System of Interstate and Defense Highways.

A few statistics are helpful for understanding the traffic growth brought on by the expanded interstate highway system. A 1904 government survey found that only 154,000 miles (248,000 kilometers) of public roads were surfaced, compared to some 2 million miles (3 million kilometers) of unpaved roads. A 1923 map of the federal-aid system showed almost 169,000 miles (272,000 kilometers) of roads. By 1998, the interstate system comprised 46,334 miles (74,567 kilometers), or about 1.2 percent of U.S. roads and streets totaling 3,920,979 miles (6,310,204 kilometers). However, the interstate system accounted for 630 billion vehicle-miles (about 1 trillion vehicle-kilometers) annually—approximately 23.9 percent of total travel for all streets and roads. The entire National Highway System (interstates plus federally aided highways) totaled 160,093 miles (257,645 kilometers) and accounted for 43.9 percent of total travel.

CULTURAL CONTEXT

The idea of a national road system was not new, but in 1919 it regained a new urgency in the mind—and body—of a young lieutenant colonel in the U.S. Army. He and his buddy Major Sereno Brett had volunteered for a transcontinental motor convoy. During the hot summer months, progress across dusty roads of gravel or broken asphalt, muddy embankments, and sand traps from which wheezing vehicles had to be pulled by wheezing soldiers turned what was expected to be a relatively simple adventure into a grueling grind that ended 62 days later when the convoy completed the trek from Washington, DC, to San Francisco, California. Little did the young lieutenant colonel know that as President Dwight D. Eisenhower, he would sign the first transcontinental highway bill in 1956.

That was not the only experience that convinced Eisenhower. He remembered as well the German *Autobahn*, which had been used by Allied forces to enter Germany in World War II, and he never forgot the expertly engineered highways of that road system. Two additional factors made such a massive project both attractive and urgent. The United States needed to provide employment for the returning soldiers, many of whom would start families and need good jobs. Unlike Franklin D. Roosevelt, whose Civilian Conservation Corps mobilized young unemployed men to build roads, dams, and parks and keep the environment clean, Eisenhower envisioned entrepreneurial opportunity as the key to prosperity. The highway bill would provide guidelines for road building, but the contracts would be let to the many small businesses across the country that could handle the construction locally. The opportunity to develop and grow small construction businesses was an attractive prospect to the president.

But why was road construction so attractive instead of other kinds of business? Because at the time, the United States was in the throes of the cold war with Soviet Russia, and many believed the country might one day be attacked by an atomic weapon. How would citizens evacuate from the major population centers if the only roads were old, dusty, gravelly, patched-up roads like the one Eisenhower remembered from his 1919 convoy? There would be disastrous consequences without roads of more suitable quality and reliability.

Careful consideration was also given to bicycle riders and their interactions with roadways. In Miyazaki, Japan, home of what was the fastest train in the world, there is a completely protected bicycle sidewalk. In Germany, there are two lanes of sidewalks: one for pedestrians, one for bicycles. Prior to World War II, in northern Germany, there were four sets of transitways in many towns: roads, sidewalks, bicycle paths, and bridle paths. Today, near Hamburg, such structures are still in use and well maintained.

When Franklin Roosevelt was president, he authorized a study of toll roads, and in 1939 the report "Tolls Roads and Free Roads" recommended that Americans would be more likely to travel on free roads. This was the first vision of what would later become the interstate highway system. The country's first limited-access highway, the Pennsylvania Turnpike, a multilane toll road largely funded by the federal government, opened in 1940.

Early American roads were built locally, and there were no national standards governing construction, no consistent signage, and no road weight restrictions. During World War I, when supplies and munitions were transported across the country, the roads buckled under the weight of the transport vehicles. World War I had focused attention on the spotty quality of roads and the wide variance in load-bearing capability. When World War II commenced, there were even more trucks hauling munitions; 13 percent of the defense plants were supplied by truck, and more than 50 percent shipped their products back out by truck. It was then that the wide variation of road quality was examined in detail.

In 1944, Congress passed the Federal-Aid Highway Act, the purpose of which was to study the criteria for common standards. Some roads could handle only light traffic, with loads up to 7,000 pounds, while other large highways could take 36,000 pounds of heavy equipment. The time had come to draw up a common framework and system, not to mention signage.

Following World War II, the first publication of *Highway Statistics* (1945) gave a complete analysis of many factors, such as number of motor vehicles registered, highway-use taxes, how highways in the states were financed, and how they were built, maintained, and repaired. Soon after publication, a synopsis of data gathered prior to 1945 was summarized, especially regarding fuel use, in *Highway Statistics—Summary to 1945*. For the first time, complete analytical planning data were readily available.

PLANNING

Planning the national highway system was a legislative maze with twists and turns, and many changes and deals. For those who wish to know more, a summary of the many aspects of financing can be found in the "For Further Reference" section.

Two figures stand out in government planning discussions: Senator Albert Gore Sr. from Tennessee and General Lucius D. Clay. Clay headed a committee that estimated highway needs would cost $101 billion. It was clear that financing would need to be addressed as a national strategic priority, and that funding would have to be reliable. On February 22, 1955, when President Eisenhower gave the report to Congress for legislative consideration, he summed up the link among free transportation, free trade, and free thought as the spark that ignited innovative America: "Our unity as a nation is sustained by free communication of thought and by easy transportation of people and goods. The ceaseless flow of information throughout the republic is matched by individual and commercial movement over a vast system of interconnected highways crisscrossing the country and joining at our national borders with friendly neighbors to the north and south. Together, the united forces of our communication and transportation systems are dynamic elements in the very name we bear—United States. Without them, we would be a mere alliance of many separate parts" (Weingroff).

Eisenhower signed the bill on June 29, 1956. He remarked, thinking back to those 62 days spent bouncing uncomfortably across the country in 1919, "That old convoy started me thinking about good, two-lane highways, but Germany had made me see the wisdom of broader ribbons across the land" (Weingroff, http://www.tfhrc.gov/pubrds/summer96/p96su10.htm).

Planning and statistics were only the first stage; the next step was financial discussions about how to fund an interstate highway system. The decision to go ahead with toll-free roads, made by Franklin Roosevelt in 1939, was a milestone. The next question was the degree to which the federal government should be involved in the financing. Wrangling over financing took many years, until Eisenhower brought it all together, mainly by stopping the financial squabbling and championing a transition from analysis and discussion to action. No doubt a general was needed to issue such a directive.

It was conceived as a "pay as you go" system that would rely primarily on federally imposed user fees on motor fuels—the federal user fee per gallon of gasoline was increased by one cent. The federal user fees would provide 90 percent of construction costs, with the balance provided primarily by state user fees. The interstate highway system also incorporated approximately 2,000 miles of already completed toll roads.

The last factor in planning the American transcontinental highway system was something not so much practical as fanciful—the World's Fair in 1939, which featured a Futurama exhibit. There, models were futuristic and not especially practical, but they made the point that a wide-ranging web of roads would be attractive, inspiring, and commercially viable.

BUILDING

As soon as the 1956 bill became law, Secretary of Commerce Sinclair Weeks signed a $1.1 billion allocation to the states, calling it "the greatest public works program in the history of the world" (Weingroff, http://www.tfhrc.gov/pubrds/summer96/p96su10.htm). But this public-works approach differed from others, in which the government or a few large companies conducted centralized hiring. The interstate highway system was built by local contractors who were responsible for building the highways in their own state.

The 1956 legislation envisioned a huge network of some 41,000 miles (66,000 kilometers). How would these roads be connected, and would they be relatively equal in quality? The federal government provided just four guidelines: (a) they must have at least two lanes in each direction (that is, four-lane highways were the minimum); (b) lanes must be 12 feet (3.7 meters) wide or more; (c) there must be a shoulder beyond the right lane at least 10 feet (3 meters) wide and paved; and (d) the speeds that could be accommodated should be between 50 and 70 miles (81 and 113 kilometers) per hour.

High standards were adopted for the interstate highway system. Access to all interstates was to be fully controlled. There would be no intersections or traffic signals. All traffic and railroad crossings would be grade separated, requiring the construction of more than 55,000 bridges. Interstates were to be divided and have at least four traffic lanes (two in each direction) and adequate shoulders. Curves were to be engineered for safe negotiation at high speed, while grades were to be moderated, eliminating blind hills. Rest areas were to be conveniently spaced. Each interstate was to be designed to handle traffic loads expected 20 years after completion.

The states were soon underway with construction. As time passed, it became clear that the goal of system completion by 1975 would not be achieved. But by 1960, more than 10,000 miles (16,000 kilometers) were opened; by 1965, 20,000 miles(32,000 kilometers) ; by 1970, 30,000 miles (48,000 kilometers); and by 1980, 40,000 miles(64,000 kilometers). The interstate highway system serves virtually all of the nation's large urban areas. Despite this broad expanse, the interstate highway system represents just over one percent of the nation's road network.

An interesting aspect of highway standardization was naming and signage. It was decided that all interstate roads would have simple one- or two-digit numbers. Roads running from north to south would have odd numbers beginning in the west; roads running east to west would have even numbers and begin in the south. Thus the road number tells the general direction and location of a road; for example, I-5 is a north–south road on the West Coast, while I-95 runs north–south on the East Coast. Connecting these major arteries are roads that loop around major cities; these bypass routes are numbered with three digits.

Building the highway system was always related to safety, both from a defense aspect and also for accident-free transport. In a country still possessing large tracts of undeveloped land that are home to creatures of the forest, collisions with deer and other animals crossing roads can be hazardous to all concerned. Directing motorists to a Web site where success stories on preserving wildlife and natural habitats are described, the Federal Highway Department of Transportation has developed four categories in which guidelines are given: "Along Roads," "On or Near Bridges," "On or Along Waterways," and "On Wetlands and Uplands" (http://www.fhwa.dot.gov/pressroom/fhwa0316.htm). It is of note that early road legislation did not take the environment into as prominent a place as it is now given.

IMPORTANCE IN HISTORY

When the network of U.S. roads is viewed from the air, the grid pattern is apparent, with north–south roads intersecting with east–west roads at 90-degree angles. Such cross sections tend to produce more accidents. Would European roundabouts be a way to avoid head-on collisions?

President Eisenhower brought to the nation's attention the tragedy of highway accidents. In 1952, the president-elect reported that 37,500 men, women, and children had died in traffic accidents in 1951, and many more (1.3 million) had been injured. He compared those terrible numbers to the loss of life in war, which he knew all too well as an army general. The interstate highway system became a mission related to his leadership as a military commander who wanted to save lives and create peace.

Should future highways take advantage of a plan developed by David Gordon Wilson? In an article on palleted automated transit (PAT) systems written for the *IATSS Journal* (the professional journal for the International Association of Traffic and Safety Sciences, founded by Soichiro Honda), Wilson describes his invention of pallets that move on independent guideways; motor vehicles ride on pallets that can exit the highway system at selected points. Wilson's research has resulted in a proposed solution for the primary cause of highway fatalities: collisions. Each pallet is equipped with a differential electric motor, and therefore cannot physically collide with another pallet.

If there were further development of a pallet system to take advantage of magnetic levitation (maglev), the pallet track could be further streamlined by being enclosed in a tube from which the forces of air resistance would be removed, thus making it possible to travel from Boston to New York in 60 minutes by car. We could suggest the auto train, running from just south of Washington, DC, to Florida as an attractive beta test site.

When considering modes of self-directed transport as opposed to group transit like rail or bus, it may be time to revisit the sportsway design proposed at the Massachusetts Institute of Technology. The sportsway includes lanes for automobiles, bicycles, pedestrians, and equestrians—reminiscent of a design developed and applied in Germany and still in use today. The difference is that the sportsway would be paid for by revenues from high-speed trains located in efficient tunnels underneath the sportsway. A precedent can be found under a state park in California, where tolls from a water-supply tunnel are dedicated to keeping the surface paths in excellent condition.

FOR FURTHER REFERENCE

Books and Articles

Gutfreund, Owen D. *Twentieth Century Sprawl: Highways and the Reshaping of the American Landscape.* New York: Oxford University Press, 2004.

Ritter, Joyce N., comp. "Development of the Interstate Program." In *America's Highways: 1776–1976.* Washington, DC: U.S. Department of Transportation, Federal Highway Administration, 1976. Can also be found, with pictures, at www.fhwa.dot.gov/byday/fhbd0511.htm.

Rose, Mark H. *Interstate Express Highway Politics 1941–1989.* Revised ed. Knoxville: University of Tennessee Press, 1990.

Seely, Bruce E. *Building the American Highway System: Engineers as Policy Makers*. Philadelphia: Temple University Press, 1987.

Von Hagen, Victor Wolfgang. *The Roads That Led to Rome*. Cleveland, OH: World Publishing Company, 1967.

Wilson, David Gordon. "Palleted Automated Transportation—A View of Developments at the Massachusetts Institute of Technology." *IATSS Research* 13, no. 1 (1989): 53–59.

Internet

Richard F. Weingraff of the Federal Highway Administration has written an history of the legislative process, with historic photos including Eisenhower standing at attention during the seminal 1919 cross-country convoy; see http://www.tfhrc. gov/pubrds/summer96/p96su10.htm.

To find information on American roads, with hyperlinks to topics from Turner-Fairbank Research Center to Winter Maintenance Virtual Clearinghouse, see http:// www.fhwa.dot.gov/fhwaweb.htm.

Joyce N. Ritter's story of how highway statistics were developed as the basis for the interstate system is available at http://www.fhwa.dot.gov/ohim/1994/text/history. htm.

For details on concrete, slag, and other road-construction techniques, see http://www. cement.org/pavements/pv_cp_highways.asp.

For animal and environmental highway safety, see http://www.fhwa.dot.gov/pressroom/ fhwa0316.htm.

For highway traffic accidents during the Eisenhower era, see http://www.fhwa.dot.gov/ infrastructure/safetyin.htm.

For Ricco Villanueva Siasoco's article "Red, White, and Blue Highways: The Story of the U.S. Interstate," which includes a quote by poet Gertrude Stein on the "American space that is filled with moving," see http://www.infoplease.com/ spot/interstate1.html.

For reference to MacAdam and Telford, see http://www.greatachievements.org/ greatachievements/ga_11_2.html.

For the relationship of bicycling to road planning and development, see http://www. itdp.org/ (includes a link to the Global Bicycle Fund for mobility in developing areas), http://www.transalt.org/, http://www.carfree.com/, and http://www. patternlanguage.com/.

For a history of the bicycle, including the first Japanese production bike (made by Eisuke Miyata, the owner of a gun factory, who believed that munitions would become obsolete but that mobility was a future business), and the Miyata Gun Factory, which in 1892 built the first bikes in Japan, see http://www.cycle-info. bpaj.or.jp/.

Film and Television

Easy Rider. Directed by Dennis Hopper. Culver City, CA: Columbia-Tristar Studios, 1969.

Thelma and Louise. Directed by Ridley Scott. Culver City, CA: MGM-UA Studios, 1991.

Music

Cole, Nat King. "Route 66." Written by Bobby Troup. Audio clip available at http://members.cox.net/jdmount/route_66.aif.

Edwards, Gus. "In My Merry Oldsmobile" Lyricist: Vincent Bryan. New York: M. Witmark & Sons, 1905. http://libraries.mit.edu/music/sheetmusic/childpages/inmymerryoldsmobile.html.

Documents of Authorization—I

I. CHAP. 241.—AN ACT TO PROVIDE THAT THE UNITED STATES SHALL AID THE STATES IN THE CONSTRUCTION OF RURAL POST ROADS, AND FOR OTHER PURPOSES.

Be it enacted by the Senate and House of Representatives of the United States of America in Congress assembled, That the Secretary of Agriculture is authorized to cooperate with the States, through their respective State highways departments, in the construction of rural post roads; but no money apportioned under this Act to any State shall be expended therein until its legislature shall have assented to the provisions of this Act, except that, until the final adjournment of the first regular session of the legislature held after the passage of this Act, the assent of the governor of the State shall be sufficient. The Secretary of Agriculture and the State highways department of each State shall agree upon the roads to be constructed therein and the character and method of construction: Provided, That all roads constructed under the provisions of this Act shall be free from tolls of all kinds.

SEC. 2. That for the purpose of this Act the term "rural post road" shall be construed to mean any public road over which the United States mails now are or may hereafter be transported, excluding every street and road in a place having a population, as shown by the latest available Federal census, of two thousand five hundred or more, except that portion of any such street or road along which the houses average more than two hundred feet apart; the term "state highway department" shall be construed to include any department of another name, or commission, or official or officials, of a State empowered, under its laws, to exercise the functions ordinarily exercised by a State highway department; the term "construction" shall be construed to include reconstruction and improvement of roads; "properly maintained" as used herein shall be construed to mean the making of needed repairs and the preservation of a reasonably smooth surface considering the type of the road; but shall not be held to include extraordinary repairs, nor reconstruction; necessary bridges and culverts shall be deemed parts of the respective roads covered by the provisions of this Act.

SEC. 3. That for the purpose of carrying out the provisions of this Act there is hereby appropriated, out of any money in the Treasury not otherwise appropriated, for the fiscal year ending June thirtieth, nineteen hundred and seventeen, the sum of $5,000,000; for the fiscal year ending June thirtieth, nineteen hundred and eighteen, the sum of $10,000,000; for the fiscal year ending June thirtieth, nineteen hundred and nineteen, the sum of $15,000,000; for the fiscal year ending June thirtieth, nineteen hundred and twenty, the sum of $20,000,000; and for the fiscal year ending June thirtieth, nineteen hundred and twenty-one, the sum of $25,000,000. So much of the appropriation apportioned to any State for any fiscal year as remains unexpended at the close thereof shall be available for expenditure in that State until the close of the succeeding fiscal year, except that amounts apportioned for any fiscal year to any State which has not a State highway department shall be available for expenditure in that State until the close of the third fiscal year succeeding the close of the fiscal year for which such apportionment was made. Any amount apportioned under the provisions of this Act unexpended at the end of the period during which it is available for expenditure under the terms of this section shall be reapportioned, within sixty days thereafter, to all the States in the same manner and on the same basis, and certified to the Secretary of the Treasury and to the State highway departments and to the governors of States having no State highway department in the same way as if it were being apportioned under this Act for the first time: Provided, That in States where the constitution prohibits the State from engaging in any work of internal improvements, then the amount of the appropriation under this Act apportioned to any such State shall be turned over to the highway department of the State or to the governor of said State to be expended under the provisions of this Act and under the rules and regulations of the Department of Agriculture, when any number of counties in any such State shall appropriate or provide the proportion or share needed to be raised in order to entitle such State to its part of the appropriation apportioned under this Act.

SEC. 4. That so much, not to exceed three per centum, of the appropriation for any fiscal year made by or under this Act as the Secretary of Agriculture may estimate to be necessary for administering the provisions of this Act shall be deducted for that purpose, available until expended. Within sixty days after the close of each fiscal year the Secretary of Agriculture shall determine what part, if any, of the sums theretofore deducted for administering the provisions of this Act will not be needed for that purpose and apportion such part, if any, for the fiscal year then current in the same manner and on the same basis, and certify it to the Secretary of the Treasury and to the State highway departments, and to the governors of States having no State highway departments, in the same way as other amounts authorized by this Act to be apportioned among all the States for such current fiscal year. The Secretary of Agriculture, after making the deduction authorized by this section, shall apportion the remainder of the appropriation for each fiscal year among the several States in the following manner: One-third in the ratio which the area of each State bears to the total area of all the States; one-third in the ratio which the population of each State bears to the total population of all the States, as shown by the latest available Federal census; one-third in the ratio which the mileage of rural delivery routes and star routes in each State bears to the total mileage or rural delivery routes and star routes in all the States, at the close of the next preceding fiscal year, as shown by the certificate of the Postmaster General, which he is directed to make and furnish annually to the Secretary of Agriculture.

SEC. 5. That within sixty days after the approval of this Act the Secretary of Agriculture shall certify to the Secretary of the Treasury and to each State highway department and to the governor of each State having no State highway department the sum which he has estimated to be deducted for administering the provisions of this Act and the sum which he has apportioned to each State for the fiscal year ending June thirtieth, nineteen hundred and seventeen, and on or before January twentieth next preceding the commencement of each succeeding fiscal year shall make like certificates for such fiscal year.

SEC. 6. That any State desiring to avail itself of the benefits of this Act shall, by its State highway department, submit to the Secretary of Agriculture project statements setting forth proposed construction of any rural post road or roads therein. If the Secretary of Agriculture approve a project, the State highway department shall furnish to him such surveys, plans, specifications, and estimates therefor as he may require: Provided, however, That the Secretary of Agriculture shall approve only such projects as may be substantial in character and the expenditure of funds hereby authorized shall be applied only to such improvements. Items included for engineering, inspection, and unforeseen contingencies shall not exceed ten per centum of the total estimated cost of the work. If the Secretary of Agriculture approve the plans, specification, and estimates, he shall notify the State highway department and immediately certify the fact to the Secretary of the Treasury. The Secretary of the Treasury shall thereupon set aside the share of the United States payable under this Act on account of such project, which shall not exceed fifty per centum of the total estimated cost thereof. No payment of any money apportioned under this Act shall be made on any project until such statement of this project, and the plans, specifications, and estimates therefor, shall have been submitted to and approved by the Secretary of Agriculture.

When the Secretary of Agriculture shall find that any project so approved by him has been constructed in compliance with said plans and specification he shall cause to be paid to the proper authority of said State the amount set aside for said project: Provided, that the Secretary of Agriculture may, in his discretion, from time to time make payments on said construction as the same progresses, but these payments including previous payments, if any, shall not be more than the United States' pro rata part of the value of the labor and materials which have been actually put into said construction in conformity to said plans and specifications; nor shall any such payment be in excess of $10,000 per mile, exclusive of the cost of bridges of more than twenty feet clear span. The construction work and labor in each State shall be done in accordance with its laws, and under the direct supervision of the State highway department, subject to the inspection and approval of the Secretary of Agriculture and in accordance with the rules and regulations made pursuant to this Act.

The Secretary of Agriculture and the State highway department of each State may jointly determine at what times, and in what amounts, payments, as work progresses, shall be made under this Act. Such payments shall be made by the Secretary of the Treasury, on warrants drawn by the Secretary of agriculture, to such official, or officials, or depository, as may be designated by the State highway department and authorized under the laws of the State to receive public funds of the State or county.

SEC. 7. To maintain the roads constructed under the provisions of this Act shall be the duty of the States, or their civil subdivisions, according to the laws of the several states. If at any time the Secretary of Agriculture shall find that any road in any State constructed under the provisions of this Act is not being properly maintained he shall give notice of such fact to the highway department of such State and if within four months from the receipt of said notice said road has not been put in a proper condition of maintenance then the Secretary of Agriculture shall thereafter refuse to approve any project for road construction in said State, or the civil subdivision thereof, as the fact may be, whose duty it is to maintain said road, until it has been put in a condition of proper maintenance.

SEC. 8. That there is hereby appropriated and made available until expended, out of any moneys in the National Treasury not otherwise appropriated, the sum of $1,000,000 for the fiscal year ending June thirtieth, nineteen hundred and seventeen, and each fiscal year thereafter, up to and including the fiscal year ending June thirtieth, nineteen hundred and twenty-six, in all $10,000,000, to be available until expended under the supervision of the Secretary of Agriculture, upon request from the proper officers of the State, Territory, or county for the survey, construction, and maintenance of roads and trails within or only partly within the national forests, when necessary for the use and development of resources upon which communities within and adjacent to the national forests are dependent: Provided, that the State, Territory, or county shall enter into a cooperative agreement with the Secretary of Agriculture for the survey, construction, and maintenance of such roads or trails upon a basis equitable to both the State, territory, or county, and the United States: And provided also, That the aggregate expenditures in any State, Territory, or county shall not exceed ten per centum of the value, as determined by the Secretary of Agriculture, of the timber and forage resources which are or will be available for income upon the national first lands within the respective county or counties wherein the roads or trails will be constructed; and the Secretary of Agriculture shall make annual report to Congress of the amounts expended hereunder.

That immediately upon the execution of any cooperative agreement hereunder the Secretary of Agriculture shall notify the Secretary of the Treasury of the amount to be expended by the United States within or adjacent to any national forest thereunder, and beginning with the next fiscal year and each fiscal year thereafter the Secretary of the Treasury shall apply from any and all revenues from such forest ten per centum thereof to reimburse the United States for expenditures made under such agreement until the whole amount advanced under such agreement shall have been returned from the receipts from such national forest.

SEC. 9. That out of the appropriations made by or under this Act, the Secretary of Agriculture is authorized to employ such assistants, clerks, and other persons in the city of Washington and elsewhere, to be taken from the eligible lists of the Civil Service Commission, to rent buildings outside of the city of Washington, to purchase such supplies, materials, equipment, office fixtures, and apparatus, and to incur such travel and other expense as he may deem necessary for carrying out the purposes of this Act.

SEC. 10. That the Secretary of Agriculture is authorized to make rules and regulations for carrying out the provisions of this Act.

SEC. 11. That this Act shall be in force from the date of its passage.

Approved, July 11, 1916.

From *U.S. Statutes at Large* 39 (1916), chap. 241, pp. 355–59.

Documents of Authorization—II

II. CHAP. 119.—AN ACT TO AMEND THE ACT ENTITLED "AN ACT TO PROVIDE THAT THE UNITED STATES SHALL AID THE STATES IN THE CONSTRUCTION OF RURAL POST ROADS, AND FOR OTHER PURPOSES," APPROVED JULY 11, 1916, AS AMENDED AND SUPPLEMENTED, AND FOR OTHER PURPOSES.

Be it enacted by the Senate and House of Representatives of the United States of America in Congress assembled, That this Act may be cited as the Federal Highway Act.

SEC. 2. That, when used in this Act, unless the context indicates otherwise—

The term "Federal Aid Act" means the Act entitled "An Act to provide that the United States shall aid the States in the construction of rural post roads, and for other purposes," approved July 11, 1916, as amended by sections 5 and 6 of an Act entitled "An Act making appropriations for the service of the Post Office Department for the fiscal year ending June 30, 1920, and for other purposes," approved February 28, 1919, and all other Acts amendatory thereof or supplementary thereto.

The term "highway" includes rights of way, bridges, drainage structures, signs, guard rails, and protective structures in connection with highways, but shall not include any highway or street in a municipality having a population of two thousand five hundred or more as shown by the last available census, except that portion of any such highway or street along which within a distance of one mile the houses average more than two hundred feet apart.

The term "State highway department" includes any State department, commission, board, or official having adequate powers and suitably equipped and organized to discharge to the satisfaction of the Secretary of Agriculture the duties herein required.

The term "maintenance" means the constant making of needed repairs to preserve a smooth surfaced highway.

The term "construction" means the supervising, inspecting, actual building, and all expenses incidental to the construction of a highway, except locating, surveying, mapping, and costs of rights of way.

The term "reconstruction" means a widening or a rebuilding of the highway or any portion thereof to make it a continuous road, and of sufficient width and strength to care adequately for traffic needs.

The term "forest roads" means roads wholly or partly within or adjacent to and serving the national forests.

The term "State funds" included for the purposes of this Act funds raised under the authority of the State, or any political or other subdivision thereof, and made available for expenditure under the direct control of the State highway department.

SEC. 3. All powers and duties of the Council of national Defense under the Act entitled "An Act making appropriations for the support of the Army for the fiscal year ending June 30, 1917, and for other purposes," approved August 29, 1916, in relation to highway or highway transport, are hereby transferred to the Secretary of Agriculture, and the Council of national Defense is directed to turn over to the Secretary of Agriculture the equipment, material, supplies, papers, maps, and documents utilized in the exercise of such powers. The powers and duties of agencies dealing with highways in the national parks or in military or naval reservations under the control of the United States Army or Navy, or with highways used principally for military or naval purposes, shall not be taken over by the Secretary of Agriculture, but such highways shall remain under the control and jurisdiction of such agencies.

The Secretary of Agriculture is authorized to cooperate with the State highway departments, and with the Department of the Interior in the construction of public highways within Indian reservations, and to pay the amount assumed therefor from the funds allotted or apportioned under this Act to the State wherein the reservation is located.

SEC. 4. That the Secretary of Agriculture shall establish an accounting division which shall devise and install a proper method of keeping the accounts.

SEC. 5. That the Secretary of War be, and he is hereby, authorized and directed to transfer to the Secretary of Agriculture, upon his request, all war material, equipment, and supplies now or hereafter declared surplus from stock now on hand and not needed for the purposes of the War Department but suitable for use in the improvement of highways, and that the same shall be distributed among the highway departments of the several States to be used in the construction, reconstruction, and maintenance of highways, such distribution to be upon the same basis as that hereinafter provided for in this Act in the distribution of Federal-aid fund: Provided, That the Secretary of Agriculture, in his discretion, may reserve from such distribution not to exceed 10 per centum of such material, equipment, and supplies for use in the construction, reconstruction, and maintenance of national forest roads or other roads constructed, reconstructed, or maintained under his direct supervision.

SEC. 6. That in approving projects to receive Federal aid under the provisions of this Act the Secretary of Agriculture shall give preference to such projects as will expedite the completion of an adequate and connected system of highways, interstate in character.

Before any projects are approved in any State, such State, through its State highway department, shall select or designate a system of highways not to exceed 7 per centum of the total highway mileage of such State as shown by the records of the State highway department at the time of the passage of this Act.

Upon this system all Federal-aid apportionments shall be expended.

Highways which may receive Federal aid shall be divided into two classes, one of which shall be known as primary or interstate highways, and shall not exceed three-sevenths of the total mileage which may receive Federal aid, and the other which shall connect or correlate therewith and be known as secondary or intercounty highways, and shall consist of the remainder of the mileage which may receive Federal Aid.

The Secretary of Agriculture shall have authority to approve in whole or in part the systems as designated or to require modification or revisions thereof: Provided, That the States shall submit to the Secretary of Agriculture for his approval any proposed revisions of the designated systems of highways above provided for.

Not more than 60 per centum of all Federal aid allotted to any State shall be expended upon the primary or interstate highways until provision has been made for the improvement of the entire system of such highways: Provided, That with the approval of any State highway department the Secretary of Agriculture may approve the expenditure of more than 60 per centum of the Federal aid apportioned to such State upon the primary or interstate highways in such State.

The Secretary of Agriculture may approve projects submitted by the State highway departments prior to the selection, designation, and approval of the system of Federal-aid highways herein provided for if he may reasonably anticipate that such projects will become a part of such system.

Whenever provision has been made by any State for the completion and maintenance of a system of primary or interstate and secondary or intercounty highways equal to 7 per centum of the total mileage of such State, as required by this Act, said State, through its State highway department, by and with the approval of the Secretary of Agriculture, is hereby authorized to add to the mileage of primary or interstate and secondary or intercounty systems as funds become available for the construction and maintenance of such additional mileage.

SEC. 7. That before any project shall be approved by the Secretary of Agriculture for any State such State shall make provisions for State funds required each year of such States by this Act for construction, reconstruction, and maintenance of all Federal-aid highways within the State, which funds shall be under the direct control of the State highway department.

SEC. 8. That only such durable types of surface and kinds of materials shall be adopted for the construction and reconstruction of any highway which is a part of the primary or interstate and secondary or intercounty systems as will adequately meet the existing and probable future traffic needs and conditions thereon. The Secretary of Agriculture shall approve the types and width of construction and reconstruction and the character of improvement, repair, and maintenance in each case, consideration being given to the type and character which shall be best suited for each locality and to the probable character and extent of the future traffic.

SEC. 9. That all highways constructed or reconstructed under the provisions of this Act shall be free from tolls of all kinds.

That all highways in the primary or interstate system constructed after the passage of this Act shall have a right of way of ample width and a wearing surface of an adequate width which shall not be less than eighteen feet, unless, in the opinion of the Secretary of Agriculture, it is rendered impracticable by physical conditions, excessive costs, probable traffic requirements, or legal obstacles.

SEC. 10. That when any State shall have met the requirements of this Act, the Secretary of the Treasury, upon receipt of certification from the governor of such State to such effect, approved by the Secretary of Agriculture, shall immediately make available to such State, for the purpose set forth in this Act, the sum apportioned to such State as herein provided.

SEC. 11. That any State having complied with the provisions of this Act, and desiring to avail itself of the benefits thereof, shall have its State highway department submit to the Secretary of Agriculture project statements setting forth proposed construction or reconstruction of any primary or interstate, or secondary or intercounty highway therein. If the Secretary of Agriculture approve the project, the State highway department shall furnish to him such surveys, plans, specifications, and estimates therefor as he may require; items including for engineering, inspection, and unforeseen contingencies shall not exceed 10 per centum of the total estimated cost of its construction.

That when the Secretary of Agriculture approves such surveys, plans, specifications, and estimates, he shall notify the State highway department and immediately certify the fact to the Secretary of the Treasury. The Secretary of the Treasury shall thereupon set aside the share of the United States payable under this Act on account of such projects, which shall not exceed 50 per centum of the total estimated cost thereof, except that in the case of any State containing unappropriated public lands exceeding 5 per centum of the total area of all lands in the State, the share of the United States payable under this Act on account of such projects shall not exceed 50 per centum of the total estimated cost thereof plus a percentage which the area of the unappropriated public lands in such State bears to the total area of such State: Provided, That the limitation of payments not to exceed $20,000 per mile, under existing law, which the Secretary of Agriculture may make be, and the same is hereby, increased in proportion to the increased percentage of Federal aid authorized by this section: Provided further, That these provisions relative to the public land States shall apply to all unobligated or unmatched funds appropriated by the Federal Aid Act and payment for approved projects upon which actual building construction work had not begun on the 30th day of June, 1921.

SEC. 12. That the construction and reconstruction of the highways or parts of highways under the provisions of this Act, and all contracts, plans, specifications, and estimates relating thereto, shall be undertaken by the State highway departments subject to the approval of the Secretary of Agriculture. The construction and reconstruction work and labor in each State shall be done in accordance with its laws and under the direct supervision of the State highway department, subject to the inspection and approval of the Secretary of Agriculture and in accordance with the rules and regulations pursuant to this Act.

SEC. 13. That when the Secretary of Agriculture shall find that any project approved by him has been constructed or reconstructed in compliance with said plans and specifications, he shall cause to be paid to the proper authorities of said State the amount set aside for said project.

That the Secretary of Agriculture may, in his discretion, from time to time, make payments on such construction or reconstruction as the work progresses, but these payments, including previous payments, if any, shall not be more than the United States pro rata part of the value of the labor and materials which have been actually put into such construction or reconstruction in conformity to said plans and specifications. The Secretary of Agriculture and the State highway department of each State may jointly determine at what time and in what amounts payments as work progresses shall be made under this Act.

Such payments shall be made by the Secretary of the Treasury, on warrants drawn by the Secretary of Agriculture, to such official or officials or depository as may be designated by the State highway department and authorized under the laws of the State to receive public funds of the State.

SEC. 14. That should any State fail to maintain any highway within its boundaries after construction or reconstruction under the provisions of this Act, the Secretary of Agriculture shall then serve notice upon the State highway department of that fact, and if within ninety days after receipt of such notice said highway has not been placed in proper condition of maintenance, the Secretary of Agriculture shall proceed immediately to have such highway placed in a proper condition of maintenance and charge the cost thereof against the Federal funds allotted to such State, and shall refuse to approve any other project in such State, except as hereinafter provided.

Upon the reimbursement by the State of the amount expended by the Federal Government for such maintenance, said amount shall be paid into the Federal highway fund for reapportionment among all the States for the construction of roads under this Act, and the Secretary of Agriculture shall then approve further projects submitted by the State as in this Act provided.

Whenever it shall become necessary for the Secretary of Agriculture under the provisions of this Act to place any highway in a proper condition of maintenance the Secretary of Agriculture shall contract with some responsible party or parties for doing such work: Provided, however, That in case he is not able to secure a satisfactory contract he may purchase, lease, hire, or otherwise obtain all necessary supplies, equipment, and labor, and may operate and maintain such motor and other equipment and facilities as in his judgment are necessary for the proper and efficient performance of his functions.

SEC. 15. That within two years after this Act takes effect the Secretary of Agriculture shall prepare, publish, and distribute a map showing the highways and forest roads that have been selected and approved as a part of the primary or interstate, and the secondary or intercounty systems, and at least annually thereafter shall publish supplementary maps showing his program and the progress made in selection, construction, and reconstruction.

SEC. 16. That for the purpose of this Act the consent of the United States is hereby given to any railroad or canal company to convey to the highway department of any State any part of its right of way or other property in that State acquired by grant from the United States.

SEC. 17. That if the Secretary of Agriculture determines that any part of the public lands or reservations of the United States is reasonably necessary for the right of way of any highway or forest road or as a source of materials for the construction or maintenance of any such highway or forest road adjacent to such lands or reservations, the Secretary of Agriculture shall file with the Secretary of the department supervising the administration of such land or reservation a map showing the portion of such lands or reservations which it is desired to appropriate.

If within a period of four months after such filing the said Secretary shall not have certified to the Secretary of Agriculture that the proposed appropriation of such land or material is contrary to the public interest or inconsistent with the purposes for which such land or materials have been reserved, or shall have agreed to the appropriation and transfer under conditions which he deems necessary for the adequate protection and utilization of the reserve, then such land and materials may be appropriated and transferred to the State highway department for such purposes and subject to the conditions so specified.

If at any time the need for any such lands or materials for such purposes shall no longer exist, notice of the fact shall be given by the State highway department to the Secretary of Agriculture, and such lands or materials shall immediately revert to the control of the Secretary of the department from which they had been appropriated.

SEC. 18. That the Secretary of Agriculture shall prescribe and promulgate all needful rules and regulations for the carrying out of the provisions of this Act, including such recommendations to the Congress and the State highway departments as he may deem necessary for preserving and protecting the highways and insuring the safety or traffic thereon.

SEC. 19. That on or before the first Monday in December of each year the Secretary of Agriculture shall make a report to Congress, which shall include a detailed statement of the work done, the status of each project undertaken, the allocation of appropriations, an itemized statement of the expenditures and receipts during the preceding fiscal year under this Act, an itemized statement of the traveling and other expenses, including a list of employees, their duties, salaries, and traveling expenses, if any, and his recommendations, if any, for new legislation amending or supplementing this Act. The Secretary of Agriculture shall also make such special reports as Congress may request.

SEC. 20. That for the purpose of carrying out the provisions of this Act there is hereby appropriated, out of the moneys in the Treasury not otherwise appropriated, $75,000,000 for the fiscal year ending June 30, 1922, $25,000,000 of which shall become immediately available, and $50,000,000 of which shall become available January 1, 1922.

SEC. 21. That so much, not to exceed 2 1/2 per centum, of all moneys hereby or hereafter appropriated for expenditure under the provisions of this Act, as the Secretary of Agriculture may deem necessary for administering the provisions of this Act and for carrying on necessary highway research and investigational studies independently or in cooperation with the State highway departments and other research agencies, and for publishing the results thereof, shall be deducted for such purposes, available until expended.

Within sixty days after the close of each fiscal year the Secretary of Agriculture shall determine what part, if any, of the sums theretofore deducted for such purposes will not be needed and apportion such part, if any, for the fiscal year then current in the same manner and on the same basis as are other amounts authorized by this Act apportioned among all the States, and shall certify such apportionment to the Secretary of the Treasury and to the State highway department.

The Secretary of Agriculture, after making the deduction authorized by this section, shall apportion the remainder of the appropriation made for expenditure under the provisions of the Act for the fiscal year among the several States in the following manner: One-third in the ratio which the area of each State bears to the total area of all the States; one-third in the ratio which the population of each State bears to the total population of all the States, as shown by the latest available Federal census; one-third in the ratio which the mileage of rural delivery routes and star routes in each State bears to the total mileage of rural delivery and star routes in all the States at the close of the next preceding fiscal year, as shown by certificate of the Postmaster General, which he is directed to make and furnish annually to the Secretary of Agriculture: Provided, That no State shall receive less than one-half of 1 per centum of each year's allotment. All moneys herein or hereafter appropriated for expenditure under the provisions of this Act shall be available until the close of the second succeeding fiscal year for which apportionment was made: Provided further, That any sums apportioned to any State under the provisions of the Act entitled "An Act to provide that the United States shall aid the States in the construction of rural post roads, and for other purposes," approved July 11, 1916, and all Acts amendatory thereof and supplemental thereto, shall be available for expenditure in that State for the purpose set forth in such Acts until two years after the close of the respective fiscal years for which any such sums become available, and any amount so apportioned remaining unexpended at the end of the period during which it is available for expenditure under the terms of such Acts shall be reapportioned according to the provisions of the Act entitled "An Act to provide that the United States shall aid the States in the construction of rural post roads, and for other purposes," approved July 11, 1916: And provided further, That any amount apportioned under the provisions of this Act unexpended at the end of the period during which it is available for expenditure under the terms of this section shall be reapportioned within sixty days thereafter to all the States in the same manner and on the same basis, and certified to the Secretary of the Treasury and the State highway departments in the same way as if it were being apportioned under this Act for the first time.

SEC. 22. That within sixty days after the approval of this Act the Secretary of Agriculture shall certify to the Secretary of the Treasury and to each of the State highway departments the sum he has estimated to be deducted for administering the provisions of this Act and the sums which he has apportioned to each State for the fiscal year ending June 30, 1922, and on or before January 20 next preceding the commencement of each succeeding fiscal year, and shall make like certificates for each fiscal year.

SEC. 23. That out of the moneys in the Treasury not otherwise appropriated, there is hereby appropriated for the survey, construction, reconstruction, and maintenance of forest roads and trails, the sum of $5,000,000 for the fiscal year ending June 30, 1922, available immediately and until expended, and $10,000,000 for the fiscal year ending June 30, 1923, available until expended.

(a) Fifty per centum, but not to exceed $3,000,000 for any one fiscal year, of the appropriation made or that may hereafter be made for expenditure under the provisions of this section shall be expended under the direct supervision of the Secretary of Agriculture in the survey, construction, reconstruction, and maintenance of roads, and trials of primary importance for the protection, administration, and utilization of the national forests, or when necessary, for the use and development of resources upon which communities within or adjacent to the national forests are dependent, and shall be apportioned among the several States, Alaska, and Puerto Rico by the Secretary of Agriculture, according to the relative needs of the various national forests, taking into consideration the existing transportation facilities, value of timber, or other resources served, relative fire danger, and comparative difficulties of road and trail construction.

The balance of such appropriations shall be expended by the Secretary of Agriculture in the survey, construction, reconstruction, and maintenance of forest roads of primary importance to the State, counties, or communities within, adjoining, or adjacent to the national forests, and shall be prorated and apportioned by the Secretary of Agriculture for expenditures in the several States, Alaska, and Puerto Rico, according to the area and value of the land owned by the Government within the national forests therein as determined by the Secretary of Agriculture from such information, investigation, sources, and departments as the Secretary of Agriculture may deem most accurate.

(b) Cooperation of Territories, States, and civil subdivisions thereof may be accepted but shall not be required by the Secretary of Agriculture.

(c) The Secretary of Agriculture may enter into contracts with any Territory, State or civil subdivision thereof for the construction, reconstruction, or maintenance of any forest road or trail or part thereof.

(d) Construction work on forest roads or trails estimated to cost $5,000 or more per mile, exclusive of bridges, shall be advertised and let to contract.

If such estimated cost is less than $5,000 per mile, or if, after proper advertising, no acceptable bid is received, or the bids are deemed excessive, the work may be done by the Secretary of Agriculture on his own account; and for such purpose the Secretary of Agriculture may purchase, lease, hire, rent, or otherwise obtain all necessary supplies, materials, tools, equipment, and facilities required to perform the work.

The appropriation made in this section or that may hereafter be made for expenditure under the provisions of this section may be expended for the purpose herein authorized and for the payment of wages, salaries, and other expenses for help employed in connection with such work.

SEC. 24. That in any State where the existing constitution or laws will not permit the State to provide revenues for the construction, reconstruction, or maintenance of highways, the Secretary of Agriculture shall continue to approve projects for said State until three years after the passage of this Act, if he shall find that said State has complied with the provisions of this Act in so far as its existing constitution and laws will permit.

SEC. 25. That if any provision of this Act, or the application thereof to any person or circumstances, shall be held invalid, the validity of the remainder of the Act and of the application of such provision to other persons or circumstances shall not be affected thereby.

SEC. 26. That all Acts or parts of Acts in any way inconsistent with the provisions of this Act are hereby repealed, and this Act shall take effect on its passage.

Approved, November 9, 1921.

From *U.S. Statutes at Large* 42 (1921), chap. 119, pp. 212–19.

Documents of Authorization—III

TITLE I – FEDERAL-AID HIGHWAY ACT OF 1956

SEC. 101. Short Title For Title I.

This title may be cited as the "Federal-Aid Highway Act of 1956".

SEC. 102. Federal-Aid Highways.

(a) (1) AUTHORIZATION OF APPROPRIATIONS.—For the purpose of carrying out the provisions of the Federal-Aid Road Act approved July 11, 1916 (39 Stat. 355), and all Acts amendatory thereof and supplementary thereto, there is hereby authorized to be appropriated for the fiscal year ended June 30, 1957, $125,000,000 in addition to any sums heretofore authorized for such fiscal year; the sum $850,000,000 for the fiscal year ending June 30, 1958; and the sum of $875,000,000 for the fiscal year ending June 30, 1959. The sums herein authorized for each fiscal year shall be available for expenditure as follows:

(A) 45 per centum for projects on the Federal-aid primary highway system.

(B) 30 per centum for projects on the Federal-aid secondary highway system.

(C) 25 per centum for projects on extensions of these systems within urban areas.

(2) APPORTIONMENTS.—The sums authorized by this section shall be apportioned among the several States in the manner now provided by law and in accordance with the formulas set forth in section 4 of the Federal-Aid Highway Act of 1944, approved December 20, 1944 (58 Stat. 838): Provided, That additional amounts herein authorized for the fiscal year ending June 30, 1957, shall be apportioned immediately upon enactment of this Act.

(b) AVAILABILITY FOR EXPENDITURE.—Any sums apportioned to any State under this section shall be available for expenditure in that State for two years after the close of the fiscal year for which such sums are authorized, and any amounts so apportioned remaining unexpended at the end of such period shall lapse: Provided, That such funds shall be deemed to have been expended if a sum equal to the total of the sums herein and heretofore apportioned to the State is covered by formal agreements with the Secretary of Commerce for construction, reconstruction, or improvement of specific projects as provided in this title and prior Acts: Provided further, That in the case of those sums heretofore, herein, or hereinafter apportioned to any State for projects on the Federal-aid secondary highway system, the Secretary of Commerce may, upon the request of any State, discharge his responsibility relative to the plans, specifications, estimates, surveys, contract awards, design, inspection, and construction of such secondary road projects by his receiving and approving a certified statement by the State highway department setting forth that the plans, design, and construction for such projects are in accord with the standards and procedures of such State applicable to projects in this category approved by him: Provided further, That such approval shall not be given unless such standards and procedures are in accordance with the objectives set forth in section 1 (b) of the Federal-Aid Highway Act of 1950: And provided further, That nothing contained in the foregoing provisos shall be construed to relieve any State of its obligation now provided by law relative to maintenance, nor to relieve the Secretary of Commerce of his obligation with respect to the selection of the secondary system or the location of projects thereon, to make a final inspection after construction of each project, and to require an adequate showing of the estimated and actual cost of construction of each project. Any Federal-aid primary, secondary, or urban funds released by the payment of the final voucher or by modification of the formal project agreement shall be credited to the same class of funds, primary, secondary, or urban, previously apportioned to the State and be immediately available for expenditure.

(c) TRANSFERS OF APPORTIONMENTS.—Not more than 20 per centum of the respective amounts apportioned to a State for any fiscal year from funds made available for expenditure under clause (A), clause (B), or clause (C) of subsection (a) (1) of this section, may be transferred to the apportionment made to such State under any other of such clauses, except that no such apportionment may be increased by more than 20 per centum by reason of transfers to it under this subsection: Provided, That such transfer is requested by the State highway department and is approved by the Governor of such State and the Secretary of Commerce as being in the public interest: Provided further, That the transfers hereinabove permitted for funds authorized to be appropriated for the fiscal years ending June 30, 1958, and June 30, 1959, shall likewise be permitted on the same basis for funds which may be hereafter authorized to be appropriated for any subsequent fiscal year: And provided further, That nothing herein contained shall be deemed to alter or impair the authority contained in the last proviso to paragraph (b) of section 3 of the Federal-Aid Highway Act of 1944.

Sec. 103 Forest Highways And Forest Development Roads And Trails.

(a) AUTHORIZATION OF APPROPRIATIONS.—For the purpose of carrying out the provisions of section 23 of the Federal Highway Act of 1921 (42 Stat. 218), as amended and supplemented, there is hereby authorized to be appropriated (1) for forest highways the sum of $30,000,000 for the fiscal year ending June 30, 1958, and a like sum for the fiscal year ending June 30, 1959; and (2) for forest development roads and trails the sum of $27,000,000 for the fiscal year ending June 30, 1958, and a like sum for the fiscal year ending June 30, 1959: Provided, That with respect to any proposed construction or reconstruction of a timber access road, advisory public hearings shall be held at a place convenient or adjacent to the area of construction or reconstruction with notice and reasonable opportunity for interested persons to present their views as to the practicability and feasibility of such construction or reconstruction: Provided further, That hereafter funds available for forest highways and forest development roads and trails shall also be available for adjacent vehicular parking areas and for sanitary, water, and fire control facilities: And provided further, That the appropriation herein authorized for forest highways shall be apportioned by the Secretary of Commerce for expenditure in the several States, Alaska, and Puerto Rico in accordance with the provisions of section 3 of the Federal-Aid Highway Act of 1950.

(b) REPEAL OF CERTAIN APPORTIONMENT PROCEDURES.—The provision of section 23 of the Federal Highway Act of 1921, as amended and supplemented, requiring apportionment of funds authorized for forest development roads and trails among the several States, Alaska, and Puerto Rico is hereby repealed.

Sec. 104. Roads And Trails In National Parks, Etc.

(a) NATIONAL PARKS, ETC.—For the construction, reconstruction, and improvement of roads and trails, inclusive of necessary bridges, in national parks, monuments, and other areas administered by the National Park Service, including areas authorized to be established as national parks and monuments, and national park and monument approach roads authorized by the Act of January 31, 1931 (46 Stat. 1053), as amended, there is hereby authorized to be appropriated the sum of $16,000,000 for the fiscal year ending June 30, 1958, and a like sum for the fiscal year ending June 30, 1959.

(b) PARKWAYS.—For the construction, reconstruction, and improvement of parkways, authorized by Acts of Congress, on lands to which title is vested in the United States, there is hereby authorized to be appropriated the sum of $16,000,000 for the fiscal year ending June 30, 1958, and a like sum for the fiscal year ending June 30, 1959.

(c) INDIAN RESERVATIONS AND LANDS.—For the construction, improvement, and maintenance of Indian reservation roads and bridges and roads and bridges to provide access to Indian reservations and Indian lands under the provisions of the Act approved May 26, 1928 (45 Stat. 750), there is hereby authorized to be appropriated the sum of $12,000,000 for the fiscal year ending June 30, 1958, and a like sum for the fiscal year ending June 30, 1959: Provided, That the location, type, and design of all roads and bridges constructed shall be approved by the Secretary of Commerce before any expenditures are made thereon, and all such construction shall be under the general supervision of the Secretary of Commerce.

Sec. 105. Public Land Highways.

For the purpose of carrying out of the provisions of section 10 of the Federal-Aid Highway Act of 1950 (64 Stat. 785), there is hereby authorized to be appropriated for the survey, construction, reconstruction, and maintenance of main roads through unappropriated or unreserved public lands, nontaxable Indian lands, or other Federal reservations the additional sum of $2,000,000 for the fiscal year ending June 30, 1957, and the sum of $2,000,000 for the fiscal year ending June 30, 1958, and a like sum for the fiscal year ending June 30, 1959.

Sec. 106. Special Provisions For Federal Domain Roads, Etc.

Any funds authorized herein for forest highways, forest development roads and trails, park roads and trails, parkways, Indian roads, and public lands highways shall be available for contract upon apportionment, or a date not earlier than one year preceding the beginning of the fiscal year for which authorized if no apportionment is required: Provided, That any amount remaining unexpended two years after the close of the fiscal year for which authorized shall lapse. The Secretary of the department charged with the administration of such funds is hereby granted authority to incur obligations, approve projects, and enter into contracts under such authorizations, and his action in doing so shall be deemed a contractual obligation of the Federal Government for the payment of the cost thereof, and such funds shall be deemed to have been expended when so obligated. Any funds heretofore, herein, or hereafter authorized for any fiscal year for forest highways, forest development roads and trails, park roads and trails, parkways, Indian roads, and public lands highways shall be deemed to have been expended if a sum equal to the total of the sums authorized for such fiscal year and previous fiscal years since and including the fiscal year ending June 30, 1955, shall have been obligated. Any of such funds released by payment of final voucher or modification of project authorizations shall be credited to the balance of unobligated authorizations and be immediately available for expenditure.

Sec. 107. Highways For Alaska.

(a) APPORTIONMENT; MATCHING; SELECTION OF SYSTEMS.— The Territory of Alaska shall be entitled to share in funds herein or hereafter authorized for expenditure for projects on the Federal-aid primary and secondary highway systems, and extensions thereof within urban areas, under the Federal Aid Road Act approved July 11, 1916 (39 Stat. 355), and Acts amendatory thereof or supplementary thereto, upon the same terms and conditions as the several States and Hawaii and Puerto Rico, and the Territory of Alaska shall be included in the calculations to determine the basis of apportionment of such funds, except that one-third only of the area of Alaska shall be used in the calculations to determine the area factor in the apportionment of such funds: Provided, That the Territory of Alaska shall contribute funds each fiscal year in an amount that shall be not less than 10 per centum of the Federal funds apportioned to it for such fiscal year, such contribution to be deposited in a special account in the Federal Treasury for use in conjunction with the Federal funds apportioned to the Territory. The system or systems of roads on which Federal-aid apportionments to the Territory of Alaska are to be expended shall be determined and agreed upon by the Governor of Alaska, the Territorial Highway Engineer of Alaska, and the Secretary of Commerce without regard to the limitations contained in section 6 of the Federal highway Act (42 Stat. 212), as amended and supplemented. The Federal funds apportioned to the Territory of Alaska and the funds contributed by such Territory in accordance herewith may be expended by the Secretary of Commerce either directly or in cooperation with the Territorial Board of Road Commissioners of Alaska, and may be so expended separately or in combination and without regard to the matching provisions of the Federal

Highway Act (42 Stat. 212); and both such funds may be expended for the maintenance of roads within the system or systems of roads agreed upon under the same terms and conditions as for the construction of such roads.

(b) TRANSFER OF FUNCTIONS.—Effective not more than ninety days after the approval of this Act, the functions, duties, and authority pertaining to the construction, repair, and maintenance of roads, tramways, ferries, bridges, trails, and other works in Alaska, conferred upon the Department of the Interior and heretofore administered by the Secretary of the Interior under the Act of June 30, 1932 (47 Stat. 446; 48 U.S.C., sec. 321a and following), are hereby transferred to the Department of Commerce, and thereafter shall be administered by the Secretary of Commerce, or under his direction, by such officer, or officers, as may be designated by him.

(c) TRANSFER OF PERSONNEL, ETC.—There are hereby transferred to the Department of Commerce, to be employed and expended in connection with the functions, duties, and authority transferred to said Department by subsection (b) hereof, all personnel employed in connection with any such functions, duties, or authority, and the unexpended balances of appropriations, allocations, or other funds now available, or that hereafter may be made available, for use in connection with such functions, duties, or authority; and the Department of the Interior is directed to turn over to the Secretary of Commerce all equipment, materials, supplies, papers, maps, and documents, or other property (real or personal, and including office equipment and records) used or held in connection with such functions, duties, and authority.

(d) EFFECTUATION OF TRANSFER.—The Secretary of the Interior and the Secretary of Commerce shall take such steps as may be necessary or appropriate to effect the transfer from the Department of the Interior to the Department of Commerce of the functions, duties, and authority, and the funds and property, as herein provided for.

(e) DISTRIBUTION OF FUNCTION.—The Secretary of Commerce shall have power, by order or regulations, to distribute the functions, duties, and authority hereby transferred, and appropriations pertaining thereto, as he may deem proper to accomplish the economical and effective organization and administration thereof.

Sec. 108. National System Of Interstate And Defense Highways.

(a) INTERSTATE SYSTEM.—It is hereby declared to be essential to the national interest to provide for the early completion of the "National System of Interstate Highways", as authorized and designated in accordance with section 7 of the Federal-Aid Highway Act of 1944 (58 Stat. 838). It is the intent of the Congress that the Interstate System be completed as nearly as practicable over a thirteen-year period and that the entire System in all the States be brought to simultaneous completion. Because of its primary importance to the national defense, the name of such system is hereby changed to the "National System of Interstate and Defense Highways". Such National System of Interstate and Defense Highways is hereinafter in this Act referred to as the "Interstate System".

(b) AUTHORIZATION OF APPROPRIATIONS.—For the purpose of expediting the construction, reconstruction, or improvement, inclusive of necessary bridges and tunnels, of the Interstate System, including expansions thereof through urban areas, designated in accordance with the provisions of section 7 of the Federal-Aid Highway Act of 1944 (58 Stat. 838), there is hereby authorized to be appropriated the additional sum of $1,000,000,000 for the fiscal year ending June 30, 1957, which sum shall be in addition to the authorization heretofore made for that year, the additional sum of $1,700,000,000 for the fiscal year ending June 30, 1958, the additional sum of $2,000,000,000 for the fiscal year ending June 30, 1959, the additional sum of $2,200,000,000 for the fiscal year ending June 30, 1960, the additional sum of $2,200,000,000 for the fiscal year ending June 30, 1961, the additional sum of $2,200,000,000 for the fiscal year ending June 30, 1962, the additional sum of $2,200,000,000 for the fiscal year ending June 30, 1963, the additional sum of $2,200,000,000 for the fiscal year ending June 30, 1964, the additional sum of $2,200,000,000 for the fiscal year ending June 30, 1965, the additional sum of $2,200,000,000 for the fiscal year ending June 30, 1966, the additional sum of $2,200,000,000 for the fiscal year ending June 30, 1967, the additional sum of $1,500,000,000 for the fiscal year ending June 30, 1968, and the additional sum of $1,025,000,000 for the fiscal year ending June 30, 1969.

(c) APPORTIONMENTS FOR 1957, 1958, AND 1959.—the additional sums herein authorized for the fiscal years ending June 30, 1957, June 30, 1958, and June 30, 1959, shall be apportioned among the several States in the following manner: one-half in the ratio which the population of each State bears to the total population of all the States, as shown by the latest available Federal census: Provided, That no State shall receive less than three-fourths of 1 per centum of the money so apportioned; and one-half in the manner now provided by law for the apportionment of funds for the Federal-aid primary system. The additional sum herein authorized for the fiscal year ending June 30, 1957, shall be apportioned immediately upon enactment of this Act. The additional sums herein authorized for the fiscal years ending June 30, 1958, and June 30, 1959, shall be apportioned on a date not less than six months and not more than twelve months in advance of the beginning of the fiscal year for which authorized.

(d) APPORTIONMENTS FOR SUBSEQUENT YEARS BASED UPON REVISED ESTIMATES OF COST.—All sums authorized by this section to be appropriated for the fiscal years 1960 through 1969, inclusive, shall be apportioned among the several States in the ratio which the estimated cost of completing the Interstate System in each State, as determined and approved in the manner provided in this subsection, bears to the sum of the estimated cost of completing the Interstate System in all of the States. Each apportionment herein authorized for the fiscal years 1960 through 1969, inclusive, shall be made on a date as far in advance of the beginning of the fiscal year for which authorized as practicable but in no case more than eighteen months prior to the beginning of the fiscal year for which authorized. As soon as the standards provided for in subsection (i) have been adopted, the Secretary of Commerce, in cooperation with the State highway departments, shall make a detailed estimate of the cost of completing the Interstate System as then designated, after taking into account all previous apportionments made under this section, based upon such standards and in accordance with rules and regulations adopted by him and applied uniformly to all of the States. The Secretary of Commerce shall transmit such estimate to the Senate and the House of Representatives within ten days subsequent to January 2, 1958. Upon approval of such estimate by the Congress by concurrent resolution, the Secretary of Commerce shall use such approved estimate in making apportionments for the fiscal years ending June 30, 1960, June 30, 1961, and June 30, 1962. The Secretary of Commerce shall make a revised estimate of the cost of completing the then designated Interstate System, after taking into account all previous apportionments made under this section, in the same manner as stated above, and transmit the same to the Senate and the House of Representatives within ten days subsequent to January 2, 1962. Upon approval of such estimate by the Congress by concurrent resolution, the Secretary of Commerce shall use such approved estimate in making apportionments for the fiscal years ending June 30, 1963, June 30, 1964, June 30, 1965, and June 30, 1966. The Secretary of Commerce shall make a revised estimate of the cost of completing the then designated Interstate System, after taking into account all previous apportionments made under this section, in the same manner as stated above, and transmit the same to the Senate and the House of Representatives within ten days subsequent to January 2, 1966, and annually thereafter through and including January 2, 1968. Upon approval of any such estimate by the Congress by concurrent resolution, the Secretary of Commerce shall use such approved estimate in making apportionments for the fiscal year which begins next following the fiscal year in which such report is transmitted to the Senate and the House of Representatives. Whenever the Secretary of Commerce, pursuant to this subsection, requests and receives estimates of cost from the State highway departments, he shall furnish copies of such estimates at the same time to the Senate and the House of Representatives.

(e) FEDERAL SHARE.—The Federal share payable on account of any project on the Interstate system provided for by funds made available under the provisions of this section shall be increased to 90 per centum of the total cost thereof, plus a percentage of the remaining 10 per centum of such cost in any State containing unappropriated and unreserved public lands and nontaxable Indian lands, individual and tribal, exceeding 5 per centum of the total area of all lands therein, equal to the percentage that the area of such lands in such State is of its total area: Provided, That such Federal share payable on any project in any State shall not exceed 95 per centum of the total cost of such project.

(f) AVAILABILITY FOR EXPENDITURE.—Any sums apportioned to any State under the provisions of this section shall be available for expenditure in that State for two years after the close of the fiscal year for which such sums are authorized: Provided, That such funds for any fiscal year shall be deemed to be expended if a sum equal to the total of the sums apportioned to the State specifically for the Interstate System for such fiscal year and previous fiscal years is covered by formal agreements with the Secretary of Commerce for the construction, reconstruction, or improvement of specific projects under this section.

(g) LAPSE OF AMOUNTS APPORTIONED.—Any amount apportioned to the States under the provisions of this section unexpended at the end of the period during which it is available for expenditure under the terms of subsection (f) of this section shall lapse, and shall immediately be reapportioned among the other States in accordance with the provisions of subsection (d) of this section: Provided, That any Interstate System funds released by the payment of the final voucher or by the modification of the formal project agreement shall be credited to the Interstate System funds previously apportioned to the State and be immediately available for expenditure.

(h) CONSTRUCTION BY STATES IN ADVANCE OF APPORTION-MENT.—In any case in which a State has obligated all funds apportioned to it under this section and proceeds, subsequent to the date of enactment of this Act, to construct (without the aid of Federal funds) any project (including one or more parts of any project) on the Interstate System, as designated at that time, in accordance with all procedures and all requirements applicable to projects financed under the provisions of this section (except insofar as such procedures and requirements limit a State to the construction of projects with the aid of Federal funds previously apportioned to it), the Secretary of Commerce, upon application by such State and his approval of such application, is authorized, whenever additional funds are apportioned to such State under this section, to pay to such State from such funds the Federal share of the costs of construction of such project: Provided, That prior to construction of any such project, the plans and specifications therefor shall have been approved by the Secretary of Commerce in the same manner as other projects on the Interstate System: Provided further, That any such project shall conform to the standards adopted under subsection (i). In determining the apportionment for any fiscal year under the provisions of subsection (d) of this section, any such project constructed by a State without the aid of Federal funds shall not be considered completed until an application under the provisions of this subsection with respect to such project has been approved by the Secretary of Commerce.

(i) STANDARDS.—The geometric and construction standards to be adopted for the Interstate System shall be those approved by the Secretary of Commerce in cooperation with the State highway departments. Such standards shall be adequate to accommodate the types and volumes of traffic forecast for the year 1975. The right-of-way width of the Interstate System shall be adequate to permit construction of projects on the Interstate System up to such standards. The Secretary of Commerce shall apply such standards uniformly throughout the States. Such standards shall be adopted by the Secretary of Commerce in cooperation with the State highway departments as soon as practicable after the enactment of this Act.

(j) MAXIMUM WEIGHT AND WIDTH LIMITATIONS.—No funds authorized to be appropriated for any fiscal year by this section shall be apportioned to any State within the boundaries of which the Interstate System may lawfully be used by vehicles with weight in excess of eighteen thousand pounds carried on any one axle, or with a tandem-axle weight in excess of thirty-two thousand pounds, or within an overall gross weight in excess of 73,280 pounds, or with a width in excess of 96 inches, or the corresponding maximum weights or maximum widths permitted for vehicles using the public highways of such State under laws or regulations established by appropriate State authority in effect on July 1, 1956, whichever is the greater. Any amount which is withheld from apportionment to any State pursuant to the foregoing provisions shall lapse: Provided however, That nothing herein shall be construed to deny apportionment to any State allowing the operation within such State of any vehicles or combinations thereof that could be lawfully operated within such State on July 1, 1956.

(k) TESTS TO DETERMINE MAXIMUM DESIRABLE DIMENSIONS AND WEIGHTS.—The Secretary of Commerce is directed to take all action possible to expedite the conduct of a series of tests now planned or being conducted by the Highway Research Board of the National Academy of Sciences, in cooperation with the Bureau of Public Roads, the several States, and other persons and organizations, for the purpose of determining the maximum desirable dimensions and weights for vehicles operated on the Federal-aid highway systems, including the Interstate System, and, after the conclusion of such tests, but not later than March 1, 1959, to made recommendations to the Congress with respect to such maximum desirable dimensions and weights.

(l) INCREASE IN MILEAGE.—Section 7 of the Federal-Aid Highways Act of 1944 (58 Stat. 838), relating to the Interstate System, is hereby amended by striking out "forty thousand", and inserting in lieu thereof "forty-one thousand": Provided, That the cost of completing any mileage designated from the one thousand additional miles authorized by this subsection shall be excluded in making the estimates of cost for completing the Interstate System as provided in subsection (d) of this section.

Sec. 109. Acquisition Of Rights-of-way For Interstate System.

(a) FEDERAL ACQUISITION FOR STATES.—In any case in which the Secretary of Commerce is requested by any State to acquire any lands or interests in lands (including within the term "interests in lands", the control of access thereto from adjoining lands) required by such State for right-of-way or other purposes in connection with the prosecution of any project for the construction, reconstruction, or improvement of any section of the Interstate System, the

Secretary of Commerce is authorized, in the name of the United States and prior to the approval of title by the Attorney General, to acquire, enter upon, and take possession of such lands or interests in lands by purchase, donation, condemnation, or otherwise in accordance with the laws of the United States (including the Act of February 26, 1931, 46 Stat. 1421), if—

(1) the Secretary of Commerce has determined either that such State is unable to acquire necessary lands or interests in lands, or is unable to acquire such lands or interests in lands with sufficient promptness; and

(2) such State has agreed with the Secretary of Commerce to pay, at such time as may be specified by the Secretary of Commerce, an amount equal to 10 per centum of the costs incurred by the Secretary of Commerce, in acquiring such lands or interests in lands, or such lesser percentage which represents the State's pro rata share of project costs as determined in accordance with section 108 (e) of this title.

The authority granted by this section shall also apply to lands and interests in lands received as grants of lands from the United States and owned or held by railroads or other corporations.

(b) COSTS OF ACQUISITION.—The costs incurred by the Secretary of Commerce in acquiring any such lands or interests in lands may include the cost of examination and abstract of title, certificate of title, advertising, and any fees incidental to such acquisition. All costs incurred by the Secretary of Commerce in connection with the acquisition of any such lands or interests in lands shall be paid from the funds for construction, reconstruction, or improvement of the Interstate System apportioned to the State upon the request of which such lands or interests in lands are acquired, and any sums paid to the Secretary of Commerce by such State as its share of the costs of acquisition of such lands or interest in lands shall be deposited in the Treasury to the credit of the appropriation for Federal-aid highways and shall be credited to the amount apportioned to such State as its apportionment of funds for construction, reconstruction, or improvement of the Intestate System, or shall be deducted from other moneys due the State for reimbursement under section 108 of this title.

(c) CONVEYANCE OF ACQUIRED LANDS TO THE STATES.—The Secretary of Commerce is further authorized and directed by proper deed, executed in the name of the United States, to convey any such lands or interest in lands acquired in any State under the provisions of this section, except the outside five feet of any such right-of-way in any State which does not provide control of access, to the State highway department of such State or such political subdivisions thereof as its laws may provide, upon such terms and conditions as to such lands or interests in lands as may be agreed upon by the Secretary of Commerce and the State highways department or political subdivisions to which the conveyance is to be made. Whenever the State makes provision for control of access satisfactory to the Secretary of Commerce, the outside five feet then shall be conveyed to the State by the Secretary of Commerce, as herein provided.

(d) RIGHTS-OF-WAY OVER PUBLIC LANDS.—Whenever rights-of-way, including control of access, on the Interstate System are required over public lands or reservations of the United States, the Secretary of Commerce may make such arrangements with the agency having jurisdiction over such lands as may be necessary to give the State or other person constructing the projects on such lands adequate rights-of-way and control of access thereto from adjoining lands, and any such agency is hereby directed to cooperate with the Secretary of Commerce in this connection.

Sec. 110. Availability Of Funds To Acquire Rights-of-way And To Make Advances To The States.

(a) ADVANCE RIGHT-OF-WAY ACQUISITIONS.—For the purpose of facilitating the acquisition of rights-of-way on any of the Federal-aid highway systems, including the Interstate System, in the most expeditious and economical manner, and recognizing that the acquisition of rights-of-ways requires lengthy planning and negotiations if it is to be done at a reasonable cost, the Secretary of Commerce is hereby authorized, upon request of a State highway department, to make available to such State for acquisition of rights-of-way, in anticipation of construction and under such rules and regulations as the Secretary of Commerce may prescribe, the funds apportioned to such State for expenditure on any of the Federal-aid highway systems, including the Interstate System: Provided, That the agreement between the Secretary of Commerce and the State highway department for the reimbursement of the cost of such rights-of-way shall provide for the actual construction of a road on such rights-of-way within a period not exceeding five years following the fiscal year in which such request is made: Provided further, That Federal participation in the cost of rights-of-way so acquired shall not exceed the Federal pro rata share applicable to the class of funds from which Federal reimbursements is made.

(b) ADVANCES TO STATES.—Section 6 of the Federal-Aid Highway Act of 1944 is hereby amended to read as follows:

"Sec. 6. If the Secretary of Commerce shall determine that it is necessary for the expeditious completion of projects on any of the Federal-aid highways systems, including the Interstate System, he may advance to any State out of any existing appropriations the Federal share of the cost of construction thereof to enable the State highway department to make prompt payments for acquisition of rights-of-way, and for construction as it progresses. The sums so advanced shall be deposited in a special revolving trust fund, by the State official authorized under the laws of the State to receive Federal-aid highway funds, to be disbursed solely upon vouchers approved by the State highway department for rights-of-way which have been or are being acquired, and for construction which has been actually performed and approved by the Secretary of Commerce. Upon determination of the Secretary of Commerce that any part of the funds advanced to any State under the provisions of this section are no longer required, the amount of the advance which is determined to be in excess of current requirements of the State shall be repaid upon his demand, and such repayments shall be returned to the credit of the appropriation from which the funds were advanced. Any sums advanced and not repaid on demand shall be deducted from sums due the State for the Federal pro rata share of the cost of construction of Federal-aid projects."

Sec. 111. Relocation Of Utility Facilities.

(a) AVAILABILITY OF FEDERAL FUNDS FOR REIMBURSEMENT TO STATES.—Subject to the conditions contained in this section, whenever a State shall pay for the cost of relocation of utility facilities necessitated by the construction of a project on the Federal-aid primary or secondary systems or on the Interstate System, including extension thereof within urban areas, Federal funds may be used to reimburse the State for such cost in the same proportion as Federal funds are expended on the project; Provided, That Federal funds shall not be apportioned to the States under this section when the payment to the utility violates the law of the State or violates a legal contract between the utility and the State.

(b) UTILITY DEFINED.—For the purposes of this section, the term "utility" shall include publicly, privately, and cooperatively owned utilities.

(c) COST OF RELOCATION DEFINED.—For the purposes of this section, the term "cost of relocation" shall include the entire amount paid by such utility properly attributable to such relocation after deducting therefrom any increase in the value of the new facility and any salvage value derived from the old facility.

Sec. 112. Agreements Relating To Use Of And Access To Rights-of-way.

All agreements between the Secretary of Commerce and the State highway department for the construction of projects on the Interstate System shall contain a clause providing that the State will not add any points of access to, or exit from, the project in addition to those approved by the Secretary in the plans for such project, without the prior approval of the Secretary. Such agreements shall also contain a clause providing that the State will not permit automotive service stations or other commercial establishments for serving motor vehicle users to be constructed or located on the rights-of-way of the Interstate System. Such agreements may, however, authorize a State or political subdivision thereof to use the air space above and below the established grade line of the highway pavement for the parking of motor vehicles provided such use does not interfere in any way with the free flow of traffic on the Interstate System.

Sec. 113. Toll Roads, Bridges, And Tunnels.

(a) APPROVAL AS PART OF INTERSTATE SYSTEM.—Upon a finding by the Secretary of Commerce that such action will promote the development of an integrated Interstate System, the Secretary is authorized to approve as part of the Interstate System any toll road, bridge, or tunnel, now or hereafter constructed which meets the standards adopted for the improvement of projects located on the Interstate System, whenever such toll road, bridge, or tunnel is located on a route heretofore or hereafter designated as a part of the Interstate System: Provided, That no Federal-aid highway funds shall be expended for the construction, reconstruction, or improvement of any such toll road except to the extent hereafter permitted by law: Provided further, That no Federal-aid highway funds shall be expended for the construction, reconstruction, or improvement of any such toll road, bridge, or tunnel except to the extent now or hereafter permitted by law.

(b) APPROACHES HAVING OTHER USE.—The funds authorized under this title, or under prior Acts, shall be available for expenditure on projects approaching any toll road, bridge, or tunnel to a point where such project will have some use irrespective of its use for such toll road, bridge, or tunnel.

(c) APPROACHES HAVING NO OTHER USE.—The funds authorized under section 108 (b) of this title, or under prior Acts, shall be available for expenditure on Interstate System projects approaching any toll road on the Interstate System, even though the project has no use other than as an approach to such toll road: Provided, That agreement satisfactory to the Secretary of Commerce has been reached with the State prior to approval of any such project (1) that the section of toll road will become free to the public upon the collection of tolls sufficient to liquidate the cost of the toll road or any bonds outstanding at the time constituting a valid lien against said section of toll road covered in the agreement and their maintenance and operation and debt service during the period of toll collections, and (2) that there is one or more reasonably satisfactory alternate free routes available to traffic by which the toll section of the System may be bypassed.

(d) EFFECT ON CERTAIN PRIOR ACTS.—Nothing in this title shall be deemed to repeal the Act approved March 3, 1927 (44 Stat. 1398), or subsection (g) of section 204 of the National Industrial Recovery Act (48 State. 200), and such Acts are hereby amended to include tunnels as well as bridges.

Sec. 114. Determination Of Policy With Respect To Reimbursement For Certain Highways.

It is hereby declared to be the intent and policy of the Congress to determine whether or not the Federal Government should equitably reimburse any State for a portion of a highway which is on the Interstate System, whether toll or free, the construction of which has been completed subsequent to August 2, 1947, or which is either in actual use or under construction by contract, for completion, awarded not later than June 30, 1957: Provided, That such highway meets the standards required by this title for the Interstate System. The time, methods, and amounts of such reimbursement, if any, shall be determined by the Congress following a study which the Secretary of Commerce is hereby authorized and directed to conduct, in cooperation with the State highways departments, and other agencies as may be required, to determine which highways in the Interstate System measure up to the standards required by this title, including all related factors of cost, depreciation, participation of Federal funds, and any other items relevant thereto. A complete report of the results of such study shall be submitted to the Congress within ten days subsequent to January 2, 1958.

Sec. 115. Prevailing Rate Of Wage.

(a) APPLICATION OF DAVIS-BACON ACT.—The Secretary of Commerce shall take such action as may be necessary to insure that all laborers and mechanics employed by contractors or subcontractors on the initial construction work performed on highway projects on the Interstate System authorized under section 108 of this title shall be paid wages at rates not less than those prevailing on the same type of work on similar construction in the immediate locality as determined by the Secretary of Labor in accordance with the Act of August 30, 1935, known as the Davis-Bacon Act (40 U.S.C., sec. 276-a).

(b) CONSULTATION WITH STATE HIGHWAY DEPARTMENT; PRE-DETERMINATION OF RATES.—In carrying out the duties of the foregoing subsection, the Secretary of Labor shall consult with the highway department of the State in which a project on the Interstate System is to be performed. After giving due regard to the information thus obtained, he shall make a predetermination of the minimum wages to be paid laborers and mechanics in accordance with the provisions of the foregoing subsection which shall be set out in each project advertisement for bids and in each bid proposal form and shall be made a part of the contract covering the project.

Sec. 116. Declarations Of Policy With Respect To Federal-aid Highway Program.

(a) ACCELERATION OF PROGRAM.—It is hereby declared to be in the national interest to accelerate the construction of the Federal-aid highway systems, including the Interstate System, since many of such highways, or portions thereof, are in fact inadequate to meet the needs of local and interstate commerce, the national and the civil defense.

(b) COMPLETION OF INTERSTATE SYSTEM; PROGRESS REPORT ON FEDERAL-AID HIGHWAY PROGRAM.—It is further declared that one of the most important objectives of this Act is the prompt completion of the Interstate System. Insofar as possible in consonance with this objective, existing highways located on an interstate route shall be used to the extent that such use is practicable, suitable, and feasible, it being the intent that local needs, to the extent practicable, suitable, and feasible, shall be given equal consideration with the needs of interstate commerce. The Secretary of Commerce is hereby directed to submit to the Congress not later than February 1, 1959, a report on the progress made in attaining the objectives set forth in this subsection and in subsection (a), together with recommendations.

(c) PUBLIC HEARINGS.—Any State highway department which submits plans for a Federal-aid highway project involving the bypassing of, or going through, any city, town, or village, either incorporated or unincorporated, shall certify to the Commissioner of Public Roads that it has had public hearings, or has afforded the opportunity for such hearings, and has considered the economic effects of such a location: Provided, That, if such hearings have been held, a copy of the transcript of said hearing shall be submitted to the Commissioner of Public Roads, together with the certification.

(d) PARTICIPATION BY SMALL BUSINESS ENTERPRISES.—It is hereby declared to be in the national interest to encourage and develop the actual and potential capacity of small business and to utilize this important segment of our economy to the fullest practicable extent in construction of the Federal-aid highway systems, including the Interstate System. In order to carry out that intent and encourage full and free competition, the Secretary of Commerce should assist, insofar as feasible, small business enterprises in obtaining contracts in connection with the prosecution of the highway program.

Sec. 117. Highway Safety Study.

Section 7 of the Federal-Aid Highway Act of 1952 (66 Stat. 158) is hereby amended to read as follows:

"SEC. 7. There is hereby authorized an emergency fund in the amount of $30,000,000 for expenditure by the Secretary of Commerce, in accordance with the provisions of the Federal-Aid Road Act approved July 11, 1916, as amended and supplemented, after receipt of an application therefor from the highway department of any State, in the repair or reconstruction of highways and bridges on the Federal-aid highway systems, including the Interstate System, which he shall find have suffered serious damage as the result of disaster over a wide area, such as by flood, hurricanes, tidal waves, earthquakes, severe storms, landslides, or other catastrophes in any part of the United States. The appropriation of such moneys as may be necessary for the establishment of the fund in accordance with the provisions of this section and for its replenishment on an annual basis is hereby authorized: Provided, That pending the appropriation of such sum, or its replenishment, the Secretary of Commerce may expend, from existing Federal-aid highway appropriations, such sums as may be necessary for the immediate prosecution of the work herein authorized, such appropriations to be reimbursed from the appropriation herein authorized when made: Provided further, That no expenditures shall be made hereunder with respect to any such catastrophe in any State unless an emergency has been declared by the Governor of such State and concurred in by the Secretary of Commerce: Provided further, That the Federal share payable on account of any repair or reconstruction project provided for by funds made available under this section shall not exceed 50 per centum of the cost thereof: And provided further, that the funds herein authorized shall be available for use on any projects programmed and approved at any time during the fiscal year ending June 30, 1956, and thereafter, which meet the provisions of this section, including projects which may have been previously approved during the fiscal year ending June 30, 1956, from any other category of funds under the Federal-Aid Road Act approved July 11, 1916, as amended and supplemented."

Sec. 118 Omitted

Sec. 119. Definition Of Construction.

The definition of the term "construction" in section 1 of the Federal-Aid Highway Act of 1944 is hereby amended by inserting after "mapping" the following: "(including the establishment of temporary and permanent geodetic markers in accordance with specifications of the Coast and Geodetic Survey in the Department of Commerce)".

Sec. 120. Archeological And Paleontological Salvage.

Funds authorized by this title to be appropriated, to the extent approved as necessary by the highway department of any State, may be used for archeological and paleontological salvage in that State in compliance with the Act entitled "An Act for the preservation of American antiquities", approved June 8, 1906 (34 Stat. 225), and State laws where applicable.

Sec. 121. Mapping.

In carrying out the provisions of this title the Secretary of Commerce may, wherever practicable, authorize the use of photogram-metric methods in mapping, and the utilization of commercial enterprise for such services.

Sec. 122. Relationship Of This Title To Other Acts; Effective Date.

All provisions of the Federal-Aid Road Act approved July 11, 1916, together with all Acts amendatory thereof or supplementary thereto, not inconsistent with this title, shall remain in full force and effect and be applicable hereto. All Acts or parts of Acts in any way inconsistent with the provisions of this title are hereby repealed. This title shall take effect on the date of the enactment of this Act.

Documents of Authorization—IV

TITLE II—HIGHWAY REVENUE ACT OF 1956

Sec. 201. Short Title For Title Ii.

(a) SHORT TITLE.—This title may be cited as the "Highway Revenue Act of 1956".

(b) AMENDMENT OF 1954 CODE.—Whenever in this title an amendment is expressed in terms of an amendment to a section or other provision, the reference shall be considered to be made to a section or other provision of the Internal Revenue Code of 1954.

Sec. 202. Increase In Taxes On Diesel Fuel And On Special Motor Fuels.

(a) DIESEL FUEL.—Subsection (a) of section 4041 (relating to tax on diesel fuel) is amended by striking out "2 cents a gallon" and inserting in lieu thereof "3 cents a gallon", and by adding after paragraph (2) the following:

"In the case of a liquid taxable under this subsection sold for use or used as a fuel in a diesel-powered highway vehicle (A) which (at the time of such sale or use) is not registered, and is not required to be registered, for highway use under the laws of any State or foreign country, or (B) which, in the case of a diesel-powered highway vehicle owned by the United States, is not used on the highway, the tax imposed by paragraph (1) or by paragraph (2) shall be 2 cents a gallon in lieu of 3 cents a gallon. If a liquid on which tax was imposed by paragraph (1) at the rate of 2 cents a gallon by reason of the preceding sentence is used as a fuel in a diesel-powered highway vehicle (A) which (at the time of such use) is registered, or is required to be registered, for highway use under the laws of any State or foreign country, or (B) which, in the case of a diesel-powered highway vehicle owned by the United States, is used on the highway, a tax of 1 cent a gallon shall be imposed under paragraph (2)."

(b) SPECIAL MOTOR FUELS.—Subsection (b) of section 4041 (relating to special motor fuels) is amended by striking out "2 cents a gallon" and inserting in lieu thereof "3 cents a gallon", and by adding after paragraph (2) the following:

"In the case of a liquid taxable under this subsection sold for use or used otherwise than as a fuel for the propulsion of a highway vehicle (A) which (at the time of such sale or use) is registered, or is required to be registered, for highway use under the laws of any State or foreign country, or (B) which, in the case of a highway vehicle owned by the United States, is used on the highway, the tax imposed by paragraph (1) or by paragraph (2) shall be 2 cents a gallon in lieu of 3 cents a gallon. If a liquid on which tax was imposed by paragraph (1) at the rate of 2 cents a gallon by reason of the preceding sentence is used as a fuel for the propulsion of a highway vehicle (A) which (at the time of such use) is registered, or is required to be registered, for highway use under the laws of any State or foreign country, or (B) which, in the case of a highway vehicle owned by the United States, is used on the highway, a tax of 1 cent a gallon shall be imposed under paragraph (2)."

(c) RATE REDUCTION.—Subsection (c) of section 4041 (relating to rate reduction) is amended to read as follows:

"(c) RATE REDUCTION.—On and after July 1, 1972—

"(1) the taxes imposed by this section shall be 1 1/2 cents a gallon; and

"(2) the second and third sentences of subsections (a) and (b) shall not apply."

Sec. 203. Increase In Tax On Trucks, Truck Trailers, Buses, Etc.

So much of paragraph (1) of section 4061 (a) (relating to tax on trucks, truck trailers, buses, etc.) as precedes "Automobile truck chassis" is amended to read as follows:

"(1) Articles taxable at 10 percent, except that on and after July 1, 1972, the rate shall be 5 percent—".

Sec. 204. Increase In Taxes On Tires Of The Type Used On HighwayVehicles; Tax On Tread Rubber, Etc.

(A) IN GENERAL.—Section 4071 (relating to tax on tires and tubes) is amended to read as follows:

"SEC. 4071. IMPOSITION OF TAX.

"(a) IMPOSITION AND RATE OF TAX.—There is hereby imposed upon the following articles, if wholly or in part of rubber, sold by the manufacturer, producer, or importer, a tax at the following rates:

"(1) Tires of the type used on highway vehicles, 8 cents a pound.

"(2) Other tires, 5 cents a pound.

"(3) Inner tubes for tires, 9 cents a pound.

"(4) Tread rubber, 3 cents a pound.

"(b) DETERMINATION OF WEIGHT.—For purposes of this section, weight shall be based on total weight, except that in the case of tires such total weight shall be exclusive of metal rims or rim bases. Total weight of the articles shall be determined under regulations prescribed by the Secretary or his delegate.

"(c) RATE REDUCTION.—On and after July 1, 1972—

"(1) the tax imposed by paragraph (1) of subsection (a) shall be 5 cents a pound; and

"(2) paragraph (4) of subsection (a) shall not apply."

(b) TREAD RUBBER DEFINED.—Section 4072 (defining the term "rubber") is amended to read as follows:

"SEC. 4072. DEFINITIONS.

"(a) RUBBER.—For purposes of this chapter, the term 'rubber' includes synthetic and substitute rubber.

"(b) TREAD RUBBER.—For purposes of this chapter, the term 'tread rubber' means any material—

"(1) which is commonly or commercially known as tread rubber or camelback; or

"(2) which is a substitute for a material described in paragraph (1) and is of a type used in recapping or retreading tires.

"(c) TIRES OF THE TYPE USED ON HIGHWAY VEHICLES. For purposes of this part, the term 'tires of the type used on highway vehicles' means tires of the type used on—

"(1) motor vehicles which are highway vehicles, or

"(2) vehicles of the type used in connection with motor vehicles which are highway vehicles."

(c) EXEMPTION OF CERTAIN TREAD RUBBER FROM TAX.—Section 4073 (relating to exemptions) is amended by adding at the end thereof the following new subsection:

"(c) EXEMPTION FROM TAX ON TREAD RUBBER IN CERTAIN CASES.—Under regulations prescribed by the Secretary or his delegate, the tax imposed by section 4071 (a) (4) shall not apply to tread rubber sold by the manufacturer, producer, or importer, to any person for use by such person otherwise than in the recapping or retreading of tires of the type used on highway vehicles."

(d) TECHNICAL AMENDMENT.—The table of sections for part II of subchapter A of chapter 32 is amended by striking out

"SEC. 4072. Definition of rubber."

and inserting in lieu thereof

"SEC. 4072. Definitions."

Sec. 205. Increase In Tax On Gasoline.

Section 4081 (relating to tax on gasoline) is amended to read as follows:
"SEC. 4081. IMPOSITION OF TAX.

"(a) IN GENERAL.—There is hereby imposed on gasoline sold by the producer or importer thereof, or by any producer of gasoline, a tax of 3 cents a gallon.

"(b) RATE REDUCTION.—On and after July 1, 1972, the tax imposed by this section shall be 1 1/2 cents a gallon."

Sec. 206. Tax On Use Of Certain Vehicles.

(a) IMPOSITION OF TAX.—Chapter 36 (relating to certain other excise taxes) is amended by adding at the end thereof the following new subchapter:
"SUBCHAPTER D—Tax on Use of Certain Vehicles

"SEC. 4481. IMPOSITION OF TAX.

"(a) IMPOSITION OF TAX. A tax is hereby imposed on the use of any highway motor vehicle which (together with the semitrailers and trailers customarily used in connection with highway motor vehicles of the same type as such highway motor vehicle) has a taxable gross weight of more than 26,000 pounds, at the rate of $1.50 a year for each 1,000 pounds of taxable gross weight or fraction thereof.

"(b) BY WHOM PAID.—The tax imposed by this section shall be paid by the person in whose name the highway motor vehicle is, or is required to be, registered under the law of the State in which such vehicle, is or is required to be, registered, or, in case the highway motor vehicle is owned by the United States, by the agency or instrumentality of the United States operating such vehicle.

"(c) PRORATION OF TAX.—If in any year the first use of the highway motor vehicle is after July 31, the tax shall be reckoned proportionately from the first day of the month in which such use occurs to and including the 30th day of June following.

"(d) ONE PAYMENT PER YEAR.—If the tax imposed by this section is paid with respect to any highway motor vehicle for any year, no further tax shall be imposed by this section for such year with respect to such vehicle.

"(e). PERIOD TAX IN EFFECT.—The tax imposed by this section shall apply only to use after June 30, 1956, and before July 1, 1972.

"SEC. 4482. DEFINITIONS.

"(a) HIGHWAY MOTOR VEHICLE.—For purposes of this subchapter, the term 'highway motor vehicle' means any motor vehicle which is a highway vehicle.

"(b) TAXABLE GROSS WEIGHT.—For purposes of this subchapter, the term 'taxable gross weight', when used with respect to any highway motor vehicle, means the sum of—

"(1) the actual unloaded weight of—

"(A) such highway motor vehicle fully equipped for service, and

"(B) the semitrailers and trailers (fully equipped for service) customarily used in connection with highway motor vehicles of the same type as such highway motor vehicle, and

"(2) the weight of the maximum load customarily carried on highway motor vehicles of the same type as such highway motor vehicle and on the semitrailers and trailers referred to in paragraph (1) (B).

Taxable gross weight shall be determined under regulations prescribed by the Secretary or his delegate (which regulations may include formulas or other methods for determining the taxable gross weight of vehicles by classes, specifications, or otherwise).

"(c) OTHER DEFINITIONS.—For purposes of this subchapter—

"(1) STATE.—The term 'State' means a State, a Territory of the United States, and the District of Columbia.

"(2) YEAR.—The term 'year' means the one-year period beginning on July 1.

"(3) USE.—The term 'use' means use in the United States on the public highways.

"SEC. 4483. EXEMPTIONS.

"(a) STATE AND LOCAL GOVERNMENTAL EXEMPTION.—Under regulations prescribed by the Secretary or his delegate, no tax shall be imposed by section 4481 on the use of any highway motor vehicle by any State or any political subdivision of a State.

"(b) EXEMPTION FOR UNITED STATES. The Secretary may authorize exemption from the tax imposed by section 4481 as to the use by the United States of any particular highway motor vehicle, or class of highway motor vehicles, if he determines that the imposition of such tax with respect to such use will cause substantial burden or expense which can be avoided by granting tax exemption and that full benefit of such exemption, if granted, will accrue to the United States.

"(c) CERTAIN TRANSIT-TYPE BUSES.—Under regulations prescribed by the Secretary or his delegate, no tax shall be imposed by section 4481 on the use of any bus which is of the transit type (rather than of the inter-city type) by a person who, for the last 3 months of the preceding year (or for such other period as the Secretary or his delegate may by regulations prescribe for purposes of this subsection), met the 60-percent passenger fare revenue test set forth in section 6421 (b) (2) as applied to the period prescribed for purposes of this subsection.

"SEC. 4484. CROSS REFERENCE.

"For penalties and administrative provisions applicable to this subchapter, see subtitle F."

(b) MODE AND TIME OF COLLECTION OF TAX.—Section 6302 (b) (relating to discretion as to method of collecting tax) is amended by inserting "section 4481 of chapter 36," after "33,".

(c) TECHNICAL AMENDMENT.—The table of subchapters for chapter 36 is amended by adding at the end thereof the following:

"Subchapter D. Tax on use of certain vehicles."

Sec. 207. Floor Stocks Taxes.

(a) IMPOSITION OF TAXES.—Subchapter F of chapter 32 (special provisions applicable to manufacturers excise taxes) is amended by renumbering section 4226 as 4227 and by inserting after section 4225 the following new section:

"SEC. 4226. FLOOR STOCKS TAXES.

"(a) IN GENERAL.—

"(1) 1956 TAX ON TRUCKS, TRUCK TRAILERS, BUSES, ETC.—On any article subject to tax under section 4061 (a) (1) (relating to tax on trucks, truck trailers, buses, etc.) which, on July 1, 1956, is held by a dealer for sale, there is hereby imposed a floor stocks tax at the rate of 2 percent of the price for which the article was purchased by such dealer. If the price for which the article was sold by the manufacturer, producer, or importer is established to the satisfaction of the Secretary or his delegate, then in lieu of the amount specified in the preceding sentence, the tax imposed by this paragraph shall be at the rate of 2 percent of the price for which the article was sold by the manufacturer, producer, or importer.

"(2) 1956 TAX ON TIRES OF THE TYPE USED ON HIGHWAY VEHICLES.—On tires subject to tax under section 4071 (a) (1) (as amended by the Highway Revenue Act of 1956) which, on July 1, 1956, are held—

"(A) by a dealer for sale,

"(B) for sale on, or in connection with, other articles held by the manufacturer, producer, or importer of such other articles, or

"(C) for use in the manufacture or production of other articles, there is hereby imposed a floor stocks tax at the rate of 3 cents a pound. The tax imposed by this paragraph shall not apply to any tire which is held for sale by the manufacturer, producer, or importer of such tire or which will be subject under section 4218 (a) (2) or 4219 to the manufacturers excise tax on tires.

"(3) 1956 TAX ON TREAD RUBBER.—On tread rubber subject to tax under section 4071 (a) (4) (as amended by the Highway Revenue Act of 1956) which, on July 1, 1956, is held by a dealer, there is hereby imposed a floor stocks tax at the rate of 3 cents a pound. The tax imposed by this paragraph shall not apply in the case of any person if such person established, to the satisfaction of the Secretary or his delegate, that all tread rubber held by him on July 1, 1956, will be used otherwise than in the recapping or retreading of tires of the type used on highway vehicles (as defined in section 4072 (c)).

"(4) 1956 TAX ON GASOLINE.—On gasoline subject to tax under section 4081 which, on July 1, 1956, is held by a dealer for sale, there is hereby imposed a floor stocks tax at the rate of 1 cent a gallon. The tax imposed by this paragraph shall not apply to gasoline in retail stocks held at the place where intended to be sold at retail, nor to gasoline held for sale by a producer or importer of gasoline.

"(b) OVERPAYMENT OF FLOOR STOCKS TAXES.—Section 6416 shall apply in respect of the floor stocks taxes imposed by this section, so as to entitle, subject to all provisions of 6416, any person paying such floor stocks taxes to a credit or refund thereof for any of the reasons specified in section 6416.

"(c) MEANING OF TERMS.—For purposes of subsection (a), the terms 'dealer' and 'held by a dealer' have the meaning assigned to them by section 6412 (a) (3).

"(d) DUE DATE OF TAXES.—The taxes imposed by subsection (a) shall be paid at such time after September 30, 1956, as may be prescribed by the Secretary or his delegate."

(b) TECHNICAL AMENDMENT.—The table of sections for subchapter F of chapter 32 is amended by striking out

"Sec. 4226. Cross references."

and inserting in lieu thereof

"Sec. 4226. Floor stocks taxes.

"Sec. 4227. Cross references."

Sec. 208. Credit Or Refund Of Tax.

(a) FLOOR STOCKS REFUNDS.—So much of section 6412 (relating to floor stocks refunds) as precedes subsection (d) is amended to read as follows:

"SEC. 6412. FLOOR STOCKS REFUNDS.

"(a) IN GENERAL.—

"(1) PASSENGER AUTOMOBILES, ETC.—Where before April 1, 1957, any article subject to the tax imposed by section 4061 (a) (2) has been sold by the manufacturer, producer, or importer and on such date is held by a dealer and has not been used and is intended for sale, there shall be credited or refunded (without interest) to the manufacturer, producer, or importer an amount equal to the difference between the tax paid by such manufacturer, producer, or importer on his sale of the article and the amount of tax made applicable to such article on and after April 1, 1957, if claim for such credit or refund is filed with the Secretary or his delegate on or before August 10, 1957, based upon a request submitted to the manufacturer, producer, or importer before July 1, 1957, by the dealer who held the article in respect of which the credit or refund is claimed, and, on or before August 10, 1957, reimbursement has been made to such dealer by such manufacturer, producer, or importer for the tax reduction on such article or written consent has been obtained from such dealer to allowance of such credit or refund.

"(2) TRUCKS AND BUSES, TIRES, TREAD RUBBER, AND GASO-LINE.—Where before July 1, 1972, any article subject to the tax imposed by section 4061 (a) (1), 4071 (a) (1) or (4), or 4081 has been sold by the manufacturer, producer, or importer and on such date is held by a dealer and has not been used and is intended for sale (or, in the case of tread rubber, is intended for sale or is held for use), there shall be credited or refunded (without interest) to the manufacturer, producer, or importer an amount equal to the difference between the tax paid by such manufacturer, producer, or importer on his sale of the article and the amount of tax made applicable to such article on and after July 1, 1972, if claim for such credit or refund is filed with the Secretary or his delegate on or before November 10, 1972, based upon a request submitted to the manufacturer, producer, or importer before October 1, 1972, by the dealer who held the article in respect of which the credit or refund is claimed, and, on or before November 10, 1972, reimbursement has been made to such dealer by such manufacturer, producer, or importer for the tax reduction on such article or written consent has been obtained from such dealer to allowance of such credit or refund. No credit or refund shall be allowable under this paragraph with respect to gasoline in retail stocks held at the place where intended to be sold at retail, nor with respect to gasoline held for sale by a producer or importer of gasoline.

"(3) DEFINITIONS.—for purposes of this section—

"(A) The term 'dealer' includes a wholesaler, jobber, distributor, or retailer, or, in the case of tread rubber subject to tax under section 4071 (a) (4), includes any person (other than the manufacturer, producer, or importer thereof) who holds such tread rubber for sale or use.

"(B) An article shall be considered as 'held by a dealer' if title thereto has passed to such dealer (whether or not delivery to him has been made), and if for purposes of consumption title to such article or possession thereof has not at any time been transferred to any person other than a dealer.

"(b) LIMITATION ON ELIGIBILITY FOR CREDIT OR REFUND.—No manufacturer, producer, or importer shall be entitled to credit or refund under subsection (a) unless he has in his possession such evidence of the inventories with respect to which the credit or refund is claimed as may be required by regulations prescribed under this section.

"(c) OTHER LAWS APPLICABLE.—All provisions of law, including penalties, applicable in respect of the taxes imposed by sections 4061, 4071, and 4081 shall, insofar as applicable and not inconsistent with subsections (a) and (b) of this section, apply in respect of the credits and refunds provided for in subsection (a) to the same extent as if such credits or refunds constituted overpayments of such taxes."

(b) SPECIAL CASES.—Section 6416 (b) (2) (special cases in which tax payments considered overpayments) is amended by striking out the period at the end of subparagraph (I) and inserting in lieu thereof a semicolon, and by adding at the end thereof the following:

"(J) In the case of a liquid in respect of which tax was paid under section 4041 (a) (1) at the rate of 3 cents a gallon, used or resold for use as a fuel in a diesel-powered highway vehicle (i) which (at the time of such use or resale) is not registered, and is not required to be registered, for highway use under the laws of any State or foreign country, or (ii) which, in the case of a diesel-powered highway vehicle owned by the United States, is not used on the highway; except that the amount of any overpayment by reason of this subparagraph shall not exceed an amount computed at the rate of 1 cent a gallon;

"(K) In the case of a liquid in respect of which tax was paid under section 4041 (b) (1) at the rate of 3 cents a gallon, used or resold for use otherwise than as a fuel for the propulsion of a highway vehicle (i) which (at the time of such use or resale) is registered, or is required to be registered, for highway use under the laws of any State or foreign country, or (ii) which, in the case of a highway vehicle owned by the United States, is used on the highway; except that the amount of any overpayment by reason of this subparagraph shall not exceed an amount computed at the rate of 1 cent a gallon;

"(L) In the case of a liquid in respect of which tax was paid under section 4041 at the rate of 3 cents a gallon, used during any calendar quarter in vehicles while engaged in furnishing scheduled common carrier public passenger land transportation service along regular routes; except that (i) this subparagraph shall apply only if the 60 percent passenger fare revenue test set forth in section 6421 (b) (2) is met with respect to such quarter, and (ii) the amount of such overpayment for such quarter shall be an amount determined by multiplying 1 cent for each gallon of liquid so used by the percentage which such person's tax-exempt passenger fare revenue (as defined in section 6421 (d) (2)) derived from such scheduled service during such quarter was of his total passenger fare revenue (not including the tax imposed by section 4261, relating to the tax on transportation of persons) derived from such scheduled service during such quarter;

"(M) In the case of tread rubber in respect of which tax was paid under section 4071 (a) (4), used or resold for use otherwise than in the recapping or retreading of tires of the type used on highway vehicles (as defined in section 4072 (c)), unless credit or refund of such tax is allowable under subsection (b) (3)."

(c) PAYMENTS TO ULTIMATE PURCHASERS.—Subchapter B of chapter 65 (relating to rules of special application for abatements, credits, and refunds) is amended by renumbering section 6421 as 6422 and by inserting after section 6420 the following new section:

"SEC. 6421. GASOLINE USED FOR CERTAIN NON-HIGHWAY PURPOSES OR BY LOCAL TRANSIT SYSTEMS.

"(a) NON-HIGHWAY USES.—If gasoline is used otherwise than as a fuel in a highway vehicle (1) which (at the time of such use) is registered, or is required to be registered, for highway use under the laws of any State or foreign country or (2) which, in the case of a highway vehicle owned by the United States, is used on the highway, the Secretary or his delegate shall pay (without interest) to the ultimate purchaser of gasoline an amount equal to 1 cent for each gallon of gasoline so used.

"(b) LOCAL TRANSIT SYSTEMS.—

"(1) ALLOWANCE.—If gasoline is used during any calendar quarter in vehicles while engaged in furnishing scheduled common carrier public passenger land transportation service along regular routes, the Secretary or his delegate shall, subject to the provisions of paragraph (2), pay (without interest) to the ultimate purchaser of such gasoline the amount determined by multiplying—

"(A) 1 cent for each gallon of gasoline so used, by

"(B) the percentage which the ultimate purchaser's tax-exempt passenger fare revenue derived from such scheduled service during such quarter was of his total passenger fare revenue (not including the tax imposed by section 4261, relating to the tax on transportation of persons) derived from such scheduled service during such quarter.

"(2) LIMITATION.—Paragraph (1) shall apply in respect of gasoline used during any calendar quarter only if at least 60 percent of the total passenger fare revenue (not including the tax imposed by section 4261, relating to the tax on transportation of persons) derived during such quarter from scheduled service described in paragraph (1) by the person filing the claim was attributable to tax-exempt passenger fare revenue derived during such quarter by such person from such scheduled service.

"(c) TIME FOR FILING CLAIM; PERIOD COVERED.—Not more than one claim may be filed under subsection (a), and not more than one claim may be filed under subsection (b), by any person with respect to gasoline used during the one-year period ending on June 30 of any year. No claim shall be allowed under this section with respect to any one-year period unless filed on or before September 30 of the year in which such one-year period ends.

"(d) DEFINITIONS.—For purposes of this section-

"(1) EXEMPT SALES.—No amount shall be paid under this section with respect to any gasoline which the Secretary or his delegate determines was exempt from the tax imposed by section 4081. The amount which (but for this sentence) would be payable under this section with respect to any gasoline shall be reduced by any other amount which the Secretary or his delegate determines is payable under this section, or is refundable under any provision of this title, to any person with respect to such gasoline.

"(2) GASOLINE USED ON FARMS.—This section shall not apply in respect of gasoline which was (within the meaning of paragraph (1), (2), and (3) of section 6420 (c)) used on a farm for farming purposes.

"(f) APPLICABLE LAWS.—

"(1) IN GENERAL.—All provisions of law, including penalties, applicable in respect of the tax imposed by section 4081 shall, insofar as applicable and not inconsistent with this section, apply in respect of the payments provided for in this section to the same extent as if such payments constituted refunds of overpayments of the tax so imposed.

"(2) EXAMINATION OF BOOKS AND WITNESSES.—For the purpose of ascertaining the correctness of any claim made under this section, or the correctness of any payment made in respect of any such claim, the Secretary or his delegate shall have the authority granted by paragraph (1), (2), and (3) of section 7602 (relating to examination of books and witnesses) as if the claimant were the person liable for tax.

"(g) REGULATIONS.—the Secretary or his delegate may by regulations prescribe the conditions, not inconsistent with the provisions of this section, under which payments may be made under this section.

"(h) EFFECTIVE DATE.—This section shall apply only with respect to gasoline purchased after June 30, 1956, and before July 1, 1972.

"(i) CROSS REFERENCES.—

"(1) For reduced rate of tax in case of diesel fuel and special motor fuels used for certain non-highway purposes, see subsections (a) and (b) of section 4041.

"(2) For partial refund of tax in case of diesel fuel and special motor fuels used for certain non-highway purposes, see section 6416 (b) (2) (J) and (K).

"(3) For partial refund of tax in case of diesel fuel and special motor fuels used by local transit systems, see section 6416 (b) (2) (L).

"(4) For civil penalty for excessive claims under this section, see section 6675.

"(5) For fraud penalties, etc., see chapter 75 (section 7201 and following relating to crimes, other offenses, and forfeitures)."

(d) TECHNICAL AMENDMENTS.—

(1) Section 6206 (relating to special rules applicable to excessive claims) is amended—

(A) by striking out "SECTION 6420" in the heading and inserting in lieu thereof "SECTIONS 6420 AND 6421";

(B) by inserting after "6420" in the first sentence thereof "or 6421"; and

(C) by inserting after "6420" in the second sentence thereof "or 6421, as the case may be".

(2) Section 6675 (relating to excessive claims for gasoline used on farms) is amended—

(A) by striking out "FOR GASOLINE USED ON FARMS" in the heading and inserting in lieu thereof "WITH RESPECT TO THE USE OF CERTAIN GASOLINE";

(B) by inserting after "6420 (relating to gasoline used on farms)" in subsection (a) thereof "or 6421 (relating to gasoline used for certain non-highway purposes or by local transit system)"; and

(C) by inserting after "6420" in subsection (b) thereof "or 6421, as the case may be,".

(3) Section 7210 (relating to failure to obey summons) is amended by inserting after "sections 6420 (e) (2)," the following: "6421 (f) (2),".

(4) Section 7603 (relating to service of summons) and 7604 (relating to enforcement of summons) and the first sentence of section 7605 (relating to time and place of examination) are each amended by inserting after "section 6420 (e) (2)" wherever it appears a comma and the following: "6421 (f) (2),". The second sentence of section 7605 is amended by inserting after "section 6420 (e) (2)" the following: "or 6421 (f) (2)".

(e) CLERICAL AMENDMENTS.—

(1) Section 4084 is amended to read as follows;

"SEC. 4084. CROSS REFERENCES.

"(1) For provisions to relieve farmers from excise tax in the case of gasoline used on the farm for farming purposes, see section 6420.

"(2) For provisions to relieve purchasers of gasoline from excise tax in the case of gasoline used for certain non-highway purposes or by local transit systems, see section 6421."

(2) The table of sections for subpart A of part III of subchapter A of chapter 32 is amended by striking out

"Sec. 4084. Relief of farmers from tax in case of gasoline used on the farm."

and inserting in lieu thereof

"Sec. 4084. Cross references."

(3) The table of sections for subchapter A of chapter 63 is amended by striking out

"Sec. 6206. Special rules applicable to excessive claims under sections 6420 and 6421."

(4) The table of sections for subchapter B of chapter 65 is amended by striking out

"Sec. 6421. Cross references."

and inserting in lieu thereof

"Sec. 6421. Gasoline used for certain non-highway purposes or by local transit systems.

"Sec. 6422. Cross references."

(5) Section 6504 is amended by adding at the end thereof the following:

"(14) Assessments to recover excessive amounts paid under section 6421 (relating to gasoline used for certain non-highway purposes or by local transit systems) and assessments of civil penalties under section 6675 for excessive claims under section 6421, see section 6206."

(6) Section 6511 (f) is amended by adding at the end thereof the following:

"(6) For limitations in case of payments under section 6421 (relating to gasoline used for certain non-highway purposes or by local transit systems), see section 6421 (c)."

(7) Section 6612 (c) is amended by striking out "and" before "6420" and by inserting before the period at the end thereof the following: ", and 6421 (relating to payments in the case of gasoline used for certain non-highway purposes or by local transit systems)".

(8) The table of sections for subchapter B of chapter 68 is amended by striking out

"Sec. 6675. Excessive claims for gasoline used on farms."

and inserting in lieu thereof

"Sec. 6675. Excessive claims with respect to the use of certain gasoline."

Sec. 209. Highway Trust Fund.

(a) CREATION OF TRUST FUND.—There is hereby established in the Treasury of the United States a trust fund to be known as the "Highway Trust Fund" (hereinafter in this section called the "Trust Fund"). The Trust Fund shall consist of such amounts as may be appropriated or credited to the Trust Fund as provided in this section.

(b) DECLARATION OF POLICY.—It is hereby declared to be the policy of the Congress that if it hereafter appears—

(1) that the total receipts of the Trust Fund (exclusive of advances under subsection (d)) will be less than the total expenditures from such Fund (exclusive of repayments of such advances); or

(2) that the distribution of the tax burden among the various classes of persons using the Federal-aid highways, or otherwise deriving benefits from such highways, is not equitable,

the Congress shall enact legislation in order to bring about a balance of total receipts and total expenditures, or such equitable distributions, as the case may be.

(c) TRANSFER TO TRUST FUND OF AMOUNTS EQUIVALENT TO CERTAIN TAXES.—

(1) IN GENERAL.—There is hereby appropriated to the Trust Fund, out of any money in the Treasury not otherwise appropriated, amounts equivalent to the following percentages of the taxes received in the Treasury before July 1, 1972, under the following provisions of the Internal Revenue Code of 1954 (or under the corresponding provisions of prior revenue laws)—

(A) 100 percent of the taxes received after June 30, 1956, under sections 4041 (taxes on diesel fuel and special motor fuels), 4071 (a) (4) (tax on tread rubber), and 4081 (tax on gasoline);

(B) 20 percent of the tax received after June 30, 1956, and before July 1, 1957, under section 4061 (a) (1) (tax on trucks, buses, etc.);

(C) 50 percent of the tax received after June 30, 1957, under section 4061 (a) (1) (tax on trucks, buses, etc.);

(D) 37 1/2 percent of the tax received after June 30, 1956, and before July 1, 1957, under section 4071 (a) (1) (tax on tires of the type used on highway vehicles);

(E) 100 percent of the tax received after June 30, 1957, under section 4071 (a) (1) (2), and (3) (taxes on tires of the type used on highway vehicles, other tires, and inner tubes);

(F) 100 percent of the tax received under section 4481 (tax on use of certain vehicles); and

(G) 100 percent of the floor stocks taxes imposed by section 4226 (a).

In the case of any tax described in subparagraph (A), (B), or (D), amounts received during the fiscal year ending June 30, 1957, shall be taken into account only to the extent attributable to liability for tax incurred after June 30, 1956.

(2) LIABILITIES INCURRED BEFORE JULY 1, 1972, FOR NEW OR INCREASED TAXES.—There is hereby appropriated to the Trust Fund, out of any money in the Treasury not otherwise appropriated, amounts equivalent to the following percentages of the taxes which are received in the Treasury after June 30, 1972, and before July 1, 1973, and which are attributable to liability for tax incurred before July 1, 1972, under the following provisions of the Internal Revenue Code of 1954—

(A) 100 percent of the taxes under sections 4041 (taxes on diesel fuel and special motor fuels), 4071 (a) (4) (tax on tread rubber), and 4081 (tax on gasoline);

(B) 20 percent of the tax under section 4061 (a) (1) (tax on trucks, buses, etc.);

(C) 37 1/2 percent of the tax under section 4071 (a) (1) (tax on tires of the type used on highway vehicles); and

(D) 100 percent of the tax under section 4481 (tax on use of certain vehicles).

(3) METHOD OF TRANSFER.—The amounts appropriated by paragraphs (1) and (2) shall be transferred at least monthly from the general fund of the Treasury to the Trust Fund on the basis of estimates by the Secretary of the Treasury of the amounts, referred to in paragraphs (1) and (2), received in the Treasury. Proper adjustments shall be made in the amounts subsequently transferred to the extent prior estimates were in excess of or less than the amounts required to be transferred.

(d) ADDITIONAL APPROPRIATIONS TO TRUST FUND.—There are hereby authorized to be appropriated to the Trust Fund, as repayable advances, such additional sums as may be required to make the expenditures referred to in subsection (f).

(e) MANAGEMENT OF TRUST FUND.—

(1) IN GENERAL.—It shall be the duty of the Secretary of the Treasury to hold the Trust Fund, and (after consultation with the Secretary of Commerce) to report to the Congress not later than the first day of March of each year on the financial condition and the results of the operations of the Trust Fund during the preceding fiscal year and on its expected condition and operations during each fiscal year thereafter up to and including the fiscal year ending June 30, 1973. Such report shall be printed as a House document of the session of the Congress to which the report is made.

(2) INVESTMENT.—It shall be the duty of the Secretary of the Treasury to invest such portion of the Trust Fund as is not, in his judgment, required to meet current withdrawals. Such investments may be made only in interest-bearing obligations of the United States or in obligations guaranteed as to both principal and interest by the United States. For such purpose such obligations may be acquired (A) on original issue at par, or (B) by purchase of outstanding obligations at the market price. The purposes for which obligations of the United States may be issued under the Second Liberty Bond Act, as amended, are hereby extended to authorize the issuance at par of special obligations exclusively to the Trust Fund. Such special obligations shall bear interest at a rate equal to the average rate of interest, computed as to the end of the calendar

month next preceding the date of such issue, borne by all marketable interest-bearing obligations of the United States then forming a part of the Public Debt; except that where such average rate is not a multiple of one-eighth of 1 percent, the rate of interest of such special obligations shall be the multiple of one-eighth of 1 percent next lower than such average rate. Such special obligations shall be issued only if the Secretary of the Treasury determines that the purchase of other interest-bearing obligations of the United States, or of obligations guaranteed as to both principal and interest by the United States on original issue or at the market price, is not in the public interest. Advances to the Trust Fund pursuant to subsection (d) shall not be invested.

(3) SALE OF OBLIGATIONS.—Any obligation acquired by the Trust Fund (except special obligations issued exclusively to the Trust Fund) may be sold by the Secretary of the Treasury at the market price, and such special obligations may be redeemed at par plus accrued interest.

(4) INTEREST AND CERTAIN PROCEEDS.—the interest on, and the proceeds from the sale or redemption of, any obligations held in the Trust Fund shall be credited to and form a part of the Trust Fund.

(f) EXPENDITURES FROM TRUST FUND.—

(1) FEDERAL-AID HIGHWAY PROGRAM.—Amounts in the Trust Fund shall be available, as provided by appropriation Acts, for making expenditures after June 30, 1956, and before July 1, 1972, to meet those obligations of the United States heretofore or hereafter incurred under the Federal-Aid Road Act approved July 11, 1916, as amended and supplemented, which are attributable to Federal-aid highways (including those portions of general administrative expenses of the Bureau of Public Roads payable from such appropriations).

(2) REPAYMENT OF ADVANCES FROM GENERAL FUND.—Advances made pursuant to subsection (d) shall be repaid, and interest on such advances shall be paid, to the general fund of the Treasury when the Secretary of the Treasury determines that moneys are available in the Trust Fund for such purposes. Such interest shall be at rates computed in the same manner as provided in subsection (e) (2) for special obligations and shall be compounded annually.

(3) TRANSFERS FROM TRUST FUND FOR GASOLINE USED ON FARMS AND FOR CERTAIN OTHER PURPOSES.—The Secretary of the Treasury shall pay from time to time from the Trust Fund into the general fund of the Treasury amounts equivalent to the amounts paid before July 1, 1973, under sections 6420 (relating to amounts paid in respect of gasoline used on farms) and 6421 (relating to amounts paid in respect of gasoline used for certain non-highway purposes or by local transit systems) of the Internal Revenue Code of 1954 on the basis of claims filed for periods beginning after June 30, 1956, and ending before July 1, 1972.

(4) FLOOR STOCKS REFUNDS.—The Secretary of the Treasury shall pay from time to time from the Trust Fund into the general fund of the Treasury amounts equivalent to the following percentages of the floor stocks refunds

made before July 1, 1973, under section 6412 (a) (2) of the Internal Revenue Code of 1954—

(A) 40 percent of the refunds in respect of articles subject to the tax imposed by section 4061 (a) (1) of such Code (trucks, buses, etc.);

(B) 100 percent of the refunds in respect of articles subject to tax under section 4071 (a) (1) or (4) of such Code (tires of the type used on highway vehicles and tread rubber); and

(C) 66 2/3 percent of the refunds in respect of gasoline subject to tax under section 4081 of such Code.

(g) ADJUSTMENTS OF APPORTIONMENTS.—The Secretary of the Treasury shall from time to time, after consultation with the Secretary of Commerce, estimate the amounts which will be available in the Highway Trust Fund (excluding repayable advances) to defray the expenditures which will be required to be made from such fund. In any case in which the Secretary of the Treasury determines that, after all other expenditures required to be made from the Highway Trust Fund have been defrayed, the amounts which will be available in such fund (excluding repayable advances) will be insufficient to defray the expenditures which will be required as a result of the apportionment to the States of the amounts authorized to be appropriated for any fiscal year for the construction, reconstruction, or improvement of the Interstate System, he shall so advise the Secretary of Commerce and shall further advise the Secretary of Commerce as to the amount which, after all other expenditures required to be made from such fund have been defrayed, will be available in such fund (excluding repayable advances) to defray the expenditures required as a result of apportionment to the States of Federal-aid highway funds for the Interstate System for such fiscal year. The Secretary of Commerce shall determine the percentage which such amount is of the amount authorized to be appropriated for such fiscal year for the construction, reconstruction, or improvement of the Interstate System and, notwithstanding any other provision of law, shall thereafter apportion to the States for such fiscal year for the construction, reconstruction, or improvement of the Interstate System, in lieu of the amount which but for the provisions of this subsection would be so apportioned, the amount obtained by multiplying the amount authorized to be appropriated for such fiscal year by such percentage. Whenever the Secretary of the Treasury determines that there will be available in the Highway Trust Fund (excluding repayable advances) amounts which, after all other expenditures required to be made from such fund have been defrayed, will be available to defray the expenditures required as a result of the apportionment of any Federal-aid highway funds for the Interstate System previously withheld from apportionment for any fiscal year, he shall so advise the Secretary of Commerce and the Secretary of Commerce shall apportion to the States such portion of the funds so withheld from apportionment as the Secretary of the Treasury has advised him may be so apportioned without causing expenditures from the Highway Trust Fund for the Interstate System to exceed amounts available in such fund (excluding repayable advances) to defray such expenditures. Any funds apportioned pursuant to the provisions of the preceding sentence shall remain available for expenditure until the close of the third fiscal year following that in which apportioned.

Sec. 210. Investigation And Report To Congress.

(a) PURPOSE.—The purpose of this section is to make available to the Congress information on the basis of which it may determine what taxes should be imposed by the United States, and in what amounts, in order to assure, insofar as practicable, an equitable distribution of the tax burden among the various classes of persons using the Federal-aid highways or otherwise deriving benefits from such highways.

(b) STUDY AND INVESTIGATION.—In order to carry out the purpose of this section, the Secretary of Commerce is hereby authorized and directed, in cooperation with other Federal officers and agencies (particularly the Interstate Commerce Commission) and with the State highway departments, to make a study and investigation of—

(1) the effects on design, construction, and maintenance of Federal-aid highways of (A) the use of vehicles of different dimensions, weights, and other specifications, and (B) the frequency of occurrences of such vehicles in the traffic stream,

(2) the proportionate share of the design, construction, and maintenance costs of the Federal-aid highways attributable to each class of persons using such highways, such proportionate share to be based on the effects referred to in paragraph (1) and the benefits derived from the use of such highways, and

(3) any direct and indirect benefits accruing to any class which derives benefits from Federal-aid highways, in addition to benefits from actual use of such highways, which are attributable to public expenditures for such highways.

(c) COORDINATION WITH OTHER STUDIES.—The Secretary of Commerce shall coordinate the study and investigation required by this section with—

(1) the research and other activities authorized by section 10 of the Federal-Aid Highway Act of 1954, and

(2) the tests referred to in section 108 (k) of this Act.

(d) REPORTS ON STUDY AND INVESTIGATION.—The Secretary of Commerce shall report to the Congress the results of the study and investigation required by this section. The final report shall be made as soon as possible but in no event later than March 1, 1959. On or before March 1, 1957, and on or before March 1, 1958, the Secretary of Commerce shall report to the Congress the progress that has been made in carrying out the study and investigation required by this section. Each such report shall be printed as a House document of the session of the Congress to which the report is made.

24

The Colorado River and Hoover Dam

United States

DID YOU KNOW . . . ?

➤ The Colorado River is 1,450 miles (2,330 kilometers) long, with 20 dams to harness the raw energy.

➤ The Hoover Dam stands 726.4 feet (221.4 meters) high.

➤ Dam construction began in 1931 and was completed in 1935—two years ahead of schedule.

➤ During peak periods of electrical demand, enough water runs through the generators to fill 15 average-size swimming pools (20,000 gallons each) in one second.

➤ There is enough concrete in Hoover Dam to build a two-lane road from Seattle, Washington, to Miami, Florida, or a four-foot-wide sidewalk around the earth at the equator.

➤ The dam is the first example in history of a comprehensive interstate compact to deal with allocation of water rights.

The engineering marvel that is the Hoover Dam evolved out of a multiparty agreement, the Colorado River Compact (reproduced here as Document I). Of course, all the parties were states; nevertheless, they had different interests, and the agreement needed to accommodate all seven Colorado River Basin states. This was a different kind of agreement. During the 1920 negotiations by the parties along the Colorado Basin, all began to realize that as California grew, that state's needs might become a problem. It was apparent to all seven parties that the growth of one state would have a major impact on the other partners.

The Hoover Dam. Courtesy of Shutterstock.

In 1922, the compact was considered a major agreement, the first time in American history that a group of states had apportioned the water of an interstate stream and the first time that more than two or three states had negotiated a treaty to settle any sort of problem among themselves (Kahrl, 41).

HISTORY

The Colorado River is impressive by any measure. It extends approximately 1,450 miles (2,330 kilometers) from northern Colorado to the Gulf of California in Mexico, and affects an area of 242,000 square miles (632,000 square kilometers)—all but a few thousand in the United States. It is harnessed by more than 20 dams that provide river and sedimentation regulation, flood control, irrigation, navigation, recreation, generation of electric power, municipal and industrial water supplies, and fish and wildlife habitats.

Of these dams, the largest, created by the first major multipurpose river-development project in the world, is the Hoover Dam, on the stretch of river running between Nevada and Arizona. Located in Black Canyon about 30 miles (50 kilometers) southeast of Las Vegas, Nevada, the dam created a massive water-storage facility known as Lake Mead, which flooded Boulder Canyon and is the largest artificial lake in the United States. When built, the Hoover Dam was the world's largest, a symbol of human power to control nature and shape destiny. The Hoover Dam was followed by several other high-profile projects involving the river, among them Glen Canyon Dam (Lake Powell), Parker Dam (Lake Havasu), and major irrigation and water-supply canals.

In the early twentieth century there was discussion of exploiting the Colorado River. Irrigation and flood control would help the region's agriculture and related industries thrive, and the river could supply power and water to growing cities like Las Vegas and Los Angeles. Several disastrous floods highlighted the need to harness the river: from 1905 to 1907, the Colorado breached weakened canal walls and tore into California's Salton Basin, eventually creating the Salton Sea, and there were several instances when it flooded Arizona's Yuma Valley. In 1918, U.S. Reclamation Director Arthur P. Davis proposed taming the Colorado River by erecting an unprecedentedly high dam at Boulder Canyon.

Full-scale development of the Colorado River dates from the Colorado Compact, which regulated use of the river's water by the seven U.S. states that abut it: Arizona, California, Colorado, Nevada, New Mexico, Utah, and Wyoming. The compact was signed in 1922 and endorsed by Congress in the 1928 Boulder Canyon Project Act (reproduced here as Document II), which launched construction of the Hoover Dam.

This immense job, carried out in wilderness, was completed by 1936. The result was a concrete arch-gravity-type dam standing 726.4 feet (221.4 meters) high; its crest measures 1,244 feet (379 meters) long, with a volume of 4,400,000 cubic yards (3,360,000 cubic meters). The power plant's rated generating capacity is over 2,000 megawatts. Congress officially named it the Hoover Dam in 1931, in tribute to the major role played by Herbert Hoover, who chaired the Colorado River Commission, presided over negotiations for the Colorado Compact, and, after becoming president in 1929, oversaw the initial years of construction. For some time it was referred to as the Boulder Dam, but the name Hoover Dam was officially "restored" in 1947 by Congress. The Hoover Dam, along with the power plant and reservoir, are run by the Bureau of Reclamation of the Interior Department.

Lake Mead, named for Elwood Mead, U.S. commissioner of reclamation from 1924 to 1936, extends 115 miles (185 kilometers) upriver behind the dam. It covers about 247 square miles (640 square kilometers), with a total capacity of 28,537,000 acre-feet (35.2 billion cubic meters). It is administered by the U.S. National Park Service as part of the Lake Mead National Recreation Area.

CULTURAL CONTEXT

Any discussion of important dams in human history must include the ancient Marib Dam, located in the southwest corner of the Arabian peninsula now known as Yemen. Called *Saba* in Arabic and *Sheba* in the west, the region around the fortified city of Marib, a center for trade coming from Asia and India, made that city comparable to a modern-day Hong Kong or New York. As camel drivers passed through the deserts of Yemen, many of them called in at

Workman using a compressed air vibrator for compacting concrete adjacent to forms, Hoover Dam, 1934. Courtesy of the Library of Congress.

Marib. Dating from at least 1050 B.C., Marib was a lush oasis filled with palm trees and exotic plants.

Saba had its own prized natural resources: frankincense and myrrh, both in high demand in the ancient world. Derived from native trees, they grew wild on lands that were irrigated by the waters of a unique dam of vast proportions. When the Sabeans built the famous dam at Marib, the resulting irrigation helped preserve the precious trees and made the land extremely fertile. In addition, the dam provided the infrastructure for one of the first water systems with wells and an extensive irrigation network.

The dam, an engineering marvel of its time, was cut through the Balaq Hills by the Wadi Adhanah. It spanned an 1,800-foot gap, and rose 15 feet (4.6 meters) above the water. Inscriptions indicate that the dam was built in the seventh century B.C. by a ruler named Sumhu' Alay Yanuf and his son Yatha'-Amar Bayyin, and was maintained for centuries thereafter by the succeeding Himyarite civilization. Sometime thereafter, the technical know-how of hydraulic engineering was lost, and after a series of breaks and repairs, the dam finally collapsed around A.D. 570.

As they viewed the majestic Colorado River, it must have occurred to many explorers, engineers, and naturalists to harness its raw power for economic and commercial benefit. "Unlike many other states, the California water system is in large part self-contained. . . . The all-important exception to this general rule is the Colorado River, which today supplies water to half the state's population, while at the same time supporting an agricultural industry which produces crops and livestock valued at many millions of dollars a year" (Kahrl, 38).

After the dam was designed from an engineering standpoint, architect Gordon B. Kaufmann came onto the project to rework the design to add an artistic element. His proposal talked about a system of plain surfaces relieved by shadows here and there (Rinehart). This gave the dam its reputation as a monument of public art. Because of Kaufmann's design, the Hoover Dam became an icon of modern art that helped usher in a new era in American design.

In addition to Kaufmann's design, Oskar J. W. Hansen was commissioned to create two large cast concrete panels depicting flood control, irrigation, power, and the history of the area. He is also responsible for a star map, set in the floor of the dam, which compares the Hoover Dam to the pyramids (Wilson, 310).

PLANNING

Before the Hoover Dam could be built, decisions had to be made about allocating the enormous quantities of water that would result from harnessing the Colorado River. It was not easy to reach agreement on a fair distribution of water resources. Tensions rose, not least being whether problems should be legislated at the federal level and handed down to the parties as a *fait accompli,* or the parties could themselves develop a cooperative agreement. Delph Carpenter, a Colorado attorney, proposed a self-determined path that included an interstate compact. In his opinion, such a preagreement would both head off a federal dictum and avoid future litigation.

In November 1922, delegates from the seven Colorado River Basin states met in New Mexico to discuss, negotiate, and work out a compact. On November 24, 1922, the Colorado River Compact was signed in the Palace of the Governors, Santa Fe, New Mexico. As a result, the seven states that were party to the compact would work together to decide how to apportion the Colorado River water between Upper and Lower Basin states.

At first, determination of water rights was linked to irrigable land within each state, but later the decision to draw a line between the Upper and Lower Basins and group the states into two camps seemed the only sensible thing to do. The Upper Basin consisted of Wyoming, Colorado, Utah, and New Mexico; the Lower Basin contained California, Arizona, and Nevada. Upper Basin states would be required to deliver 75 maf (million acre feet, a water-measurement term) each 10 years, a period long enough to even out the natural variances caused by drought and flood. Calculations by the Reclamation Bureau showed that flows could range from 4.4 maf to 22 maf, with 16.4 maf being the norm.

In 1928, legislation designated as the Boulder Canyon Project Aid was enacted by Congress. It provided for construction of works to protect and develop the Colorado River Basin, and it approved the Colorado River Compact of 1922.

As the population of California and Arizona grew, squabbling increased, and the parties to the compact had to reassess the situation. In the 1940s, Arizona began to question apportionments, eventually bringing the matter to the U.S. Supreme Court for adjudication. The case of *Arizona v. California* lasted 11 years and cost $5 million, one of the most contentious and expensive litigation cases of the time.

Carpenter's other hope, self-determination, was also dashed when the secretary of the interior was made water master of the Lower Basin by the U.S. Supreme Court, thereby investing that office with the power to decide future apportionments. Although the Lower Basin states could not settle their differences cooperatively, the Upper Basin states fared better, and in 1948 they redesigned and redesignated their apportionments, wisely choosing percentages of the whole rather than specific maf amounts that would be subject to natural variation and therefore would always be off the mark.

The compact left out two parties: Native Americans and Mexico. Regarding Native Americans, Herbert Hoover made certain they were mentioned. Article VII of the compact stipulates, "Nothing in this compact shall be construed as affecting the obligations of the United States of America to Indian Tribes." It is possible that this phrase refers to the 1903 Supreme Court decision in *Winters v. United States*, which stated that Native American tribes had water rights whether

The first bucket of concrete being poured in the Hoover Dam, 1933. Courtesy of the Library of Congress.

they were using the water or not, and that their rights were related to the placement of their reservations. Sixty years later, in 1963, the Supreme Court intervened to define the rights of five Native American reservations in the Lower Basin.

The requirement to delineate Native American water rights highlights the importance of their stake in this issue. Three other tribes are major players: the Walapai and Havasupai, who are partially within compact boundaries, and the Navajo, whose 25,000-square-mile reservation is entirely within the territory of the compact. All three tribes make claims on the water rights of both Upper and Lower Basin apportionments. At a 1997 Colorado River Symposium, an attorney for the Navajo observed that Navajo water rights are probably at the five-maf level. Nothing prevents the Navajo from selling their water. While states might be subject to federal law, Native American tribes are sovereign entities protected by the Winters law and therefore able to use their land and rights as they deem most appropriate.

If Native Americans were mentioned only briefly, even less attention was paid to Mexico. The compact mandated that surpluses not used by the states could be given to Mexico, and if there were deficiencies, then the Upper and Lower Basin states would make them up. In 1944, clarification was achieved in a treaty that apportioned 1.5 maf annually to Mexico.

BUILDING

Just as the planning process involved the multiparty agreement among the states to which the waters of the Colorado River belonged, building the Hoover Dam also required multiparty authorization. Six companies (Morrison-Knudsen Co.; Utah Construction Co.; J. F. Shea Co.; Pacific Bridge Co.; MacDonald & Kahn Ltd.; and a joint venture of W. A. Bechtel Co., Henry J. Kaiser, and Warren Brothers) formed a cooperative league known as Six Companies, Inc., headquartered in Las Vegas; the joint venture won the bid to build the dam.

While the Colorado River Compact had flaws that resulted in litigation, construction of the Hoover Dam went so well that it finished two years ahead of schedule. The contract included bonuses for completion and threatened fines for running beyond the construction deadline; the contractors took these clauses seriously, working day and night.

Why was this project so successful? One reason was the availability of legions of recently unemployed laborers. The prospect of work on the Hoover Dam project offered relief from the catastrophe of the Great Depression. Whole families moved to Las Vegas, and workers made the 30-mile trip from there to the job site each day.

Another community known as "Ragtown" developed near the work site, populated by families that had arrived long before construction began in the hope of being hired. Ragtown quickly grew from 1,400 people to more than 5,000. During the summer of 1931, the first year of construction, the weather was so hot that 25 women and children died from heat-related illness.

Before construction began, there were few other employment opportunities, and people became desperate for food and essentials. An unrecognized hero named Murl Emery, a merchant, opened a small store to serve Ragtown. Although workers received only 50 cents per hour, Emery allowed families to pay what they could and put the rest of their bill on credit. Testimony to the entire community is the fact that no one ever failed to pay their debt—except one person who died. There was no government help, and no help from dam contractors—just Murl Emery and his generous family.

For the first time in public construction history, the federal government mandated a policy of diversity, requiring that African Americans be hired as laborers. There were no problems finding candidates; many African Americans had come to find work.

Construction of the Hoover Dam was difficult because the powerful Colorado River had to be diverted and made dry before work on the dam could begin. Four diversion tunnels were dug through the walls of the canyon. As they were built, blasting and piping teams were formed into competing squads that participated in an intramural competition. People were so excited to win the contest that the tunnels were completed early.

Construction of the dam itself required 3.25 million cubic yards of concrete, and more for the power plant. Therefore, before building anything, the workers built several concrete plants, which worked so constantly that a bucket of concrete went out on the overhead cable delivery system every 78 seconds.

One problem seemed insurmountable: it had been calculated that it would take 100 years for the concrete to cool. The brilliant solution came from engineers who devised a system for cooling the concrete by pouring it in rows and columns that were cooled by 582 miles of steel pipes pumped with icy water. Afterward, the pipes themselves were emptied of water and filled with concrete, thus further strengthening the structure, reminiscent of the techniques used in the building of both the London Bridge and Brooklyn Bridge.

The iconic arch curve of the Hoover Dam was carefully engineered so the pressure of the water would hold the structure in place by interacting with gravity. Gustave Eiffel comes to mind; he designed the curve of his base pylons to reverse the air pressure and thus strengthen the tower. "The Hoover Dam is a curved gravity dam. Lake Mead pushes against the dam, creating compressive forces that travel along the great curved wall. The canyon walls push back, counteracting these forces. This action squeezes the concrete in the arch together, making the dam very rigid. This way, Lake Mead can't push it over" (http://www.pbs.org/wgbh/buildingbig/wonder/structure/hoover.html).

Management produced an innovation that became known as the *jumbo truck*—a truck fitted with layers of platforms on which 20–30 workers armed with drills could stand. The truck was driven right up the face of a canyon cliff and everyone drilled at once, making holes that were then stuffed with blasting powder. With a single blast, an entire section was demolished at once. With eight jumbo trucks working simultaneously, the work proceeded quickly. Electric lights were added to the trucks so drilling could be done around the clock. As a result, a ton of dynamite was planted every 14 feet, and this portion of the job was quickly accomplished.

Clearly, the management techniques and crew groups that were used in the construction of the Hoover Dam indicated that Americans were learning how to handle large-scale jobs, not only with appropriate technology but with innovative human-resource management and motivation.

No discussion of Hoover Dam construction would be complete without noting the high-scalers, who performed acrobatic feats from aloft. Mainly Native Americans familiar with the intricacies of the canyons, the high-scalers hung from ropes and scraped off loose rocks that had accumulated from blasting. Paid a relatively high wage, the high-scalers were veritable daredevils, whose lofty movements were recorded by photographers who immortalized their agility. Some high-scalers lived in Ragtown, others in Boulder City, Nevada, a new town completed in 1932 as an attractive community for workers, especially when Lake Mead was completed.

The Hoover Dam construction was begun in 1931 and completed in 1935, two years ahead of schedule. Power generation began in 1936, and improvements were continually added up to 1961. The dam was dedicated on September 30, 1935, by President Franklin D. Roosevelt.

IMPORTANCE IN HISTORY

The Colorado River Compact and Hoover Dam highlight the critical strategic issues—today and in the future—of water rights, water marketing, and water-resource policies. In this vein, the concept of an ever-normal water supply is timely. In the 1930s, Henry Wallace proposed an "ever-normal granary," advising that if grain were stored during years of surplus, it would be available for use in years of deficit. Perhaps Wallace took his idea from the biblical story in which Joseph advised the Pharaoh about the impending "seven lean years and seven fat years."

While grain has been successfully stored, less has been done to store water, despite reservoirs in many locations. If water is transferred around the planet, there should be a sufficient supply, although experts warn there may be a global water shortage.

Looking down the side of the Hoover Dam at the Colorado River. Courtesy of Shutterstock.

For instance, the Amazon River in South America debouches into the south Atlantic; from there it could be used to supply all of Africa's needs if properly harnessed and delivered. The Amazon releases 6.5 million cubic feet (184,000 cubic meters) per second into the Atlantic during the rainy season—56 times as much water as the Nile. The Amazon yields one-fifth of all the fresh water entering oceans; offshore near the mouth of the enormous river, the water is drinkable right in the midst of the ocean. At a 1978 Rensselaerville, New York, conference, J. Vincent Harrington suggested a pipeline that could be attached to the coastal mouth of the Amazon to capture potable water. However, in 2006, Ernst G. Frankel, MIT Professor Emeritus of Ocean Engineering, advised that proper coordination of under-utilized African rivers might accomplish the same results at lower costs. Robert Cathcart also found that nineteenth-century German scholars regarded the Amazon water surplus as a world resource.

Considering the agreements reached in the Colorado River Compact and the building of the Hoover Dam, as well as solutions that worked and omissions later regretted, perhaps these agreements would be helpful in conceptualizing and phrasing future authorizations for shared water on a wider global basis.

Perhaps it is time to found a college for water study. Private colleges train people for many purposes; one example is the former Hawthorne College in New Hampshire, which specialized in airline-industry studies. Another early example is the first of the *Grandes Ecoles* in France, formed in 1753 under Louis XV, the *Ecole des Ponts et Chaussées*. This "School of Bridges and Roads" still exists and holds a deserved place of primacy. But there are gaps in the world's education system, and water management and planning might be one. China could be a suitable location as that country reworks the Grand Canal toward an expected completion date of 2050. Or could a water college be founded near the Amazon? Or the Sahara?

Finally, the Hoover Dam illustrates an observation made by Eugen Rosenstock-Huessey that the unemployed should be viewed, in part, as a potential asset, and should be so valued. One advantage to use of the unemployed is their ability to live on-site, which has been the key to so many successful large-scale projects.

The Colorado River Compact could be studied as a precedent for decisions about water marketing, such as water from the Amazon River as it flows into the south Atlantic. Is there not a public need for a series of agreements between the United States and Canada in order to better control the use of rivers shared by the two countries? The St. Lawrence Seaway has long been the subject of Canadian-American compact. Another situation that appears to call for similar joint consideration is the Red River, which intermittently ravages North Dakota and Manitoba.

FOR FURTHER REFERENCE

Books and Articles

Bergman, Charles. *Red Delta: Fighting for Life at the End of the Colorado River.* Golden, CO: Fulcrum Publishing, 2002.

Cathcart, Robert. *Planet Earth Renewed: Macroprojects and Geopolitics*. Monograph. Fifth revision, 1984. Contact author at: rbcathcart@msn.com or http://geophos. blogspot.com.

Ward, Evan Ray. *Border Oasis: Water and the Political Ecology of the Colorado River Delta, 1940–1975*. Tucson: University of Arizona Press, 2003.

Wilson, Richard Guy. "American Modernism in the West: Hoover Dam." In *Images of an American Land*, edited by Thomas Carter. Albuquerque: University of New Mexico Press, 1997.

Internet

For the Hoover Dam's official website, see http://www.hooverdam.com.

For information on the Hoover Dam and its construction, see http://www.arizona-leisure.com/hoover-dam-building.html.

For Ragtown and the story of Murl Emery, see http://www.arizona-leisure.com/hoover-dam-men.html.

For Julian Rinehart's article about Oskar Hansen and Gordon Kaufmann, see http://www.usbr.gov/lc/hooverdam/history/articles/rhinehart1.html.

For more on the Hoover Dam as a curved gravity structure, see http://www.pbs.org/wgbh/buildingbig/wonder/structure/hoover.html.

For more information on Six Companies, Inc, see: http://www.constructioncompany/historic-construction-projects/hoover-dam/.

For further information on the Marib Dam, see http://www.yementimes.com/98/iss52/lastpage.htm.

Film and Television

Modern Marvels: Hoover Dam.. A&E Home Video. Orland Park, IL: M.P.I. Media Group, 1999, http://www.mpimedia.com/contact.htm.

Documents of Authorization—I

COLORADO RIVER COMPACT

SIGNED AT SANTA FE, NEW MEXICO,

November 24, 1922

The States of Arizona, California, Colorado, Nevada, New Mexico, Utah, and Wyoming, having resolved to enter into a compact under the act of the Congress of the United States of America approved August 19, 1921, (42 Stat. L., p. 171), and the acts of the legislatures of the said States, have through their governors appointed as their commissioners: W. S. Norviel for the State of Arizona, W. F. McClure for the State of California, Delph E. Carpenter for the State of Colorado, J. G. Scrugham for the State of Nevada, Stephen B. Davis, Jr. for the State of New Mexico, R. E. Caldwell for the State of Utah, Frank C. Emerson for the State of Wyoming, who, after negotiations participated in by Herbert Hoover, appointed by the President as the representative of the United States of America, have agreed upon the following articles.

ARTICLE I

The major purposes of this compact are to provide for the equitable division and apportionment of the use of the waters of the Colorado River system; to establish the relative importance of different beneficial uses of water; to promote interstate comity; to remove causes of present and future controversies and to secure the expeditious agricultural and industrial development of the Colorado River Basin, the storage of its waters, and the protection of life and property from floods. To these ends the Colorado River Basin is divided into two basins, and an apportionment of the use of part of the water of the Colorado River system is made to each of them with the provision that further equitable apportionment may be made.

ARTICLE II

As used in this compact:

(a) The term "Colorado River system" means that portion of the Colorado River and its tributaries within the United States of America.

(b) The term "Colorado River Basin" means all of the drainage area of the Colorado River system and all other territory within the United States of America to which the waters of the Colorado River system shall be beneficially applied.

(c) The term "States of the upper division" means the States of Colorado, New Mexico, Utah, and Wyoming.

(d) The term "States of the lower division" means the States of Arizona, California, and Nevada.

(e) The term "Lee Ferry" means a point in the main stream of the Colorado River 1 mile below the mouth of the Paria River.

(f) The term "Upper Basin" means those parts of the States of Arizona, Colorado, New Mexico, Utah, and Wyoming within and from which waters naturally drain into the Colorado River system above Lee Ferry, and also all parts of said States located without the drainage area of the Colorado River system which are now or shall hereafter be beneficially served by waters diverted from the system above Lee Ferry.

(g) The term "Lower Basin" means those parts of the States of Arizona, California, Nevada, New Mexico, and Utah within and from which waters naturally drain into the Colorado River system below Lee Ferry, and also all parts of said States located without the drainage area of the Colorado River system which are now or shall hereafter be beneficially served by waters diverted from the system below Lee Ferry.

(h) The term "domestic use" shall include the use of water for household, stock, municipal, mining, milling, industrial, and other like purposes, but shall exclude the generation of electrical power.

ARTICLE III

(a)   There is hereby apportioned from the Colorado River system in perpetuity to the upper basin and to the lower basin, respectively, the exclusive beneficial consumptive use of 7,500,000 acre-feet of water per annum, which shall include all water necessary for the supply of any rights which may now exist.

(b)   In addition to the apportionment in paragraph (a), the lower basin is hereby given the right to increase its beneficial consumptive use of such waters by 1,000,000 acre-feet per annum.

(c)   If, as a matter of international comity, the United States of America shall hereafter recognize in the United States of Mexico any right to the use of any waters of the Colorado River system, such waters shall be supplied first from the waters which are surplus over and above the aggregate of the quantities specified in paragraphs (a) and (b); and if such surplus shall prove insufficient for this purpose, then the burden of such deficiency shall be equally borne by the upper basin and the lower basin, and whenever necessary the States of the upper division shall deliver at Lee Ferry water to supply one-half of the deficiency so recognized in addition to that provided in paragraph (d).

(d)   The States of the upper division will not cause the flow of the river at Lee Ferry to be depleted below an aggregate of 75,000,000 acre-feet for any period of 10 consecutive years reckoned in continuing progressive series beginning with the 1st day of October next succeeding the ratification of this compact.

(e)   The States of the upper division shall not withhold water, and the States of the lower division shall not require the delivery of water, which can not reasonably be applied to domestic and agricultural uses.

(f)   Further equitable apportionment of the beneficial uses of the waters of the Colorado River system unapportioned by paragraphs (a), (b), and (c) may be made in the manner provided in paragraph (g) at any time after October 1, 1963, if and when either basin shall have reached its total beneficial consumptive use as set out in paragraphs (a) and (b).

(g)   In the event of a desire for further apportionment as provided in paragraph (f) any two signatory States, acting through their governors, may give joint notice of such desire to the governors of the other signatory States and to the President of the United States of America, and it shall be the duty of the governors of the signatory States and of the President of the United States of America forthwith to appoint representatives, whose duty it shall be to divide and apportion equitably between the upper basin and lower basin the beneficial use of the unapportioned water of the Colorado River system as mentioned in paragraph (f), subject to the legislative ratification of the signatory States and the Congress of the United States of America.

ARTICLE IV

(a) Inasmuch as the Colorado River has ceased to be navigable for commerce and the reservation of its waters for navigation would seriously limit the development of its basin, the use of its waters for purposes of navigation shall be subservient to the uses of such waters for domestic, agricultural, and power purposes. If the Congress shall not consent to this paragraph, the other provisions of this compact shall nevertheless remain binding.

(b) Subject to the provisions of this compact, water of the Colorado River system may be impounded and used for the generation of electrical power, but such impounding and use shall be subservient to the use and consumption of such water for agricultural and domestic purposes and shall not interfere with or prevent use for such dominant purposes.

(c) The provisions of this article shall not apply to or interfere with the regulation and control by any State within its boundaries of the appropriation, use, and distribution of water.

ARTICLE V

The chief official of each signatory State charged with the administration of water rights, together with the Director of the United States Reclamation Service and the Director of the United States Geological Survey, shall cooperate, ex officio.

(a) To promote the systematic determination and coordination of the facts as to flow, appropriation, consumption, and use of water in the Colorado River Basin, and the interchange of available information in such matters.

(b) To secure the ascertainment and publication of the annual flow of the Colorado River at Lee Ferry.

(c) To perform such other duties as may be assigned by mutual consent of the signatories from time to time.

ARTICLE VI

Should any claim or controversy arise between any two or more of the signatory States: (a) With respect to the waters of the Colorado River system not covered by the terms of this compact; (b) over the meaning or performance of any of the terms of this compact; (c) as to the allocation of the burdens incident to the performance of any article of this compact or the delivery of waters as herein provided; (d) as to the construction or operation of works within the Colorado River Basin to be situated in two or more States, or to be constructed in one State for the benefit of another State; or (e) as to the diversion of water in one State for the benefit of another State, the governors of the States affected upon the request of one of them, shall forthwith appoint commissioners with power to consider and adjust such claim or controversy, subject to ratification by the legislatures of the States so affected.

Nothing herein contained shall prevent the adjustment of any such claim or controversy by any present method or by direct future legislative action of the interested States.

ARTICLE VII

Nothing in this compact shall be construed as affecting the obligations of the United States of America to Indian tribes.

ARTICLE VIII

Present perfected rights to the beneficial use of waters of the Colorado River system are unimpaired by this compact. Whenever storage capacity of 5,000,000 acre-feet shall have been provided on the Main Colorado River within or for the benefit of the lower basin, then claims of such rights, if any, by appropriators or users of water in the lower basin against appropriators or users of water in the upper basin shall attach to and be satisfied from water that may be stored not in conflict with Article III.

All other rights to beneficial use of waters of the Colorado River system shall be satisfied solely from the water apportioned to that basin in which they are situated.

ARTICLE IX

Nothing in this compact shall be construed to limit or prevent any State from instituting or maintaining any action or proceeding, legal or equitable, for the protection of any right under this compact or the enforcement of any of its provisions.

ARTICLE X

This compact may be terminated at any time by the unanimous agreement of the signatory States. In the event of such termination, all rights established under it shall continue unimpaired.

ARTICLE XI

This compact shall become binding and obligatory when it shall have been approved by the legislatures of each of the signatory States and by the Congress of the United States. Notice of approval by the legislatures shall be given by the governor of each signatory State to the governors of the other signatory States and to the President of the United States, and the President of the United States is requested to give notice to the governors of the signatory States of approval by the Congress of the United States.

In witness whereof the commissioners have signed this compact in a single original, which shall be deposited in the archives of the Department of State of the United States of America and of which a duly certified copy shall be forwarded to the governor of each of the signatory States.

Done at the city of Santa Fe, New Mexico, this twenty-fourth day of November, A.D. one thousand nine hundred and twenty-two.

W. S. Norviel
W. F. McClure
Delph E. Carpenter
J. G. Scrugham
Stephen B. Davis, Jr.
R. E. Caldwell
Frank C. Emerson
Approved: Herbert Hoover

Documents of Authorization—II

AN ACT TO PROVIDE FOR THE CONSTRUCTION OF WORKS FOR THE PROTECTION AND DEVELOPMENT OF THE COLORADO RIVER BASIN, FOR THE APPROVAL OF THE COLORADO RIVER COMPACT, AND FOR OTHER PURPOSES.

Be it enacted by the Senate and House of Representatives of the United States of America in Congress assembled, That for the purpose of controlling the floods, improving navigation and regulating the flow of the Colorado River, providing for storage and for the delivery of the stored waters thereof for reclamation of public lands and other beneficial uses exclusively within the United States, and for the generation of electrical energy as a means of making the project herein authorized a self-supporting and financially solvent undertaking, the Secretary of the Interior, subject to the terms of the Colorado River compact hereinafter mentioned, is hereby authorized to construct, operate, and maintain a dam and incidental works in the main stream of the Colorado River at Black Canyon or Boulder Canyon adequate to create a storage reservoir of a capacity of not less than twenty million acre-feet of water and a main canal and appurtenant structures located entirely within the United States connecting the Laguna Dam, or other suitable diversion dam, which the Secretary of the Interior is hereby authorized to construct if deemed necessary or advisable by him upon engineering or economic considerations, with the Imperial and Coachella Valleys in California, the expenditures for said main canal and appurtenant structures to be reimbursable, as provided in the reclamation law, and shall not be paid out of revenues derived from the sale or disposal of water power or electric energy at the dam authorized to be constructed at said Black Canyon or Boulder Canyon, or for water for potable purposes outside of the Imperial and Coachella Valleys: *Provided, however,* That no charge shall be made for water or for the use, storage, or delivery of water for irrigation or water for potable purposes in the Imperial or Coachella Valleys; also to construct and equip, operate, and maintain at or near said dam, or cause to be constructed, a complete plant and incidental structures suitable for the fullest economic development of electrical energy from the water discharged from said reservoir; and to acquire by proceedings in eminent domain, or otherwise, all lands, rights of way, and other property necessary for said purposes.

SEC. 2. (a) There is hereby established a special fund, to be known as the "Colorado River Dam fund" (hereinafter referred to as the "fund"), and to be available, as hereafter provided, only for carrying out the provisions of this Act. All revenues received in carrying out the provisions of this Act shall be paid into and expenditures shall be made out of the fund, under the direction of the Secretary of the Interior.

(b) The Secretary of the Treasury is authorized to advance to the fund, from time to time and within the appropriations therefor, such amounts as the Secretary of the Interior deems necessary of carrying out the provisions of this Act, except that the aggregate amount of such advances shall not exceed the sum of $165,000,000. Of this amount the sum of $25,000,000 shall be allocated to flood control and shall be repaid to the United States out of 62 1/2 per centum of revenues, if any, in excess of the amount necessary to meet periodical payments during the period of amortization, as provided in section 4 of this Act. If said sum of $25,000,000 is not repaid in full during the period of amortization, then 62 1/2 per centum of all net revenues shall be applied to payment of the remainder. Interest at the rate of 4 per centum per annum accruing during the years upon the amounts so advanced and remaining unpaid shall be paid annually out of the fund, except as herein otherwise provided.

(c) Moneys in the fund advanced under subdivision (b) shall be available only for expenditures for construction and the payment of interest, during construction, upon the amounts so advanced. No expenditures out of the fund shall be made for operation and maintenance except from the appropriations therefor.

(d) The Secretary of the Treasury shall charge the fund as of June 30 in each year with such amount as may be necessary for the payment of interest on advances made under subdivision (b) at the rate of 4 per centum per annum accrued during the year upon the amounts so advanced and remaining unpaid, except that if the fund is insufficient to meet the payment of interest the Secretary of the Treasury may, in his discretion, defer any part of such payment, and the amount so deferred shall bear interest at the rate of 4 per centum per annum until paid.

(e) The Secretary of the Interior shall certify to the Secretary of the Treasury, at the close of each fiscal year, the amount of money in the fund in excess of the amount necessary for construction, operation, and maintenance, and payment of interest. Upon receipt of each such certificate the Secretary of the Treasury is authorized and directed to charge the fund with the amount so certified as repayment of the advances made under subdivision (b), which amount shall be covered into the Treasury to the credit of miscellaneous receipts.

SEC. 3. There is hereby authorized to be appropriated from time to time, out of any money in the Treasury not otherwise appropriated, such sums of money as may be necessary to carry out the purposes of this Act, not exceeding in the aggregate of $165,000,000.

SEC. 4. (a). This Act shall not take effect and no authority shall be exercised hereunder and no work shall be begun and no moneys expended on or in connection with the works or structures provided for in this Act, and no water rights shall be claimed or initiated hereunder, and no steps shall be taken by the United States or by others to initiate or perfect any claims to the use of water pertinent to such works or structures unless and until (1) the States of Arizona, California, Colorado, Nevada, New Mexico, Utah, and Wyoming shall have ratified the Colorado River compact, mentioned in section 13 hereof, and the President by public proclamation shall have so declared, or (2) if said States fail to ratify the said compact within six months from the date of the passage of this Act then, until six of the said States, including the State of California, shall ratify said compact and shall consent to waive the provisions of the first paragraph of Article XI of said compact, which makes the same binding and obligatory only when approved by each of the seven States signatory thereto, and shall have approved said compact without conditions, save that of such six-State approval, and the President by public proclamation shall have so declared, and, further, until the State of California, by act of its legislature, shall agree irrevocably and unconditionally with the United States and for the benefit of the States of Arizona, Colorado, Nevada, New Mexico, Utah and Wyoming, as an express covenant and in consideration of the passage of this Act, that the aggregate annual consumptive use (diversions less returns to the river) of water of and from the Colorado River for use in the State of California, including all uses under contracts made under the provisions of this Act and all water necessary for the supply of any rights which may now exist, and shall not exceed four million four hundred thousand acre-feet of the waters apportioned to the lower basin States by paragraph (a) of Article III of the Colorado River compact, plus not more than one-half of any excess or surplus waters unapportioned by said company, such uses always to be subject to the terms of said compact.

The States of Arizona, California, and Nevada are authorized to enter into an agreement which shall provide (1) that of the 7,500,000 acre-feet annually apportioned to the lower basin by paragraph (a) of Article III of the Colorado River compact, there shall be apportioned to the State of Nevada 300,000 acre-feet and to the State or Arizona 2,800,000 acre-feet for exclusive beneficial consumptive use in perpetuity, and (2) that the State of Arizona may annually use one-half of the excess or surplus waters unapportioned by the Colorado River compact, and (3) that the State of Arizona shall have the exclusive beneficial consumptive use of the Gila River and its tributaries within the boundaries of said State, and (4) that the waters of the Gila River and its tributaries, except return flow after the same enters the Colorado River, shall never be subject to any diminution whatever by any allowance of water which may be made by treaty or otherwise to the United States of Mexico but if, as provided in paragraph (c) of Article III of the Colorado River compact, it shall become necessary to supply water to the United States of Mexico from waters over and above the quantities which are surplus as defined by said compact, then the State of California shall and will mutually agree with the State of Arizona to supply, out of the main stream of the Colorado River, one-half of any deficiency which must be supplied to Mexico by the lower basin, and (5) that the State of California shall and will further mutually agree with the States of Arizona and Nevada that none of said three States shall withhold water and none shall require the delivery of water, which can not reasonably be applied to domestic and agricultural uses, and (6) that all of the provisions of said tri-State agreement shall be subject in all particulars to the provisions of the Colorado River compact, and (7) said agreement to take effect upon the ratification of the Colorado River compact by Arizona, California, and Nevada.

(b) Before any money is appropriated for the construction of said dam or power plant, or any construction work done or contracted for, the Secretary of the Interior shall make provision for revenues by contract, in accordance with the provisions of this Act, adequate in his judgment to insure payment of all expenses of operation and maintenance of said works incurred by the United States and the repayment, within fifty years from the date of the completion of said works, of all amounts advanced to the fund under subdivision (b) of section 2 for such works, together with interest thereon made reimbursable under this Act.

Before any money is appropriated for the construction of said main canal and appurtenant structures to connect the Laguna Dam with the Imperial and Coachella Valleys in California, or any construction work is done upon said canal or contracted for, the Secretary of the Interior shall make provision for revenues, by contract or otherwise, adequate in his judgment to insure payment of all expenses of construction, operation, and maintenance of said main canal and appurtenant structures in the manner provided in the reclamation law.

If during the period of amortization the Secretary of the Interior shall receive revenues in excess of the amount necessary to meet the periodical payments to the United States as provided in the contract, or contracts, executed under this Act, then, immediately after the settlement of such periodical payments, he shall pay to the State of Arizona 18 3/4 per centum of such excess revenues and to the State of Nevada 18 3/4 per centum of such excess revenues.

SEC. 5. That the Secretary of the Interior is hereby authorized, under such general regulations as he may prescribe, to contract for the storage of water in said reservoir and for the delivery thereof at such points on the river and on said canal as may be agreed upon, for irrigation and domestic uses, and generation of electrical energy and delivery at the switchboard to States, municipal corporations, political subdivisions, and private corporations of electrical energy generated at said dam, upon charges that will provide revenue which, in addition to other revenue accruing under the reclamation law and under this Act, will in his judgment cover all expenses of operations and maintenance incurred by the United States on account of works constructed under this Act and the payments to the United States under subdivision (b) of section 4. Contracts respecting water for irrigation and domestic uses shall be for permanent service and shall conform to paragraph (a) section 4 of this Act. No person shall have or be entitled to have the use for any purpose of the water stored as aforesaid except by contract made as herein stated.

After the repayments to the United States of all money advanced with interest, charges shall be on such basis and the revenues derived therefrom shall be kept in a separate fund to be expended within the Colorado River Basin as may hereafter by prescribed by the Congress. General and uniform regulations shall be prescribed by the said Secretary for the awarding of contracts for the sale and delivery of electrical energy, and for renewals under subdivision (b) of this section, and in making such contracts the following shall govern:

(a) No contract for electrical energy or for generation of electrical energy shall be of longer duration than fifty years from the date at which such energy is ready for delivery.

Contracts made pursuant to subdivision (a) of this section shall be made with a view to obtaining reasonable returns and shall contain provisions whereby at the end of fifteen years from the date of their execution and every ten years thereafter, there shall be readjustment of the contract, upon the demand of either party thereto, either upward or downward as to price, as the Secretary of the Interior may find to be justified by competitive conditions at distributing points or competitive centers, and with provisions under which disputes or disagreements as to interpretation or performance of such contract shall be determined either by arbitration or court proceedings, the Secretary of the Interior being authorized to act for the United States in such readjustments or proceedings.

(b) The holder of any contract for electrical energy not in default thereunder shall be entitled to a renewal thereof upon such terms and conditions as may be authorized or required under the then existing laws and regulations, unless the property of such holder dependent for its usefulness on a continuation of the contract be purchased or acquired and such holder be compensated for such damages to its property, used and useful in the transmission and distribution of such electrical energy and not taken, resulting from the termination of the supply.

(c) Contracts for the use of water and necessary privileges for the generation and distribution of hydroelectric energy or for the sale and delivery of electrical energy shall be made with responsible applicants therefor who will pay the price fixed by the said Secretary with a view to meeting the revenue requirements herein provided for. In case of conflicting applications, if any, such conflicts shall be resolved by the said Secretary, after hearing, with due regard to the public interest, and in conformity with the policy expressed in the Federal Water Power Act as to conflicting applications for permits and licenses, except that preference to applicants for the use of water and appurtenance works and privileges necessary for the generation and distribution of hydroelectric energy, or for delivery at the switchboard of a hydroelectric plant, shall be given, first, to a State for the generation or purchase of electric energy for use in the State, and the States of Arizona, California, and Nevada shall be given equal opportunity as such applicants.

The rights covered by such preference shall be contracted for by such State within six months after notice by the Secretary of the Interior and to be paid for on the same terms and conditions as may be provided in other similar contracts made by said Secretary: *Provided, however,* That no application of a State or a political subdivision for an allocation of water for power purposes or of electrical energy shall be denied or another application in conflict therewith be granted on the ground that the bond issue of such State or political subdivision, necessary to enable the applicant to utilize such water and appurtenant works and privileges necessary for the generation and distribution of hydroelectric energy or the electrical energy applied for, has not been authorized or marketed, until after a reasonable time, to be determined by the said Secretary, has been given to such applicant to have such bond issue authorized and marketed.

(d) Any agency receiving a contract for electrical energy equivalent to one hundred thousand firm horsepower, or more, may, when deemed feasible by the said Secretary, from engineering and economic considerations and under general regulations prescribed by him, be required to permit any other agency having contracts hereunder for less than the equivalent of twenty-five thousand firm horsepower, upon application to the Secretary of the Interior made within sixty days from the execution of the contract of the agency the use of whose transmission line is applied for, to participate in the benefits and use of any main transmission line constructed or to be constructed by the former for carrying such energy (not exceeding, however, one-fourth the capacity of such line), upon payment by such other agencies of a reasonable share of the cost of construction, operation, and maintenance thereof.

The use is hereby authorized of such public and reserved lands of the United States as may be necessary or convenient for the construction, operation, and maintenance of main transmission lines to transmit said electrical energy.

SEC. 6. That the dam and reservoir provided for by section 1 hereof shall be used; First, for river regulation, improvement of navigation and flood control; second, for irrigation and domestic uses and satisfaction of present perfected rights in pursuance of Article VIII of said Colorado River compact; and third, for power. The title to said dam, reservoir, plant, and incidental works shall forever remain in the United States, and the United States shall, until otherwise provided by Congress, control, manage, and operate the same, except as herein otherwise provided: *Provided, however,* That the Secretary of the Interior may, in his discretion, enter into contracts of lease of a unit or units of any Government-built plant, with right to generate electrical energy, or alternatively, to enter into contracts of lease for the use of water for the generation of electrical energy as herein provided, in either of which events the provisions of section 5 of this Act relating to revenue, term, renewals, determination of conflicting applications, and joint use of transmission lines under contracts for the sale of electrical energy, shall apply.

The Secretary of the Interior shall prescribe and enforce rules and regulations conforming with the requirements of the Federal Water Power Act, so far as applicable, respecting maintenance of works in condition of repair adequate for their efficient operation, maintenance of a system of accounting, control of rates and service in the absence of State regulation or interstate agreement, valuation for rate-making purposes, transfers of contracts, contracts extending beyond the lease period, expropriation of excessive profits, recapture and/or emergency use by the United States of property of lessees, and penalties for enforcing regulations made under this Act or penalizing failure to comply with such regulations or with the provisions of this Act. He shall also conform with other provisions of the Federal Water Power Act and of the rules and regulations of the Federal Power Commission, which have been devised or which may be hereafter devised, for the protection of the investor and consumer.

The Federal Power Commission is hereby directed not to issue or approve any permits or licenses under said Federal Water Power Act upon or affecting the Colorado River or any of its tributaries, except the Gila River, in the States of Colorado, Wyoming, Utah, New Mexico, Nevada, Arizona, and California until this Act shall become effective as provided in section 4 herein.

SEC. 7. That the Secretary of the Interior may, in his discretion, when repayments to the United States of all money advanced, with interest, reimbursable hereunder, shall have been made, transfer the title to said canal and appurtenant structures, except the Laguna Dam and the main canal and appurtenance structures down to and including Syphon Drop, to the districts or other agencies of the United States having a beneficial interest therein in proportion to their respective capital investments under such form of organization as may be acceptable to him. The said districts or other agencies shall have the privilege at any time of utilizing by contract or otherwise such power possibilities as may exist upon said canal, in proportion to their respective contributions or obligations toward the capital cost of said canal and appurtenant structures from and including the diversion works to the point where each respective power plant may be located. The net proceeds from any power development on said canal shall be paid into the fund and credited to said districts or other agencies on their said contracts, in proportion to their rights to develop power, until the districts or other agencies using said canal shall have paid thereby and under any contract or otherwise an amount of money equivalent to the operation and maintenance expense and cost of construction thereof.

SEC. 8. (a) The United States, its permittees, licensees, and contractees, and all users and appropriators of water stored, diverted, carried, and/or distributed by the reservoir, canals, and other works herein authorized, shall observe and be subject to and controlled by said Colorado River compact in the construction, management, and operation of said reservoir, canals, and other works and the storage, diversion, delivery, and use of water for the generation of power, irrigation, and other purposes, anything in this Act to the contrary notwithstanding, and all permits, licenses, and contracts shall so provide.

(b) Also the United States, in constructing, managing, and operating the dam, reservoir, canals, and other works herein authorized, including the appropriation, delivery, and use of water for the generation of power, irrigation, or other uses, and all users of water thus delivered and all users and appropriators of waters stored by said reservoir and/or carried by said canal, including all permittees and licensees of the United States or any of its agencies, shall observe and be subject to and controlled, anything to the contrary herein notwithstanding, by the terms of such compact, if any, between the States of Arizona, California, and Nevada, or any two thereof, for the equitable division of the benefits, including power, arising from the use of water accruing to said States, subsidiary to and consistent with said Colorado River compact, which may be negotiated and approved by said States and to which Congress shall give its consent and approval on or before January 1, 1929; and the terms of any such compact concluded between said States and approved and consented to by Congress after said date: *Provided,* That in the latter case such compact shall be subject to all contracts, if any, made by the Secretary of the Interior under section 5 hereof prior to the date of such approval and consent by Congress.

SEC. 9. That all lands of the United States found by the Secretary of the Interior to be practicable of irrigation and reclamation by the irrigation works authorized herein shall be withdrawn from public entry. Thereafter, at the direction of the Secretary of the Interior, such lands shall be opened for entry, in tracts varying in size but not exceeding one hundred and sixty acres, as may be determined by the Secretary of the Interior, in accordance with the provisions of the reclamation law, and any such entryman shall pay an equitable share in accordance with the benefits received, as determined by the said Secretary, of the construction cost of said canal and appurtenance structures; said payments to be made in such installments and at such times as may be specified by the Secretary of the Interior, in accordance with the provisions of the said reclamation law, and shall constitute revenue from said project and be covered into the fund herein provided for: *Provided,* That all persons who have served in the United States Army, Navy, or Marine Corps during the war with Germany, the war with Spain, or in the suppression of the insurrection in the Philippines, and who have been honorably separated or discharged therefrom or placed in the Regular Army or Navy Reserve, shall have the exclusive preference right for a period of three months to enter said lands, subject, however, to the provisions of subsection (c) of section 4, Act of December 5, 1924 (Forty-third Statutes at Large, page 702); and also, so far as practicable, preference shall be given to said persons in all construction work authorized by this Act: *Provided further,* That in the event such an entry shall be relinquished at any time prior to actual residence upon the land by the entryman for not less than one year, lands so relinquished shall be subject to entry for a period of sixty days after the filing and notation of the relinquishment of the local land office, and after the expiration of said sixty-day period such lands shall be open to entry, subject to the preference in this section provided.

SEC. 10. That nothing in this Act shall be construed as modifying in any manner the existing contract, dated October 23, 1918, between the United States and the Imperial Irrigation District, providing for a connection with Laguna Dam; but the Secretary of the Interior or other districts, persons, or agencies for the construction, in accordance with this Act, of said canal and appurtenant structures, and also for the operation and maintenance thereof, with the consent of the other users.

SEC. 11. That the Secretary of the Interior is hereby authorized to make such studies, surveys, investigations, and do such engineering as may be necessary to determine the lands in the State of Arizona that should be embraced within the boundaries of a reclamation project, heretofore commonly known and hereafter to be known as the Parker-Gila Valley reclamation project, and to recommend the most practicable and feasible method of irrigating lands within said project, or units thereof, and the cost of the same; and the appropriation of such sums of money as may be necessary for the aforesaid purposes from time to time is hereby authorized. The Secretary shall report to Congress as soon as practicable, and not later than December 10, 1931, his findings, conclusions, and recommendations regarding such project.

SEC. 12. "Political subdivision" or "political subdivisions" as used in this Act shall be understood to include any State, irrigation or other district, municipality, or other governmental organization.

"Reclamation law" as used in this Act shall be understood to mean that certain Act of the Congress of the United States approved June 17, 1902, entitled "An Act appropriating the receipts from the sale and disposal of public land in certain States and Territories to the construction of irrigation works for the reclamation of arid lands," and the Acts amendatory thereof and supplemental thereto.

"Maintenance" as used herein shall be deemed to include in each instance provision for keeping the works in good operating condition.

"The Federal Water Power Act," as used in this Act, shall be understood to mean that certain Act of Congress of the United States approved June 10, 1920, entitled "An Act to create a Federal Power Commission; to provide for the improvement of navigation; the development of water power; the use of the public lands in relation thereto; and to repeal section 18 of the River and Harbor Appropriation Act, approved August 8, 1917, and for other purposes," and the Acts amendatory thereof and supplemental thereto.

"Domestic" whenever employed in this Act shall include water uses defined as "domestic" in said Colorado River compact.

SEC. 13. (a) the Colorado River compact signed at Santa Fe, New Mexico, November 24, 1922, pursuant to Act of Congress approved August 19, 1921,